METALS
IN
GROUNDWATER

Edited by
Herbert E. Allen
E. Michael Perdue
David S. Brown

CRC Press
Taylor & Francis Group
Boca Raton London New York

CRC Press is an imprint of the
Taylor & Francis Group, an **informa** business

First published 1993 by Lewis Publishers

Published 2019 by CRC Press
Taylor & Francis Group
6000 Broken Sound Parkway NW, Suite 300
Boca Raton, FL 33487-2742

© 1993 by Taylor & Francis Group, LLC
CRC Press is an imprint of Taylor & Francis Group, an Informa business

First issued in paperback 2019

No claim to original U.S. Government works

ISBN 13: 978-0-367-44981-0 (pbk)
ISBN 13: 978-0-87371-277-4 (hbk)

Visit the Taylor & Francis Web site at
http://www.taylorandfrancis.com

and the CRC Press Web site at
http://www.crcpress.com

Library of Congress Cataloging-in-Publication Data

Metals in Groundwater / Herbert E. Allen, E. Michael Perdue, David S.
        Brown, editors.
                p. cm.
        Includes bibliographical references and index.
                1. Metals--Environmental aspects--Congresses. 2. Groundwater--Pollution-
-Congresses. 3. Soil pollution--Congresses. 4. Metals--Speciation--Congresses. I.
Allen, Herbert E. (Herbert Ellis), 1939-   . II. Perdue, Edward M. III. Brown,
David S.
TD427.M44M45 1993
628.1'683--dc20   93-2362
ISBN 0-87371-277-3

The significance of groundwater as an important and potentially vulnerable resource has become widely recognized within the last several years. Aquifers have become contaminated with organic and inorganic substances. A significant amount of research has been directed toward an understanding of the transport of metals to groundwaters. A key factor in this transport is partitioning of the metal between the solid phase material and water.

This book is the outgrowth of a workshop on metal speciation and transport in groundwaters, which was held at Jekyll Island, Georgia on May 24-26, 1989. The workshop brought together a group of experts in the fields of agricultural chemistry, analytical chemistry, aquatic chemistry, environmental engineering, hydrology, oceanography, and risk assessment. This book is comprised of most of the invited contributions, which collectively provide a realistic overview of metal speciation and transport in groundwaters, from theoretical, experimental, and "real world" perspectives.

Four interrelated questions influenced the organizational phase of the workshop. These questions are:

- What environmental factors enhance and suppress metal transport in groundwaters?

- How are local environmental factors that potentially affect metal transport altered by land-use practices, especially hazardous waste disposal?

- What land or water treatment processes could effectively and economically counteract the effects of these land-use practices?

- What essential features must be present in mathematical models of metal transport in groundwaters, and how are these models validated?

To provide an overview of current research approaches that may ultimately provide answers to such questions, we have chosen authors who have expertise in metal sorption by soils and soil components, mechanisms of metal-soil association, kinetic vs. equilibrium factors, transport phenomena, measurement methods for solid phases and adsorbed metals, and assessment of bioavailable material. Several papers deal with mining, municipal landfill, and agricultural sites.

Several authors have addressed experimental and modeling approaches to the study of metal adsorption to heterogeneous surfaces, trace metal behavior in agricultural soils, sorption of both cationic and anionic metal ions on calcium carbonate (in multiphase systems containing calcite, smectite, and amorphous iron oxide), complexation of cationic metal ions by dissolved humic substances, and metal complexation by organic ligands in landfill leachates. All these chapters have as a common theme the application of fundamental chemical approaches in highly complex heterogeneous systems. The current state of knowledge of redox chemistry in the sub-surface environment is reviewed in another chapter, with special emphasis on the chemical aspects of the problem.

Solute transport in groundwater is the focus of three chapters, with emphasis on field tracer studies, modeling techniques, and coupling of chemical speciation and transport models. Another chapter describes the approaches taken by the U. S. Environmental Protection Agency to quantify the potential health effects associated with chronic chemical exposure to metals in water. The remaining chapters present case studies of radionuclide transport from a uranium ore deposit and a study of the effects of landfill disposal practices on the speciation and transport of metals in landfill leachates.

We wish to thank all authors for their contributions and for their consistent cooperation during the time required to produce this book. The expert technical assistance of Ms. June Olson and Ms. Dana M. Crumety, who provided expert secretarial support, is gratefully acknowledged. Finally, we wish to thank the U. S. Environmental Protection Agency's Office of Exploratory Research for their financial support.

Herbert E. Allen	Newark, Delaware
E. Michael Perdue	Atlanta, Georgia
David S. Brown	Atlanta, Georgia
May 1993	

Herbert E. Allen is Professor of Civil Engineering at the University of Delaware, Newark, Delaware, U.S.A. Dr. Allen received his Ph.D. in Environmental Health Chemistry from the University of Michigan in 1974, his M.S. in Analytical Chemistry from Wayne State University in 1967, and his B.S. in Chemistry from the University of Michigan in 1962. He served on the faculty of the Department of Environmental Engineering at the Illinois Institute of Technology from 1974 to 1983. From 1983 to 1989 he was Professor of Chemistry and Director of the Environmental Studies Institute at Drexel University.

Dr. Allen has published more than 100 papers and chapters in books. His research interests concern the chemistry of trace metals and organics in contaminated and natural environments. He conducted research directed toward the development of standards for metals in soil, sediment and water that take into account metal speciation and bioavailability.

Dr. Allen is past-chairman of the Division of Environmental Chemistry of the American Chemical Society. He has been a frequent advisor to the World Health Organization, the Environmental Protection Agency and industry.

E. Michael Perdue is Professor of Earth and Atmospheric Science at the Georgia Institute of Technology, Atlanta, Georgia, U.S.A. Dr. Perdue received his Ph.D. in Chemistry from the Georgia Institute of Technology in 1973 and he had received his B.S. in Chemistry from the same institution in 1969. For ten years Dr. Perdue was on the Chemistry faculty at Portland State University. He joined the faculty of the Georgia Institute of Technology in 1983.

For twenty years, Dr. Perdue and his graduate students have studied the chemical compositions and properties of naturally occurring organic matter in soils and natural waters. Emphasis has been placed on the measurement and interpretation of average chemical and physical properties of humic substances and other unresolvable complex mixtures. He has published over 40 papers dealing with such topics as the thermodynamic description of acid-base and metal-ligand equilibria of humic substances, analytical constraints on the structural features of humic substances, and environmental chemical kineics.

Dr. Perdue chaired a Dahlem Workshop in 1989 and co-edited the resulting book "Organic Acids in Aquatic Ecosystems" published by John Wiley & Sons.

David S. Brown is Acting Chief of the Assessment Branch at the U.S. Environmental Protection Agency's Environmental Research Laboratory in Athens, Georgia. Dr. Brown received his Ph.D. in Soil Physical Chemistry from the University of Illinois in 1971, his M.S. in Soil Chemistry from the University of Illinois in 1968, and his B.S.A. in Soil Science from Purdue University in 1966. While at the Environmental Research Laboratory, he has held several positions including: Soil Scientist, Agricultural and Industrial Water Pollution Control Branch, 1971-1977, Soil Scientist, Processes Branch, 1977-1983, Soil Scientist, Assessment Branch, 1984-1986, and Pesticides and Metals Team Leader, Assessment Branch, 1987-1990.

Dr. Brown has devoted much of his career to the development, application and testing of exposure assessment strategies, modeling tools and applications methodologies. The primary areas of emphasis have been in pesticide transport and metals speciation processes. He received the U.S. EPA Bronze Medal Award in 1974, 1985, and 1992 for Pesticide Runoff Field Research, Development of a Technical Support Strategy for Regulating Land Disposal of Solid Waste, and Special Work Group on Sewage Sludge Use and Disposal Regulations, respectively. Dr. Brown also received the U.S. EPA/Soil Science Society of America Scientific Achievement Award in Earth Sciences in 1990, and the U.S. EPA, Office of Research and Development Award for Excellence in Management in 1990.

CONTENTS

3

SUB-SURFACE REDOX CHEMISTRY: A COMPARISON OF EQUILIBRIUM AND REACTION-BASED APPROACHES

William Fish

4

LANTHANIDE ION PROBE SPECTROSCOPY FOR METAL ION SPECIATION

Wisnu Susetyo, Lionel A. Carreira,
Leo V. Azarraga and David M. Grimm

5

AN OVERVIEW OF MODELING TECHNIQUES FOR SOLUTE TRANSPORT IN GROUNDWATER

Peter S. Huyakorn, Jan B. Kool and T. Neil Blandford

6

---

xiii

9

METAL-ORGANIC INTERACTIONS IN SUBTITLE D LANDFILL LEACHATES AND ASSOCIATED GROUNDWATERS

Peter A. Gintautas, Kristina A. Huyck,
Stephen R. Daniel and Donald L. Macalady

12

METAL SPECIATION AND MOBILITY AS INFLUENCED BY LANDFILL DISPOSAL PRACTICES

Frederick G. Pohland, Wendall H. Cross and Joseph P. Gould

METALS
IN
GROUNDWATER

ADSORPTION TO HETEROGENEOUS SURFACES

Willem H. van Riemsdijk and Tjisse Hiemstra
Department of Soil Science & Plant Nutrition
Wageningen Agricultural University
Wageningen, The Netherlands

1.   INTRODUCTION

Adsorption of metals in soils, sediments and aquifers is a complex phenomenon. Metal ions are charged species which may form various soluble complexes both with inorganic and organic ligands. The formation constants for the inorganic ligand-metal complexes are well known for many complexes and can be used to calculate the distribution of the metal over its various forms in solution. However, for complexes between metals and natural dissolved organic ligands, like fulvic acids, it is not possible to calculate the speciation. This is due to the absence of reliable physical chemical models that can account for metal binding to natural organic matter. This seriously hinders the advance of research related to metal speciation in the environment.

The presence of soluble metal complexes often reduces the metal adsorption compared with the absence of these dissolved complexes. This is the case when the affinity of the metal complexes for reactive sites on the solid matrix is much less than that of the free metal ion. Examples are metal complexes with ligands like chloride or citrate [1].

0-87371-277-3/93/$0.00 + $.50

Another important factor influencing metal adsorption to natural sorbents is the competition between a metal ion of interest and other species for the same adsorption sites. An example is the influence of zinc and calcium on the cadmium adsorption to soils [1]. In practical pollution situations a mixture of metals and other ions is often involved. Competition may, in such cases, notably reduce the adsorption of a weakly adsorbed metal ion like cadmium, causing enhanced mobility and bioavailability [2]. The retardation of a metal in groundwater is thus strongly a factor of the solution phase composition, which will vary with respect to, for example, the position in the pollutant plume.

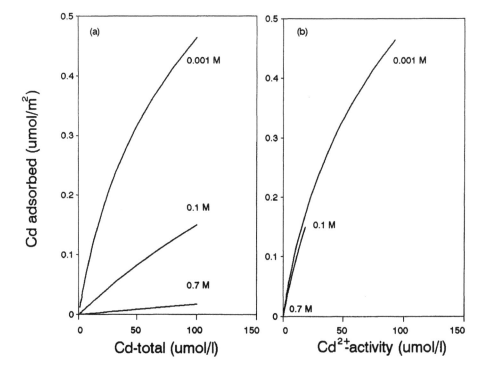

FIGURE 1. *Calculated Adsorption Isotherms of Cadmium on Iron Oxides for Various Cl⁻ Levels, Represented with a Total Cadmium Solution Concentration (a) and a Cd²⁺ Activity Scale (b).* The cadmium adsorption constant (log K = 6.8) has been taken from Van Riemsdijk et al. [24]. The Stern layer capacitance is set at 2.23 and only one protonation reaction is assumed (log $K_{1,2}$ = 10.7).

In Figure 1, the effect of chloride on cadmium adsorption is illustrated. When the adsorption is plotted as a function of total metal concentration in solution a whole range of curves results, the actual shape depending on the chloride concentration in the solution. These curves coalesce approximately into one curve when the adsorption is plotted as a function of free metal ion activity in solution, (Figure 1b), provided that the adsorption of the chloride complexes is negligible and that "competing" species are present at constant solution activity. The behavior shown in Figure 1 has been observed for cadmium adsorption for a whole range of different soils [1].

It is well known that pH strongly influences the adsorption of metal ions in natural systems. This effect is caused firstly by a change of the surface potential of the reactive solid phases and secondly by a change in the competition between protons and metal ions for the same adsorption sites. A change in the surface potential affects metal adsorption through a change in electrostatic repulsion or attraction.

Apart from changes in solution composition, changes in solid phase composition will also strongly influence the interaction of metals with the solid matrix. The question arises how to deal with the above described complexity. A 'practical' approach would be to measure the interaction with a natural solid matrix of interest for a range of solution conditions that are considered relevant for the situation that is to be dealt with. One can then try to model the obtained result with a chosen empirical model using curve fitting techniques to obtain the model parameters. Although this approach may give a good description for the situation studied, nothing can be said with any confidence for a situation that is outside the range of conditions that have been used in the experiments. Nor can the information obtained be transferred to a solid matrix with a different composition of reactive phases. For the purpose of environmental policy, soil chemical knowledge is often required that is of more general validity.

2.   CRITICAL EVALUATION OF EXISTENT (CHEMICAL) METHODOLOGIES DEALING WITH THE COMPLEXITY OF SOIL SYSTEMS

The above indicates that natural variation in both the solution composition and the composition of the solid matrix may lead to a very large variation in the distribution behavior of one and the same chemical for different locations. It is thus far from simple to obtain information that has general validity. Several approaches can be found in the soil chemical literature that are attempts to obtain this type of information.

One approach is to discriminate the total content of a chemical over various operationally defined fractions in the solid phase. This solid phase "speciation" is obtained by employing sequential extraction procedures. For metal speciation in sediments, the following fractions are "defined" by sequential extraction [3-7]:

1)   exchangeable metals plus metals associated with calcium carbonate,
2)   metals present in easily reduced phases, mainly metals associated with manganese oxides, and partly also associated with amorphous iron oxides,
3)   metals associated with amorphous and poorly crystalline iron oxides,
4)   metals associated with organic matter or present as metal sulphides, and
5)   a residual fraction.

Sequential extraction with different extraction agents by definition leads to various metal fractions. Whether the fractions obtained indeed represent the various pools is questionable. Gruebel *et al.* [8] tested a sequential extraction procedure for arsenic and selenium using well characterized minerals. Their work shows that the various phases may not always be identified correctly.

Identifying operationally defined fractions in itself does not lead to the possibility of predicting the interaction between a chemical and various soils, sediments and aquifer materials. One step in that direction is the work of Unger and Allen [7] who combined the measurement of the partitioning of a metal over the solution phase and the sediment with a sequential extraction procedure. In this way, they were able to "speciate" the amount (ad)sorbed as a function of the loading of the sediment with the metal. They estimated a binding capacity and an average equilibrium "constant" for each of the five fractions that were discriminated in this study. How this binding capacity is related to soil or sediment type was not a subject of their study.

The relation between soil characteristics and the observed overall interaction behavior of a chemical can be studied by measuring the partitioning of a chemical over solution and solid phase for a whole range of different soils and sediments [9,10]. Although the methodology in itself is promising, there are a few serious problems that may explain why the results published so far are often disappointing. The first problem is that the soil characteristics measured are poorly (operationally) defined and are not necessarily good estimators for the various types of reactive surfaces present.

The second, very important, problem is that the chemical composition of the aqueous phase from which the adsorption is studied generally varies for the different soils. Although the same electrolyte may be added to all soils, the resulting aqueous medium is **not** the same. An obvious and very important factor is the pH of the aqueous phase, which is buffered by the soil particles, resulting in different pH values of the electrolyte solution for different soils. Because the pH strongly influences most chemical interactions in soil, the result of such a study is sometimes that the pH is the soil characteristic responsible for differences in behavior between various soils [10]. This suggests that different soils, when brought to the same pH, should show the same adsorption behavior. This is obviously not true. Another example of unintentional differences in the resulting composition of the aqueous phase is the presence of a variable amount of competitor species. Studying cadmium adsorption for a range of soils will, in general, lead to a release of a variable amount of zinc ions. The resulting cadmium adsorption is not only a function of the various reactive surfaces that are present and the resulting pH, but also depends on the degree of competition with zinc ions [11].

The third problem is that soils as sampled may already contain a varying amount of the adsorbed species of interest. To be able to make a good comparison between the different soils, the desorption behavior of the species of interest should be studied. Studying desorption is often extremely difficult, especially in the case of a high affinity adsorption, i.e. a steep ad(de)sorption isotherm. Recently a promising new methodology for studying phosphate desorption has been developed [12].

Another approach in trying to obtain results that are of a more general validity for a wide range of soils is to study the interaction of a chemical with various individual soil components as a means to interpret the difference in behavior between various soils or sediments. An example of this approach is the work of McLaren et al. [9] who studied adsorption of cobalt with different soil components and also with different soils. At present, this procedure is also not very successful mainly because for two reasons.

The first reason is that the fundamental knowledge about the physicochemical interaction mechanisms between species and the various soil components is, at present, not well enough established. McLaren et al. [9] used empirical equations to describe the measured interaction with soil components. The measurements with the soil components were done for a very limited range of solution conditions, at one pH only, and the empirical model used does not allow for estimating the interaction behavior for solution conditions other than those of the measurements. The adsorption measurements with the different soils were done with one electrolyte solution without control of pH leading to different pH values for the different soils studied. Comparison between these adsorption measurements and those of soil components based on one pH measurement is bound to fail. The second important problem with this methodology is the need to establish the content and nature of the various reactive surfaces present in a soil sample.

For the development of general applicable soil chemical knowledge that can be used to interpret the bio-availablity, toxicity and mobility of pollutants, it is essential to have insight in the fundamental physical chemical interaction of species with soil components. Metal (hydr)oxides of aluminum, iron and manganese and amorphous aluminum silicates, together with natural organic matter, are very important reactive surfaces with respect to metal adsorption in soils, lakes and aquifers. In the following, we will discuss physical chemical adsorption models for metal oxides and organic matter together with a discussion on how this information can be transferred to natural systems.

3. ADSORPTION MODELS FOR METAL OXIDES

Metal (hydr)oxides of iron, aluminum and manganese are important reactive surfaces in soils, aquifers and sediments with respect to interaction with positively charged species like, H^+, Al^{3+}, Cd^{2+}, Zn^{2+}, Pb^{2+}, and Cu^{2+} and with negatively charged species like phosphate, arsenate, sulphate, selenite, borate, bicarbonate and fluoride. Organic species may also adsorb on metal oxides [13]. The adsorption of species on oxides is strongly dependent on the pH due to the variable, pH dependent, surface charge and potential of metal (hydr)oxides and to a pH dependent speciation of the adsorbate.

4. RELATIONSHIP BETWEEN SURFACE CHARGE (POTENTIAL) AND pH

Metal oxides are amphoteric surfaces which may be either positively or negatively charged depending on the pH of the system. The pH at which the metal(hydr)oxide is uncharged is called the point of zero charge, pzc. The pzc of metal (hydr)oxides differs considerably, ranging from 2 for silica to around 10 for some iron oxides. The primary surface charge, in the absence of specific adsorptions of ions other than H^+, depends on the pH and on the electrolyte concentration or ionic strength. The most widely used model to interpret the variable charge characteristics is the so called two-pK model [13-18]. In this model, it is assumed that there is only one type of reactive group present on the surface of a metal (hydr)oxide that may react with protons according to the following two reactions,

$$SO^- + H_s^+ \rightleftharpoons SOH^o \; ; \; K_{H1} \tag{1}$$

$$SOH^o + H_s^+ \rightleftharpoons SOH_2^+ \; ; \; K_{H2} \tag{2}$$

where SO^-, SOH^o, SOH_2^+ represent three different surface species belonging to one and the same reactive surface oxygen group, and $[H_s^+]$ is the proton concentration near the plane of adsorption which is related to the concentration in solution, $[H^+]$, via,

$$[H_s^+] \; = \; [H^+] \exp(-F \; \psi_o \; / \; RT) \tag{3}$$

where ψ_o is the surface potential. In order to be able to model the variable charge characteristics, an additional expression is required that relates the surface charge and the surface potential. This relationship is provided by assuming a certain double layer model. In the "oxide" literature, different approaches can be found. Sometimes only a diffuse Gouy Chapman double layer is assumed [19], or it is assumed that ratio between surface charge and potential is constant, leading to the so called constant capacitance model [13,14,17], or a combination of a Stern layer with a Gouy Chapman double layer often indicated as the Basic Stern model is used [20-25], or a model where three planes near the surface plus a diffuse double layer model is used which is called the triple layer model [15,18,26]. The basic charging reactions in all these approaches are equal. They only differ in the relation(s) used to relate the surface charge and the surface potential. A disadvantage of the constant capacitance model is that it cannot be used to describe the charging at variable salt level because the "constant" capacitance is in practice a function of the salt concentration. A disadvantage of the triple layer model, as used in literature is that in addition to the reactions with protons, pair formation with simple electrolyte ions, such as sodium and chloride, has to also be included in the model, and this increases the number of adjustable parameters. These variable charge, variable potential models have been successfully used to describe surface charge-pH curves as a function of salt level for various metal (hydr)oxides. The model parameters, e.g. the two proton affinity constants and the Stern layer capacitance are treated as adjustable parameters. The proton affinity constants may differ considerably for various metal (hydr)oxides.

An alternative for the two-pK model is the one-pK model [20-25,27]. This assumes that the basic charging reaction can be described with the following reaction,

$$SOH^{1/2-} + H_s^+ \rightleftharpoons SOH_2^{1/2+} \quad ; \quad K_H \qquad (4)$$

The surface charge is dependent on the ratio between the two different species, belonging to the same surface group. It follows from this model that the log K_H equals the pzc of the metal (hydr)oxide. The advantage of the one-pK model is its extreme simplicity, the fact that it has one parameter less than the two-pK model and that the proton affinity constant follows directly from the pzc. This one-pK model gives as good a description of the available data as the two-pK model for most metal oxides, with the exception of silica.

5. THE MUSIC MODEL

Both the one and the two-pK modeling approaches presume the presence of only one type of reactive group. However, it is well known from infrared analysis that different types of surface oxygen groups are present on metal (hydr)oxides [28-30]. Surface oxygen may be singly, doubly or triply coordinated to an underlying metal ion. It is to be expected that the reactivity of these oxygen groups towards proton adsorption differs considerably. Recently a new model, the MUSIC (MUti SIte Complexation) model, has been developed that takes these three different types of surface groups into account [21,22]. Each type of surface group may react with two protons. The affinity constants for the proton adsorption reactions for the various groups of a range of

different metal (hydr)oxides are **predicted** by the new model, leading
to the possibility of **predicting** the variable charge behavior of metal
oxides. It predicts that the difference between the log K values for two
consecutive proton adsorption reactions for one type of surface group
is very large (around 14 log K units). The one and two-pK models are
special cases of this more general new model. The predictions are in
good agreement with existing experimental data [22]. Another result
is that the quite different shape of the charging curve for silica
compared to other oxides is predicted by the MUSIC model. It also
follows from the model that different crystal planes of the same metal
(hydr)oxide may exhibit a quite different charging behavior, due to a
different composition with respect to the various types of reactive
surface groups. This may have important implications for the specific
adsorption of species other than protons on metal hydr(oxides). For
gibbsite $(Al(OH)_3)$ it is predicted that the dominant crystal plane (001)
is **uncharged** for pH values below ten, whereas a negative surface
charge develops for pH values higher than ten. In the normal pH range,
only the edge faces will develop a (positive) surface charge. These
predictions are also in accordance with experimental information [22].
Specific adsorption of anions on gibbsite will thus probably only occur
on the edge faces.

The prediction of the proton affinity constants is based on the
following formula [21]:

$$\log K = A - B n \frac{v}{L} \tag{5}$$

where A and B are constants, n is the number of metal ions that
coordinate with a surface oxygen group (n = 1, 2 or 3), L is the
distance between an adsorbed proton and the metal center and the
formal bond valence v, as introduced by Pauling [31] equals,

$$v = \frac{Z}{CN} \tag{6}$$

where Z is the charge of the metal in the metal (hydr)oxide and CN is the coordination number of this metal. The constant A differs whether a proton reacts with a 'naked' surface oxygen or with a surface hydroxyl. The constants A and B are obtained by considering the protonation of solution monomers for which log K and v/L are known. For these monomers, n is always one. The model is further calibrated based on the protonation of a singly coordinated surface hydroxyl as present on gibbsite ($Al(OH)_3$). The log K value for this reaction can be derived directly from experimental data because the charging behavior of gibbsite is determined by only this one reactive group over a wide pH range [21,22]. It turns out that the affinity for proton adsorption on a surface is approximately 4 log K units higher than the corresponding reaction for the dissolved monomer. This difference may arise because the spatial configuration of the oxygen ligands on a surface is in general quite different from the situation for a dissolved monomer. This difference in configuration will lead to a different enthalpy and entropy of adsorption for the surface compared with the reactions in solution.

The formal bond valence v can be seen as the positive charge of the metal center that is contributed to a ligand. For iron and aluminum ($v = 3/6 = 1/2$). The concept of v also facilitates a proper bookkeeping of surface charges. This is illustrated in Figure 2, where the concept is applied to three different dissolved iron species. The positive charge of the central metal ion has been equally distributed over the surrounding ligands. Note that in this case a positive, neutral or negative species can be formed by simply changing the ratio of the coordinated hydroxyl and water molecules. This is the basis of the one-pK metal oxide model.

The proton affinity constants for the three different types of surface oxygen groups that may be present on the surface of iron and aluminum (hydr)oxide as predicted by the MUSIC model are given in Table 1. Assuming that the entropy contribution to the protonation reaction is mainly derived from the change in hydration entropy, and

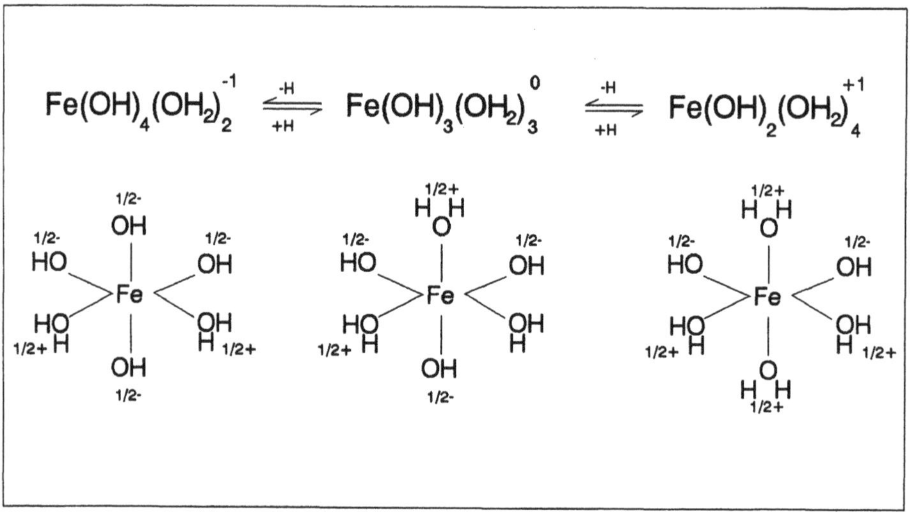

FIGURE 2.   *Schematic Representation of Three Dissolved Fe-hydroxy Complexes.*

applying the Born treatment of hydration energy [32] to the protonation of solution monomers, and using the formal bond valence in case of surface reactions, the entropy and thus also the enthalpy (because the free energy can be derived from the log $K_{intr.}$) of the surface protonation reactions can be estimated [32]. The results are given in Table 1. It follows from Table 1 that the difference in log K between the first and the second proton adsorbing on the same type of oxygen group is approximately 14 log K units. The difference between corresponding protonation reactions for different types of groups (singly, doubly, triply coordinated) is also very large. Note that the protonation for a doubly coordinated surface oxygen for iron and aluminum (hydr)oxides is equivalent with the formulation of the classical two-pK model. The difference being that in the two-pK model the adsorption constants are treated as fitting parameters, whereas here they are *a priori* predicted.

The protonation of a singly coordinated surface hydroxyl for these types of metal (hydr)oxides is equivalent with the formulation of the classical one-pK model.

TABLE 1. *Predicted log K for Surface Protonation Reactions at 298 °K [21] and for Related Thermodynamic Quantities [32].*

Reactant	v	log K	G $kJ\ mol^{-1}$	H $kJ\ mol^{-1}$	S $J\ mol^{-1}K^{-1}$
AlO	-3/2	23.8	-135.7	-106±2	101±6
AlOH	-1/2	10.0	-57.0	-42±2	50±6
Al_2O	-1	12.3	-70.1	-47±2	77±6
Al_2OH	0	-1.5	8.6	15±2	22±6
Al_3O	-1/2	2.2	-12.5	2±2	50±6
Al_3OH	+1/2	-11.6	66.1	65±2	-4±6
FeO	-3/2	24.5	-139.7	-116±4	79±4
FeOH	-1/2	10.7	-61.0	-52±4	30±13
Fe_2O	-1	13.7	-78.1	-61±4	56±13
Fe_2OH	0	-0.10	0.6	2±4	4±13
Fe_3O	-1/2	4.3	-24.5	-16±4	30±13
Fe_3OH	+1/2	-9.5	54.2	48±4	-20±13

The log K for the first protonation step of a singly coordinated oxygen, formal charge -1.5, is so high that a fully dissociated singly coordinated group will not be present to any significant extent on the surface of an iron or aluminum (hydr)oxide in contact with water. The log K for the protonation of a triply coordinated hydroxyl group is so small that a triply coordinated surface water molecule, formal charge +1.5, can also be considered to be nonexistent for these (hydr)oxides.

The outcome of the MUSIC model implies that different crystal planes may exhibit completely different properties. For clays it is common practice to consider that different planes may have quite different properties. However for metal oxides up till now it has been common practice to treat a metal oxide as if it has only one type of reactive group on the surface. The difference in properties of different faces that are predicted by the MUSIC model are first illustrated for gibbsite. Gibbsite crystals are thin hexagonal platelets. The dominant 001 plane exhibits only doubly coordinated groups on the surface. It is predicted that this plane is essentially uncharged below pH 9.0. Above pH 9.0, a negative surface charge will develop on this plane. Both singly and doubly coordinated groups occur on the edge faces. The singly coordinated groups are on these faces responsible for the charging behavior over the whole pH range. The doubly coordinated groups hardly contribute to the charging of the edge faces even at high pH, because the dissociation of the doubly coordinated surface hydroxyls is strongly suppressed by the negative surface potential that is caused by the dissociation of the singly coordinated groups. The effective charging behavior of a group on a surface is thus not only a function of its proton affinity constants, but depends also on the presence of other types of reactive groups on the same mineral face. The overall particle surface charge above pH 9.0 is determined by contributions of both the edge faces and the planar surfaces. These predictions are in very good agreement with experimental data [22]. These observations also explain why the slope of the charging curve of gibbsite below pH 9.0 at a certain salt level is relatively low when the charge is expressed per unit area BET surface.

Another metal oxide for which the surface structure is well established is goethite [33] as prepared according to Atkinson's method [34]. Three different crystal planes are present [33]. The total surface area, for particles with a specific surface area of around 100 m^2/g, is mainly determined by the 100 and the 010 face, the contribution of the two faces to the total surface area being roughly equal. The composition of these two faces with respect to the reactive groups is quite different [35,22]. The 010 face exhibits only singly and doubly coordinated groups. The outcome of the MUSIC model predicts that the charging behavior of this face is almost exclusively determined by the singly coordinated groups. The charging of this face can essentially be calculated with the classical one-pK model. The pzc of this face is predicted to be 10.7. At the 100 face singly, doubly and triply coordinated groups are present in equal densities, 3.3 sites/nm^2 each. The singly and triply coordinated groups are the reactive groups. The pzc of a hypothetical face with only triply coordinated would be at pH=4.3. The combination of the two reactive groups, where the charging of each can be described with a single affinity constant, leads to a predicted pzc of 7.5 for the 100 face. Around the pzc of the 100 face of goethite, the singly and triply coordinated groups are nearly fully but oppositely charged. This may strongly influence specific adsorption of ions at this face.

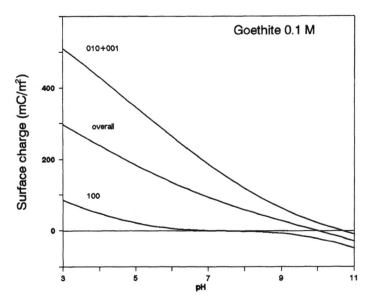

FIGURE 3. *Calculated σ_0(pH) Curves for the 100 and 010+001 Face of Goethite in 0.1 M NaNO$_3$, Using a Stern Layer Capacitance of 2.23 F/m^2 (BS Approximation).* The site density is set at 8 nm^{-2} for singly coordinated and doubly coordinated surface groups. The calculated overall σ_0(pH) curve is expressed in mC/m^2 BET surface area. The ratio of the specific surface area of the two types of faces is about 1.

In Figure 3, the predicted surface charge on the 100 and the 010 face is given as a function of pH together with the calculated overall particle charge. Note that the pzc of the particle is predicted to be around pH 10.0. Another interesting aspect of the predicted charging behavior is that, in the region between pH 7.5 and 10.7, the two dominant crystal faces are oppositely charged! The difference in charging behavior of the two planes is expected to have an influence on the colloid stability of these particles and also on the specific adsorption of various ions on these particles.

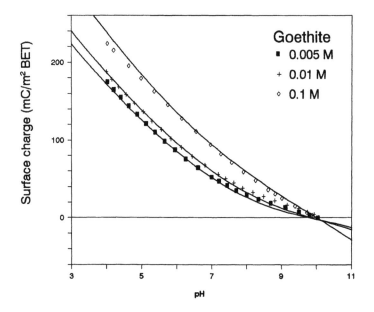

FIGURE 4.   *Predicted Overall σ_0(pH) Curves of Goethite (mC/m²BET) at Three Different NaNO₃ Concentrations (Full Lines), Taking the Weighted Sum of the σ_0-pH Curves of the Individual Planes Given in Figure 3.* The experimental information has been indicated with markers.

In Figure 4, the predicted charging behavior for goethite at three salt levels is compared with experimental data of Hiemstra *et al.* [22]. For each plane, a basic Stern type double layer has been assumed with a capacitance of 2.2 F/m². The Stern layer capacitance was treated as the only adjustable parameter. A very good description of the experimental data is obtained using the proton affinity constants as predicted by the MUSIC model. Note that the overall pzc is slightly dependent on the salt level. This effect is caused by the presence of different faces, each with its own double layer, that respond differently on a change in salt level. The goethite particle is an example of so-called patchwise heterogeneity. The effect of patchwise heterogeneity coupled with variable charge surfaces is further discussed by Bolt and Van Riemsdijk [36], Van Riemsdijk *et al.* [37] and Koopal and Van Riemsdijk [38]. The high pzc reported here for goethite is in accord with other recent findings [39-41]. The pzc reported in literature for iron oxides is often in the range from 7 to 10. A pzc lower than 10 for a certain iron oxide is not necessarily in conflict with the proton affinity constants as predicted by the MUSIC model, because the particle pzc depends on the faces being developed and the ratio of the reactive groups. Also contamination by CO_2 may cause the experimental pzc to be too low. It is, therefore, of interest to analyze the effect of the presence of CO_2.

6. ADSORPTION OF CARBONATE SPECIES AND ITS EFFECT ON THE CHARGING OF IRON OXIDES

Dissolved carbon dioxide, carbonic acid and/or (bi)carbonate are generally present in natural systems. In laboratory experiments one usually tries to exclude carbonate from the system because of the possible interference with the measurements. However, for a transfer of the laboratory results to the field, it is necessary to know the effect of the presence of CO_2. Only Zachara *et al.* [18] have reported experimentally determined adsorption of carbonate species on amorphous iron hydroxide using labelled ^{14}C, but satisfactory modeling was not obtained. The detailed surface structure of amorphous iron hydroxide is unknown. Amorphous $Fe(OH)_3$ can be regarded as water rich iron oxide in which the number of oxygens coordinated with iron (n) is lower than that of, for instance, hematite (n = 4) and probably goethite (n = 3). The coordination number of the surface groups will generally be even lower due to the presence of broken bonds. This indicates that the surface groups will be dominated by doubly ($n = 2$) and singly ($n = 1$) coordinated surface hydroxyls, Fe_2OH and $FeOH(H)$ respectively. It is assumed here that the surface structure of amorphous Fe hydroxide is comparable with that of the reactive faces of goethite (010 and 001) which only possess singly and doubly coordinated groups, each with a site density of about 8 nm^{-2} and that the Stern layer capacitance equals that of goethite ($C = 2.23$ F/m^2).

The relevant protonation reaction for the description of the surface charge [22] is given by:

$$Fe\text{-}OH^{1/2-} + H_s^+ \rightleftharpoons Fe\text{-}OH_2^{1/2+}; \quad \log K_{1.2} = 10.7 \qquad (7)$$

A species of carbon dioxide is thought to react specifically with singly coordinated surface groups [42]. The adsorption reaction can be represented as:

$$Fe\text{-}OH^{1/2-} + H_2CO_3^o \rightleftharpoons Fe\text{-}O^{-1/6}\text{-}C\text{-}OOH^{-2/6} + H_2O \qquad (8)$$
$$\log K_{Carb}$$

or equivalently as:

$$Fe\text{-}OH^{1/2+} + HCO_3^- \rightleftharpoons Fe\text{-}O^{-1/6-}\text{-}C\text{-}OOH^{-2/6} + H_2O \qquad (9)$$
$$\log K_{Carb} = -4.35$$

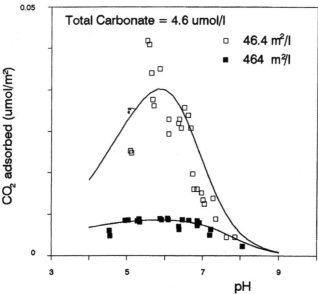

FIGURE 5.   *Adsorption of CO_2 on Amorphous Iron Hydroxide for Two Different Suspension Concentrations at a Total Carbonate Concentration in the System of 4.6 μmol/L.* Model calculations: solid lines; for parameters see text. Data are taken from Zachara *et al.* [18].

It is assumed that the positive charge of the central ions can be attributed to one or two of the planes distinguished in the Basic Stern model. The charge distribution can be described using the reduced valence principle of Pauling [31]. The carbon centre ($z = 4+$) distributes the positive charge over three surrounding ligands, attributing to each ligand on the average a valence v of $+4/3$. The Fe center attributes $1/2+$ to the surrounding ligands. This implies that the overall charge of the surface oxygen will be ($1/2 + 4/3 -2 = -1/6$) while the combined charge of the oxygen and hydroxyl situated at the second plane of the BS model equals $-2/6$.

The value of the intrinsic log K has been determined using data of Zachara *et al.* [18]. The data could be described rather well with log $K_{Carb} = 4.4$ (Figure 5). A more traditional formulation of the adsorption reaction would be:

$$Fe-OH_2^{1/2+} + HCO_3^- \leftrightarrows Fe-OH_2^{1/2+}-HCO_3^- \qquad (10)$$

or

$$Fe-OH_2^{1/2+} + HCO_3^- \leftrightarrows Fe^{1/2+}-HCO_3^- + H_2O \qquad (11)$$

With this formulation it is not possible to get a satisfactory description of the data.

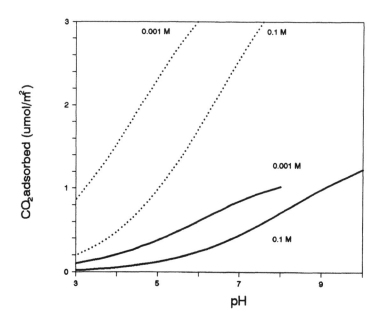

FIGURE 6.   *Calculated pH Dependency of the Adsorption of CO_2 on Amorphous Iron (Hydr)oxide for Two Electrolyte Levels (0.001 and 0.1 M of a 1:1 Electrolyte) at a Constant CO_2 Pressure.*   Solid line: $-\log P_{CO2} = 3.5$ and dotted line: $-\log P_{CO2} = 2.5$.

Knowing the affinity of carbon dioxide for singly coordinated surface species of iron (hydr)oxides, the expected level of CO_2 adsorption on the surface of iron (hydr)oxides in natural systems can be calculated, for example, as function of pH and salt level at various levels of a constant P_{CO2} (Figure 6). It is evident that surfaces in contact with the atmosphere already may adsorb relatively high amounts of carbonic acid, especially at low electrolyte levels. Such a calculation indicates that one should be aware of the possibility of a high adsorption of carbon dioxide in suspensions used in the laboratory.

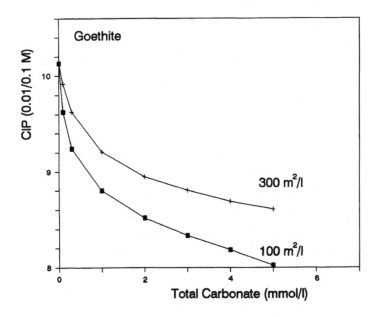

FIGURE 7. *The Influence of the Presence of Various Levels of Total Carbonate in the System on Common Intersection Point (CIP) of the Apparent Charging Curves of Goethite Suspensions (for 0.01 and 0.1 M) at Two Different Suspension Densities.*

As noted before, the experimentally determined apparent pzc of goethites may differ from the theoretical ones. Lower values can be due to the influence of CO_2. Evans *et al.* [39] and Zeltner and Anderson [41] reported that the careful removal of CO_2 by purging suspensions for quite a long time (2-4 months) increased their common intersection point (CIP) by about one pH unit. The influence of CO_2 on the CIP of goethite can now be evaluated. Calculations show that only adsorption at the 001 and 010 faces occur. This is in conflict with the suggestion of adsorption in the "grooves" at the 100 face as proposed by Russell *et al.* [43] and Zeltner and Anderson [41]. The CIP of the apparent charging curves of goethite suspensions at the 0.01 and 0.1 M electrolyte level have been calculated for various levels of total carbonate in the system (Figure 7).

The calculations indicate that very small amounts of carbonate strongly influence the CIP. A goethite suspension stored at pH 8.5 in contact with the atmosphere will have a carbonate level of about 3 mmol/l. Titration of this suspension will give a CIP of about 8.3-8.7, depending on the suspension concentration, whereas the theoretical CIP without any adsorption of CO_2 will give a CIP of about 10.1. This indicates that the CO_2 should be removed by purging the suspension. However complete removal of CO_2 by purging is a very slow process if a suspension is only purged without any acidification, as is frequently done. This is illustrated in Figures 8 and 9. A volume of 50 ml of a 10^{-3} and 10^{-4} M $NaHCO_3$ solution in a closed vessel (100 ml) was purged with purified moist N_2 gas (100 ml/min).

Figure 8 shows that the pH rises very quickly during the first hours of the experiment, and Figure 9 shows that the corresponding removal of CO_2 is rather fast. However, increasing pH decreases the P_{CO2} strongly, causing a strong reduction of the flux of removed CO_2. It can be concluded that with such a purging process even after two days less than 50 % can be removed. The resulting CO_2 levels may still reduce the CIP by about 0.5-1 pH unit. The removal of CO_2 can be enhanced by keeping the pH constant during the process. From a 10^{-3} M $NaHCO_3$ solution (pH 8.1) all CO_2 can be removed in this manner in approximately 24 hours (Figure 9). Figure 7 indicates that

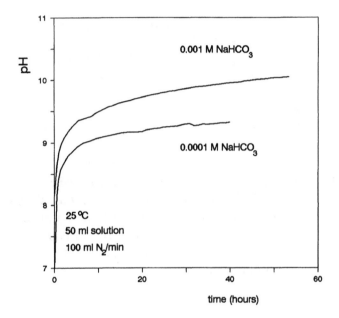

FIGURE 8. *Experimentally Determined pH Development During the Removal of CO_2 from a 10^{-4} and 10^{-3} M $NaHCO_3$ (in 0.05 M $NaNO_3$) Solution by Purging It with Moist Purified N_2.*

even a very small amount of carbon dioxide will change the CIP strongly. It implies also that all solutions added to purged suspensions should be free of CO_2, which is quite difficult for NaOH solutions.

The shift of the CIP of goethite from 10.1 to lower values is mainly due to the titration of free carbonate species in solution. The adsorption of CO_2 as such hardly influences the surface charge. Calculations show that the electric potential in the surface layer and the second adsorption layer is hardly influenced by the adsorption. It

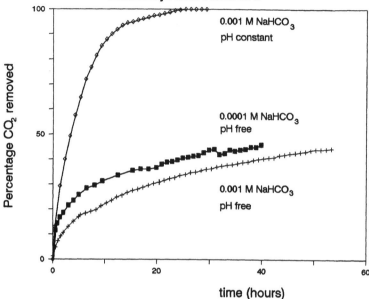

FIGURE 9. *The Removal of CO_2 from a N_2 Purged 50ml 10^{-4} and 10^{-3} M $NaHCO_3$ (in 0.05 M $NaNO_3$) Solution Without pH Control (pH Free).* Calculated from the pH change given in Figure 8 and from 50 ml of 10^{-3} M $NaHCO_3$ (in 0.05 M $NaNO_3$) solution in which the pH was kept constant at 8.15 by titrating acid.

implies that the isoelectric point (IEP) of the goethite is not changed upon adsorption of moderate amounts of carbonic acid. This prediction is in accord with the experiments of Zeltner and Anderson [41] who found a CO_2 independent IEP whereas their CIP in the titration curves was strongly CO_2 dependent. It can be concluded that the pzc of metal oxides with a high pzc like goethite can only be accurately established with very careful experimental work.

In several cases the reported values of the pzc of goethites are below pH 8.0. This cannot be completely understood from the above given analysis of the influence of CO_2 on titration curves. Other factors which may influence the CIP are the presence of silica [44] or other impurities at the surface [45] and also a variation in the relative presence of crystal planes or surface composition.

7.   ADSORPTION MODELS FOR SOIL ORGANIC MATTER

Humified organic matter, e.g. humic acids, fulvic acids, and humin, in the natural environment may be a very important sorbent for metals in soils, sediments and aquifers.   These materials are characterized by a pH dependent negative surface charge and surface potential. The surface charge is caused by dissociation of various types of functional groups. There is a wide spread in intrinsic proton affinity constants for the different functional groups, ranging from around log $K = 2$ to log $K = 11$. The interaction of ions with humified organic matter is influenced by the chemical heterogeneity and by the variable surface potential.   The variable charge character of these materials shows up clearly in the salt level dependency of the charging curves. The charging curves are equivalent with plots of the degree of proton dissociation as a function of pH. For a proper physical chemical ion adsorption model, it is necessary to account separately for the effect of the chemical heterogeneity and the effect of the variable surface potential on ion adsorption. For the assessment of the surface potential a double layer model is required. It is thus necessary to establish an appropriate double layer model and assess the chemical heterogeneity before one can construct a realistic adsorption model. All models that have been put forward so far make several *a priori* assumptions both with respect to the electrostatic part and with respect to the treatment of the chemical heterogeneity. Most often the electrostatic part is not even dealt with explicitly, i.e., that the electrostatic effects are incorporated in the chemical heterogeneity. Recently, De Wit *et al.* [46,47] have proposed a procedure that allows for the assessment of the double layer model and its parameters independently from the estimation of the intrinsic chemical affinity distribution. The principle of the method is that it is possible to filter the electrostatic effect out of the data by plotting the charging curve not as a function of pH but as a function of pH_s. It follows from Equation (3) that H_s is equal to the proton concentration in the bulk of the solution times a Boltzmann factor. The Boltzmann factor can be calculated using a double layer model. When a proper double layer model is used, the charging curves at different ionic strength values should all coincide into one curve when plotted as a function of pH_s. We have called this curve the master curve. From this master curve the intrinsic chemical heterogeneity, or the intrinsic affinity distribution, can be estimated using special techniques [48-51]. A serious problem in the affinity analysis is the treatment of experimental error in the data. Advanced methods are necessary to obtain sensible results [52]. It should be realized that information is lost when experimental error is present. When the data handling is inappropriate many spurious peaks will appear in the affinity spectrum [48,52], whereas, with a proper data treatment, the fine structure of the affinity spectrum cannot be recovered due to the uncertainty in the data. Applying the master curve concept to charging curves of humic and fulvic acids [47] shows that a spherical double layer model is in some cases appropriate, whereas in other cases, a simple Gouy Chapman double layer model suffices. For small spherical entities the salt effect on the charging curves is less than for larger entities. If the radius is larger than ca. 10 nm the spherical model becomes equivalent to the Gouy Chapman model.

8.   APPLICATION TO NATURAL SYSTEMS

Both the model development and the collection of experimental data are, at present, not far enough advanced to allow for a direct estimation of the behavior of metals in natural systems based on physical chemical models only. The knowledge with respect to adsorption phenomena on metal (hydr)oxides is far more advanced than the same for humic materials.

The variability in the extent of sorption for different soils, sediments, or aquifer materials originates from differences in solid phase characteristics and from differences in parameters like pH and redox potential. The physical chemical models allow the construction of a framework that allows for the "normalization" of the wide variability that is observed in nature. This is a very important aspect in the development of quality criteria for soil and sediment [53,54]. The possibilities for transfer of knowledge from relatively pure model systems to natural systems will be illustrated using orthophosphate as an example. Phosphate has the advantage that there is no direct interaction between this ion and humic materials. In many natural systems the iron and aluminum (hydr)oxides are the dominant reactive surfaces for phosphate sorption. Another big difference between metal ion adsorption and phosphate adsorption is that the effect of pH on phosphate adsorption is relatively small.

From adsorption studies with pure crystalline metal (hydr)oxides, e.g. goethite, it is observed that the adsorption isotherm is very flat above a few tenths of a millimole phosphate [55]. With gibbsite as sorbent it is found that the sorption reaction may proceed far beyond the adsorption maximum [56]. This was interpreted as the formation of coatings of metal phosphate at the expense of the underlying metal hydroxide. Based on these ideas, a kinetic diffusion-precipitation model was developed that is applicable for an arbitrary mixture of reactive particles that may differ in size, shape and chemical composition [57,58]. The total phosphate sorption in soil is thus a combination of reversible adsorption and diffusion-precipitation. The suggestion that phosphate sorption in soil may occur through the formation of metal phosphate coatings on metal (hydr)oxides has recently been confirmed for a wide range of soils using surface analysis techniques [59]. For both processes the amount of iron plus aluminum extractable with ammonium oxalate has been used to estimate the (ad)sorption capacity [60,61]. The reversible part of the total sorption in a given soil sample can be estimated using a newly developed desorption technique [62]. It is interesting to note that the shape of the reversible adsorption isotherms for the soils considered is quite similar to the shape of the phosphate adsorption isotherm for metal (hydr)oxides as discussed above. In contrast, the shape of the total sorption versus the phosphate concentration shows a considerable slope above a few tenths of a millimole. Transport of phosphate through soil-filled columns could be satisfactorily predicted [63] using a reversible adsorption model in combination with the diffusion-precipitation model with independently estimated model parameters.

The example discussed above illustrates that the physical chemical frame-work that originated from work with model systems was extremely powerful as a basis for the development of models that describe the behavior of phosphate in soils. It also illustrates that, apart from reversible adsorption on reactive surfaces, other processes may be of importance in determining the fate of chemicals in the environment.

9.   SUMMARY

Adsorption of ions in natural systems is quite complex and at present not completely understood. The complexity is due to the fact that natural systems are mixtures of various reactive surfaces. This in itself leads to chemical heterogeneity that affects ion adsorption. An extra complicating effect is that the sorbent components present in nature, like metal oxides or natural organic matter are heterogenous in itself.

The combination of various types of heterogeneity, with variable charge characteristics and competition between various ions for the same sorption sites are the main ingredients for the complexity of the situation. Some mechanistic ion adsorption models for metal(hydr)oxides and natural organic matter are discussed. It is shown that carbonate contamination may easily lead to a too low value of the experimentally determined pzc of metal(hydr)oxides with a high pzc. It is shown why this error may still be significant even when the system is flushed with nitrogen gas. In conclusion, it is discussed how insights derived from fundamental studies can be applied to natural systems.

REFERENCES

1.   Chardon, W. *Mobility of Cadmium in Soils.* (In Dutch) Serie Bodembescherming No. 36, (Staatsuitgeverij,'s-Gravenhage, 1984) 200 pp.

2.   Schmitt, H.W. and Sticher, H. "Long term trend analysis of heavy metal content and translocation in soils," *Geoderma* 38:195-207 (1986).

3.   Chao, L.L. "Selective dissolution of manganese oxides from soils and sediments with acidified hydroxylamine hydrochloride," *Soil Sci. Soc. Am. Proc.* 36:764-768 (1972).

4.   Förstner, U., W. Calmano, K. Conradt, H. Jaksch, C. Schimkus, and J. Schoer. "Chemical speciation of heavy metals in solid waste materials (sewage sludge, mining wastes, dredged materials, polluted sediments) by sequential extraction," in, *Proceedings of the Conference Heavy Metals in the Environment* (Amsterdam, 1979), pp. 698-704.

5.   Salomons, W. and U. Förstner. *Metals in the Hydrocycle.* (Berlin:Springer Verlag,1984), 344 pp.

6.   Tessier, A., P.G.C. Campbell, and M. Bisson. "Sequential Extraction Procedure for the Speciation of Particulate Trace Metals," *Anal. Chem.* 51:844-851 (1979).

7.   Unger, M.T. and H.E. Allen. "Distribution model of metal binding to natural sediments," in *Heavy Metals in the Hydrological Cycle*, M. Astruc and J.N. Lester, Eds. (London:Selper, Ltd., 1988) pp. 481-488.

8.   Gruebel, K.A., J.A. Davis, and J.O. Leckie. "The feasibility of using sequential extraction techniques for arsenic and selenium in soils and sediments," *Soil Sci. Soc. Am. J.* 52:390-397 (1988).

9.   McLaren, R.G., D.M. Lawson, and R.S. Swift. "Sorption and desorption of cobalt by soils and soil components," *J. Soil Sci.* 37:413-426 (1986).

10. Gerritse, R.G., R. Vriesema, J.W. Dalenburg, and H.P. De Roos. "Effect of sewage sludge on trace element mobility in soils," *J. Environ. Qual.* 11:359-364 (1982).

11.   Christensen, T.H.   "Cadmium soil sorption at low concentrations: VI. A model for zinc competition," *Water, Air Soil Pollut.* 34:305-314 (1987).

12.   Van der Zee, S.E.A.T.M., L.G.J. Fokkink and W.H. Van Riemsdijk. "A new technique for assessment of reversibly adsorbed phosphate," *Soil Sci. Soc. Am. J.* 51:599-604 (1987).

13.   Kummert, R., and W. Stumm. "Surface complexation of organic acids on hydrous γ-Al$_2$O$_3$," *J. Colloid Interface Sci.* 75:373-385 (1980).

14.   Schindler, P.W., B. Fürst, R. Dick, and P.U. Wolf. "Ligand properties of surface silanol groups. I. Surface complex formation with Fe^{3+}, Cu^{2+}, Pd^{2+} and Cd^{2+}," *J. Colloid Interface Sci.* 55:469-475 (1976).

15.   Davis, J.A., R.O. James, and J.O. Leckie. "Surface ionization and complexation at the oxide/water interface. I. Computation of electrical double layer properties in simple electrolytes," *J. Colloid Interface Sci.* 63:480-499 (1978).

16.   Westall, J. and H. Hohl. "A comparison of electrostatic models for the oxide/solution interface," *Adv. Colloid Interface Sci.* 12:265-294 (1980).

17.   Sigg, L.M. and W. Stumm. "The interaction of anions and weak acids with hydrous goethite (α-FeOOH) surface," *Colloids and Surfaces* 2:101-117 (1981).

18.   Zachara, J.M., D.C. Girvin, L. Schmidt and C.T. Resch. "Chromate adsorption on amorphous iron oxyhdroxide in the presence of major groundwater ions," *Environ. Sci. Technol.* 21:589-594 (1987).

19. Healy, T.W. and C.R. White. "Ionizable surface group models of aqueous interfaces," *Adv. Colloid Interface Sci.* 9:309-345 (1978).

20. Hiemstra, T., W.H. Van Riemsdijk and M.G.M. Bruggenwert. "Proton adsorption mechanism at the gibbsite and aluminum oxide solid/solution interface," *Neth. J. Agric. Sci.* 35:281-294 (1987).

21. Hiemstra, T., W.H. Van Riemsdijk and G.H. Bolt. "Multisite proton adsorption modelling at the solid/solution interface of (hydr)oxides: A new approach. I. Model description and evaluation of intrinsic reaction constants," *J. Colloid Interface Sci.* 133:91-104 (1989).

22. Hiemstra T., J.C.M. De Wit and W.H. Van Riemsdijk. "Multisite proton adsorption modelling at the solid/solution interface of (hydr)oxides: A new approach. II. Application to various important (hydr)oxides," *J. Colloid Interface Sci.* 133:105-117 (1989).

23. Van Riemsdijk, W.H., G.H. Bolt, L.K. Koopal, and J. Blaakmeer. "Electrolyte adsorption on heterogeneous surfaces: Adsorption models," *J. Colloid Interface Sci.* 109:219-228 (1986).

24. Van Riemsdijk, W.H., J.C.M. De Wit, L.K. Koopal, and G.H. Bolt. "Metal ion adsorption on heterogeneous surfaces. Adsorption models," *J. Colloid Interface Sci.* 116:511-522 (1987).

25. Van Riemsdijk, W.H., L.K. Koopal, and J.C.M. De Wit. "Heterogeneity and electrolyte adsorption: intrinsic and electrostatic effects," *Neth. J. Agric. Sci.* 35:241-257 (1987).

32 Riemsdijk and Hiemstra

26. Benjamin, M.M. and J.O. Leckie. "Multiple-site adsorption of Cd, Cu, Zn, and Pb on amorphous iron oxyhydroxide," *J. Colloid Interface Sci.* 79:209-221 (1981).

27. Van Riemsdijk, W.H. "Electrochemical model for adsorption of ions at the Al(OH)$_3$-electrolyte interface," Internal Report. Dept. Soil Science and Plant Nutrition, Wageningen (1979) pp. 1-25.

28. Jones, P. and J.A. Hockey. "Infra-red studies of rutile surfaces. Part 2. Hydroxylation, hydration and structure of rutile surfaces," *Trans. Faraday Soc.* 67:2679-2685 (1971).

29. Lewis, D.G. and V.C. Farmer. "Infrared adsorption of surface hydroxyl groups and lattice vibrations an lepidocrocite (γ-FeOOH) and boehmite (γ-AlOOH)," *Clay Minerals* 21:93-100 (1986).

30. Parfitt, L.R., R.J. Atkinson, and R.St.C. Smart. "The mechanism of phosphate fixation by iron oxides," *Soil Sci. Soc. Amer. Proc.* 39:837-841 (1975).

31. Pauling, L. "The principles determining the structure of complex ionic crystals," *J. Amer. Chem. Soc.* 51:1010-1026 (1929).

32. Hiemstra, T. and W.H. Van Riemsdijk. "Multi-activated-complex dissolution of metal (hydr)oxides: a thermodynamic approach applied to quartz," *J. Colloid Interface Sci.* 136:132-150 (1990).

33. Cornell, R.M., A.M. Posner, and J.P. Quirk. "Crystal morphology and the dissolution of goethite," *J. Inorg. Nucl. Chem.* 36:1937-1946 (1974).

34. Atkinson, R.J., A.M. Posner, and J.P. Quirk. "Adsorption of potential determining ions at the ferric oxide-aqueous electrolyte interface," *J. Phys. Chem.* 71:550-558 (1967).

35. Yates, D.E. "The structure of the oxide/aqueous electrolyte interface," Ph.D. Thesis, University of Melbourne, Melbourne (1975).

36. Bolt, G.H. and W.H. Van Riemsdijk. "Surface chemical processes in soil," in *Aquatic Surface Chemistry*, W. Stumm, Ed. (New York: John Wiley & Sons, Inc., 1987), pp. 127-164.

37. Van Riemsdijk, W.H., L.K. Koopal, and J.C.M. De Wit. "Heterogeneity and electrolyte adsorption: Intrinsic and electrostatic effects," *Neth. J. Agric. Sci.* 35:281-294 (1987).

38. Koopal, L.K. and W.H. Van Riemsdijk. "Electrosorption on random and patchwise heterogeneous surfaces. Electrical double layer effects," *J. Colloid Interface Sci.* 128:188-200 (1989).

39. Evans, T.D., J.R. Leal, and P.W. Arnold. "The interfacial electrochemistry of goethite (α-FeOOH), especially the effect of CO_2 contamination," *J. Electroanal. Chem.* 105:161-167 (1979).

40. Bloesch, P.M., L.C. Bell, and J.D. Hughes. "Adsorption and desorption of boron by goethite," *Aust. J. Soil Res.* 25:377-390 (1987).

41. Zeltner, W.A. and M.A. Anderson. "Surface charge development at the goethite/aqueous solution interface: Effects of CO_2 adsorption," *Langmuir* 4:469-474 (1988).

42. Rochester, C.H. and S.A. Topsham. "Infrared studies of the adsorption of probe molecules onto the surface of goethite," *J. Chem. Soc. Faraday Trans. I* 75:872-882 (1979).

43.   Russell, J.D., E. Paterson, A.R. Fraser, and V.C. Farmer. "Adsorption of carbon dioxide on goethite (α-FeOOH) surfaces and its implication for the anion adsorption," *J. Chem. Soc. Faraday Trans.* I 71: 1623-1630 (1975).

44.   Hingston, F.J., A.M. Posner, and J.P. Quirk. "Anion adsorption by goethite and gibbsite. I. The role of protons in determining adsorption envelopes," *J. Soil Sci.* 23:177-191 (1972).

45.   Penners, N.H.G. "The preparation and stability of homodisperse colloidal hematite", Ph.D. thesis, Wageningen Agricultural University, The Netherlands (1985).

46.   De Wit, J.C.M., W.H. Van Riemsdijk, and L.K. Koopal. "Heavy metal ion binding on homogeneous and heterogeneous colloids," in *Heavy Metals in the Hydrological Cycle*, M. Astruc and J.N. Lester, Eds. (London: Selper, Ltd., 1988) pp. 369-376.

47.   De Wit, J.C.M., W.H. Van Riemsdijk, and L.K. Koopal. "Proton and metal ion binding on humic substances," in *Metals Speciation, Separation and Recovery*, Vol.II, J. W. Patterson and R. Passino, Eds. (Chelsea, MI: Lewis Publishers, Inc., 1990) pp. 329-353.

48.   Nederlof, M.M., W.H. Van Riemsdijk, and L.K. Koopal. "Methods to determine affinity distributions for metal ion binding in heterogeneous systems," in *Heavy Metals in the Hydrological Cycle*, M. Astruc and J.N. Lester, Eds. (London: Selper, Ltd., 1988) pp. 361-368.

49.   Nederlof, M.M., W.H. Van Riemsdijk, and L.K. Koopal. "Determination of adsorption affinity distributions: A comparison of various methods related to local isotherm approximations," *J. Colloid Interface Sci.* 135:410-426 (1990).

50.	Sacher, R.S. and I.D. Morrison. "An improved CAEDMON program for the adsorption isotherms of heterogeneous substrates," *J. Colloid Interface Sci.* 70:153-166 (1979).

51.	Vos, C.H. and L.K. Koopal. "Surface heterogeneity analysis by gas adsorption: Improved calculation of the adsorption energy distribution using an algorithm named CAESAR," *J. Colloid Interface Sci.* 105:183-196 (1985).

52.	Nederlof, M.M., W.H. Van Riemsdijk, and L.K. Koopal. "Analysis of the binding heterogeneity of natural ligands using adsorption data," in *Heavy Metals in the Environment*, J. P. Vernet, Ed. (Amsterdam, Elsevier, 1989) pp. 365-396.

53.	Jenne, E.A., D.M. DiToro, H.E. Allen, and C.S. Zarba. "An activity-based model for developing sediment criteria for metals. Part I. A new approach," in *Proceedings of the International Conference Chemicals in the Environment*, J.N. Lester, R. Perry, and R.M. Sterrit, Eds. (London: Selper, Ltd, 1986), pp. 560-568.

54.	Shea, D. "Developing national sediment quality criteria, equilibrium partitioning of contaminants as a means of evaluating sediment criteria," *Environ. Sci. Technol.* 22:1256-1261 (1988).

55.	Bowden, J.W., S. Nagarajah, N.J. Barrow, A.M. Posner, and J.P. Quirk. "Describing the adsorption of phosphate, citrate, and selenite on a variable charge mineral surface," *Austr. J. Soil Res.* 18:49-60 (1980).

56.	Van Riemsdijk, W.H. and J. Lyklema. "Reaction of phosphate with gibbsite $(Al(OH)_3)$ beyond the adsorption maximum," *J. Colloid Interface Sci.* 76:55-66 (1980).

57. Van Riemsdijk, W.H., L.J.M. Boumans, and F.A.M. De Haan. "Phosphate sorption by soils. I. A diffusion-precipitation model for the reaction of phosphate with metal-oxides in soil," *Soil Sci. Soc. Am. J.* 48:537-540 (1984).

58. Van Riemsdijk, W.H., A.M.A. Van der Linden, and L.J.M. Boumans. "Phosphate sorption by soils. III. The diffusion-precipitation model tested for three acid sandy soils," *Soil Sci. Soc. Am. J.* 48:545-548 (1984).

59. Logan, T. Personal communication (1989).

60. Van der Zee, S.E.A.T.M., M.M. Nederlof, W.H. Van Riemsdijk, and F.A.M. De Haan (1988). "Spatial variability of phosphate adsorption parameters," *J. Environ. Qual.* 17:682-688 (1988).

61. Van der Zee, S.E.A.T.M. and W.H. Van Riemsdijk. "Model for long-term phosphate reaction kinetics in soil," *J. Environ. Qual.* 17:35-41 (1988).

62. Van der Zee, S.E.A.T.M., L.G.J. Fokkink, and W.H. Van Riemsdijk. "A new technique for assessment of reversibly adsorbed phosphate," *Soil Sci. Soc. Am. J.* 51:599-604 (1987).

63. Van der Zee, S.E.A.T.M. and W.H. Van Riemsdijk. "Sorption kinetics and transport of phosphate in sandy soil," *Geoderma* 38:293-309 (1986).

2

METAL CATION/ANION ADSORPTION ON CALCIUM CARBONATE:
Implications to Metal Ion Concentrations in Groundwater

John M. Zachara, Christina E. Cowan and Charles T. Resch
Environmental Sciences Department
Pacific Northwest Laboratory
Richland, WA 99352

1.   INTRODUCTION

Calcite ($CaCO_3(s)$) is a common mineral phase in groundwater zones. Calcite is a matrix constituent of limestone aquifers such as the Floridian Aquifer [1] and is a secondary phase found in large regional aquifers, such as the Fox Hills-Basal Hell Creek Aquifer in the Dakotas [2] the Snake River Aquifer in Idaho and Oregon [3] and the Columbia Plateau Aquifer in Washington State [4]. Calcite is also a common constituent of small near-surface aquifers that often receive metallic and organic pollutants. [5-9]. Calcite functions as a buffer for groundwater pH and influences groundwater composition via its precipitation/dissolution behavior [1,7,10,11].

Metal concentrations in calcareous groundwaters may be controlled by solubility or sorption processes (see, for example, [6,7,12,13]. Groundwaters in contact with calcite, typically in near equilibrium with this solid, are elevated in pH (pH 6.5 to 10.0 depending on the $CO_2(g)$ partial pressure), and contain significant concentrations of bicarbonate and carbonate ions [1,10]. The

0-87371-277-3/93/$0.00 + $.50
©1993 by Lewis Publishers

concentrations of metal ions in these waters may be controlled by solubility equilibria with metal carbonate, hydroxy-carbonate, and hydroxide solid phases, because many of these solids exhibit low solubility and rapid precipitation-dissolution kinetics. Examples of such solid phases include otavite, $CdCO_3$ [6,12] and hydrozincite, $Zn_5(CO_3)_2(OH)_6$ [14]. When the total metal concentration is insufficient to form a discrete metal-ligand solid phase, or when the aqueous metal concentrations are below those maintained by these discrete solids, sorption to aquifer solids, including calcite, may regulate metal concentrations. Such conditions often exist in calcareous groundwaters that are downgradient or far-field from metal-containing waste sites.

Although calcite exhibits appreciable surface area in limestone and calcareous sediments [15] its importance as a sorbent for metal ions in groundwater is not well recognized nor established through experiment. In this chapter, we review what is known about metal sorption reactions on calcite, develop and evaluate a generalized sorption model for metal cations and anions on calcite using recently measured sorption data, and perform calculations to evaluate the importance of calcite in hypothetical mixed sorbent materials containing calcite, smectite (a 2:1 clay), and amorphous iron oxide.

2.   METAL SORPTION ON CALCITE

2.1   Surface Properties of Calcite

The surface of calcite develops charge in response to the surface excess of the potential determining ions, Ca^{2+} and CO_3^{2-} [16,17,18]. The point-of-zero charge (pzc) of calcite has been reported to occur at pCa = 4.4 [17] with the surface exhibiting positive charge at Ca concentrations above this value and negative charge at Ca concentrations below this value. Calcite, therefore, carries predominantly positive charge below pH 9.0 in saturated calcium carbonate solutions in contact with atmospheric $CO_2(g)$ (pCO_2 = 3.5, where the CO_2 pressure is in atm.). Some researchers have speculated that HCO_3^-, OH^-, $CaOH^+$, and $CaHCO_3^+$ exist as surface species on calcite and influence surface charge [16,19,20]. But recent studies, using a streaming potential method, suggest that Ca^{2+} and CO_3^{2-} alone are the dominant surface species and that other solution species, including H^+ and OH^-, have no significant effect on the surface charge [18].

2.2 Sorption Behavior of Metallic Cations

A wide range of metallic cations have been shown to be sorbed onto calcite (Table 1). Sorption of metallic cations, Me^{2+}, on calcite (Figure 1) increases with increasing pH, paralleling a decrease in the aqueous concentration of calcite (Figure 2). The initial sorption of metallic cations on calcite appears, therefore, to occur by exchange with surface-associated Ca [21,22] as depicted by the following reaction:

$$X-Ca + Me^{2+} \rightleftharpoons X-Me + Ca^{2+} \qquad (1)$$

and shown conceptually in Figure 2. X is considered to be a cation-specific surface site quantifiable by isotopic exchange with ^{45}Ca. Metal cation adsorption in equilibrium $CaCO_3(aq)$ suspensions is influenced by: 1) pH and pCO_2, which control aqueous Ca concentrations (Figure 2, [23]) and 2) calcite surface area, which determines the concentration of cation-specific surface sites (X, Figure 2). Low aqueous concentrations of Ca promote Me surface exchange in accordance with the mass action expression of Equation (1).

While much of the data in Table 1 support the occurrence of a surface-exchange reaction (Equation (1)) for metallic cations, the molecular aspects of this surface reaction remain unresolved. For example, it is not known if the ion exchange occurs between dehydrated lattice ions or within a hydrated layer on the calcite surface with properties transitional between those of the crystalline lattice and the aqueous phase [22,24]. The adsorption data of Figure 1 indicate that metallic cations adsorb to calcite at pH and pCa levels where the surface carries positive charge, indicating that the $X-Me^{2+}$ surface complex is not simply electrostatic in nature, but is stabilized through chemical interaction [25].

TABLE 1.   *Metal Cation and Anion Sorption on Calcite.*

Sorbate Cations	Reference
Ba^{2+}	[74, 28]
Cd^{2+}	[27, 24, 66, 28]
Co^{2+}	[21, 28]
Cu^{2+}	[75, 76]
Mg^{2+}	[32, 31, 77, 78]
Mn^{2+}	[79, 80, 28]
Ni^{2+}	[28]
$Np(V)O_2^+$	[81]
$Pu(V)O_2^+$	[82]
Pu^{2+}	[83]
Sr^{2+}	[28]
Zn^{2+}	[84, 75, 22, 26]
Sorbate Anions	
PO_4^{3-}	[85, 86, 30]
SeO_3^{2-}	[64, 29]

Following the surface-exchange reaction, which is usually completed within time periods of seconds to hours [22,24] other phenomena specific to the metal cation sorbate and to the particular specimen of calcite have been observed to occur over time periods of hours to days. These latter phenomena involve changes in the chemical bonding of the Me^{2+} ion on the calcite surface, which influence the lability and reversibility of the surface complex (X-Me). Calcites with unstable particle energies and rapid recrystallization kinetics have been observed to trap the sorbed ions in a recrystallized surface phase or solid-solution [14,24,26] limiting desorption and isotopic exchange [22]. In the absence of significant recrystallization, Cd and, possibly, Mn surface complexes appear to dehydrate forming a discrete Me^{2+} surface precipitate (e.g., $MeCO_3$) or Me^{2+} solid solution with $CaCO_3$ (e.g., $Me_xCa_{1-x}CO_3$) [24,27,28]. Other ions that do not precipitate readily as $MeCO_3$ solids, including Zn, Co, and Ni, appear to remain as hydrated species on the calcite surface and are desorbable in the absence of recrystallization [22,28].

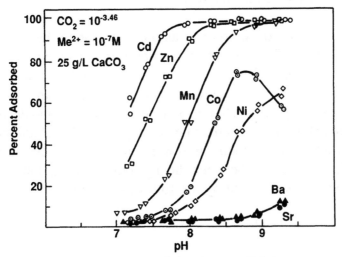

FIGURE 1.   *Metal Cation Adsorption on Calcite.*

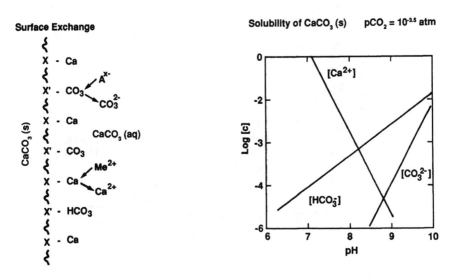

FIGURE 2.   *Conceptual Model of Solution Sorption on Calcite.*

2.3 Sorption of Inorganic Anions

The sorption of inorganic anions on calcite has not been extensively studied (Table 1). The sorption trend of the anion, SeO_3^{2-}, on calcite (Figure 3, [29]) is inverse to that observed for cations (Figure 1), with the greatest sorption observed at lower pH. Anion sorption parallels increasing positive charge on the calcite sorbent and decreasing solution concentrations of carbonate ions (Figure 2). Analogous to metal cation adsorption, anion sorption is, therefore, consistent with a surface-exchange reaction between the aqueous anion (A^{2-}) and surface carbonate species as depicted by the following reaction:

$$X'-CO_3 + A^{2-} \rightleftharpoons X'-A + CO_3^{2-} \qquad (2)$$

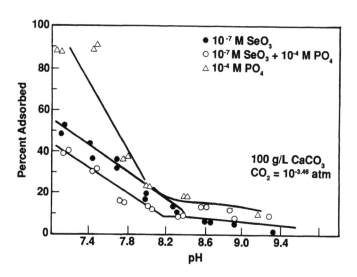

FIGURE 3. *Selenite (10^7 M) Adsorption on Calcite (100 g/L).*

and shown conceptually in Figure 2. X' is an anion-specific surface site quantifiable by isotopic exchange measurements with $H^{14}CO_3^-$. Like metal cations, sorbed anions can also be incorporated into recrystallized surface layers on calcite [30]. Observations with phosphate, however, indicate that only a small fraction of the surface-associated anion that is associated with active crystal growth sites can be incorporated into the dehydrated crystalline structure [30].

2.4   Surface Exchange on Calcite and Other Sorbents

The surface-exchange reactions postulated for metal cations and anions on calcite (Equations (1) and (2), Figure 2) are consistent with many experimental observations made using calcite and other salt-type solid phases. Auger analyses and isotopic exchange measurements have shown that the outer several atomic layers on the surface of calcite are in exchange equilibrium with the electrolyte solution with which it is in contact [31-34]. Isotopic exchange measurements using lattice constituents are commonly made to estimate surface area of salt-type minerals [35,36] suggesting that surface ions are exchangeable. Selective ion exchange reactions have been reported between the surfaces of sulfate and phosphate minerals and aqueous solutes in both trace and major concentrations [37-40].

3.   GENERALIZED SURFACE EXCHANGE MODEL

Although metal cation and anion sorption on calcite is consistent with a surface-exchange reaction between aqueous species and surface-associated Ca^{2+} and CO_3^{2-}, this hypothesis has not been evaluated using modeling techniques where the joint effects of solution speciation and the surface reaction could be assessed in relation to pH and Ca^{2+} and CO_3^{2-} concentrations. To further evaluate the plausibility of the exchange process, as depicted conceptually in Figure 2, metal cation and anion sorption data on calcite over a range in pH were modeled using a computer code incorporating hydrolysis, acid dissociation, and aqueous complexation reactions as well as surface-exchange half-reactions. The objective of this modeling was to describe the initial reaction involved in the sorption of solute ions on the calcite surface. No attempt was made, at this point, to include reactions that may follow surface exchange, such as surface precipitation or solid-solution formation. With the exception of House and Donaldson [30], Davis *et al.* [24], and Comans and Middelburg [41], there have been no comprehensive attempts to model solute sorption reactions on calcite.

3.1   Metal Cation-Exchange Model

The cation-exchange model is based on exchange half-reactions that describe the surface reaction of the metal cation (Me^{2+}) or Ca^{2+} with the cation-specific surface sites, X, on the calcite. The exchange half-reactions for Ca^{2+} and Me^{2+} are given as:

$$Ca^{2+} \rightleftharpoons X\text{-}Ca ; \quad K_{Ca}$$

and

$$Me^{2+} + X \rightleftharpoons X\text{-}Me ; \quad K_{Me}$$

respectively. The overall surface-exchange constant between Ca and the metal is calculated from the difference in the half-reaction constants; that is, $\log K_{ex} = \log K_{Me} - \log K_{Ca}$ in accordance with Equation (1). No attempt was made to charge balance the half-reactions because the charge characteristics of the surface-exchange sites, X, and the surface-exchange complexes (X-Ca, X-Me) were not known. The overall exchange reaction is, however, charge balanced.

The details of the method used to determine the half-reaction constants for the metal cations shown in Figure 1 are given in Zachara *et al.* [28]. The half-reaction constants were determined using the FITEQL program [42,43] and the number of cation-specific exchange sites, X^t (3.42×10^{-6} mol/g), was estimated by isotopic dilution using isotopic exchange data of ^{45}Ca on $CaCO_3(s)$ over a range in pH values [28]. The effects of pH and solution speciation of Ca and the metal cation are accounted for by including solution complexation reactions and associated constants like those shown in Table 2 for Ca and Zn. The solution composition for each of the sorption experiments was assumed equal to that measured in calcite-saturated solutions of approximately the same pH. The activity of the Ca ion in solution was fixed (based on the measured aqueous Ca concentrations), and the ionic strength was set at its approximate analytical value (0.1 M). The half-reaction constant for Ca (i.e., log K_{Ca} = 15.5) was fixed to ensure that essentially all the surface sites were occupied by Ca, or the metal cation, and to ensure that the concentration of "free" or unoccupied sites (X) was very small.

TABLE 2. *Aqueous Speciation Reactions with Associated Equilibrium Constants for Ca and Zn Experiments.*

Reaction		log K_r
Calcium:		
$H_2O \rightleftarrows H^+ + OH^-$		-14.0[a]
$Ca^{2+} + H_2O \rightleftarrows CaOH^+ + H^+$		-12.6[a]
$Ca^{2+} + CO_3^{2-} \rightleftarrows CaCO_3^\circ$	3.15[b]	
$Ca^{2+} + H^+ + CO_3^{2-} \rightleftarrows CaHCO_3^+$		11.3[b]
$H^+ + CO_3^{2-} \rightleftarrows HCO_3^-$		10.3[b]
$2H^+ + CO_3^{2-} \rightleftarrows H_2CO_3^\circ$	16.65[b]	
$Na^+ + H^+ + CO_3^{2-} \rightleftarrows NaHCO_3^\circ$		10.08[a]
$Na^+ + CO_3^{2-} \rightleftarrows NaCO_3^-$	1.26[a]	
Zinc:		
$Zn^{2+} + H_2O \rightleftarrows ZnOH^+ + H^+$		-8.96[a]
$Zn^{2+} + 2H_2O \rightleftarrows Zn(OH)_2^\circ + 2H^+$		-16.9[a]
$Zn^{2+} + 3H_2O \rightleftarrows Zn(OH)_3^- + 3H^+$		-28.4[a]
$Zn^{2+} + 4H_2O \rightleftarrows Zn(OH)_4^{2-} + 4H^+$		-41.4[a]
$Zn^{2+} + H^+ + CO_3^{2-} \rightleftarrows ZnHCO_3^+$		12.4[c]
$Zn^{2+} + CO_3^{2-} \rightleftarrows ZnCO_3^\circ$	4.76[d]	

(a): [87], (b): [88]; (c): [89]; (d): [90]

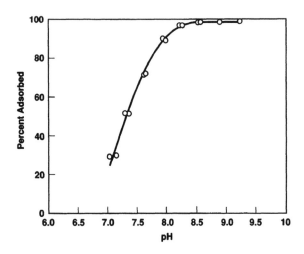

FIGURE 4.   Modeling (10^{-7} M) Adsorption on Calcite (100 g/L).

The half-reaction model using Me^{2+} as the only sorbing species (e.g., reaction of the form X-Ca + Me^{2+} ⇌ X-Me + Ca^{2+}) was able to quantitatively describe the pH-dependent sorption of all the metallic cations in Figure 1 (Ba, Sr, Cd, Mn, Co, and Zn) except Ni [28]. An example is shown in Figure 4 for 10^{-7} M Zn (pCO_2 = 3.5). The overall exchange constant for the reaction at 10^{-7} M Zn (log K_{Me} - log K_{Ca}) was 2.45, and its positive value signifies strong preference of the $CaCO_3$ surface for Zn. In contrast, quantitative description of the sorption edge data of Ni required the use of two sorbing species (Ni^{2+} and $NiOH^+$), each with a different exchange constant [28].

The robustness of the surface-exchange model was first evaluated by testing the ability of the exchange constant developed using the data in Figure 4 to predict the measured adsorption of 10^{-7} M Zn at 1) different partial pressures of $CO_2(g)$ and 2) different initial concentrations of X-Ca (e.g., $CaCO_3$ solids concentration). The exchange constant from Figure 4 was able to closely predict the measured sorption of 10^{-7} M Zn on a second calcite specimen at three different levels of pCO_2 after the different surface area of the two calcite sorbents was taken into consideration (Figures 5a,b,c).

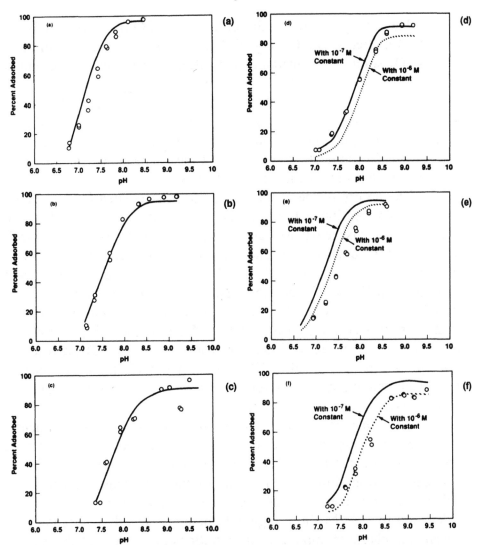

FIGURE 5. *Comparison of Adsorption on Calcite (25 g/L) in Variable P_{CO_2} (g) with Half-reaction Exchange Model Predictions.*
Lines indicate half-reaction exchange model predictions.
a) $Zn = 10^{-7}$ M, $P_{CO_2} = 10^{-2.46}$ (10^{-7} M constants); b) $Zn = 10^{-7}$ M, $P_{CO_2} = 10^{-3.46}$ (10^{-7} M constants); c) $Zn = 10^{-7}$ M, $P_{CO_2} = 10^{-4.465}$, (10^{-7} M constants); d) $Zn = 10^{-6}$ M, $P_{CO_2} = 10^{-2.46}$ (10^{-7} and 10^{-6} M constants); e) $Zn = 10^{-6}$ M, $P_{CO_2} = 10^{-3.46}$ (10^{-7} M and 10^{-6} M constants); f) $Zn = 10^{-6}$ M, $P_{CO_2} = 10^{-4.46}$ (10^{-7} and 10^{-6} M constants).

The ability to predict the influence of $CO_2(g)$ based solely on mass action effects with Ca (e.g., $CO_2(g)$ effects on Ca activity through the $CaCO_3$ solubility constraint) and differing Zn aqueous speciation indicates that Equation (1) is an accurate, first approximation of the exchange process. Electrostatic effects related to charge on the calcite surface appear to be second order. The exchange constant from Figure 4 also provided good predictions of Zn sorption at lower X-Ca (i.e., 10 g/L, Figure 6a), indicating constancy of the exchange constant over at least a limited range in Zn sorption density.

The ability of the exchange constants calculated for 10^{-7} M Zn to predict Zn sorption at 10^{-6} M initial concentration on a second calcite specimen was also evaluated as a further test of the exchange model. The 10^{-7} M Zn exchange constants accurately predicted the placement of the sorption edge at pCO_2 = 3.5 (Figure 5d), but overpredicted sorption at pCO_2 = 2.5 and 4.5 (Figures 5e,f). Zn sorption at 10^{-6} M initial concentrations in a low surface area calcite suspension was also significantly overpredicted by the 10^{-7} M Zn constant (Figure 6b). A new exchange constant of lower intensity (log K_{ex} = 2.2) fitted to the 10^{-6} M sorption edge with pCO_2 = 3.5 gave much-improved but still imperfect predictions of Zn sorption at different $pCO_2(g)$ levels and solids concentration (Figures 5e,f,g).

The discrepancies between predictions and experimental observations in Figures 5 and 6 are caused by decreasing selectivity of the $CaCO_3$ surface for Zn with surface loading and are not indicative of a departure from Equation (1). Zn sorption/exchange isotherms on calcite are nonlinear [22,28] and conditional exchange constants calculated for 10^{-7} M Zn in Figure 4, therefore, greatly overpredict the sorption isotherm at initial Zn concentrations of 10^{-6} M and above [28]. A power-exchange model requiring two constants is needed to describe Zn sorption over a range in surface loading [22]. This selectivity decrease is specific to Zn and is not observed for other metallic cations [28].

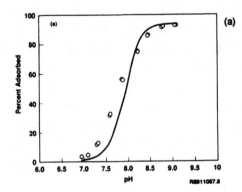
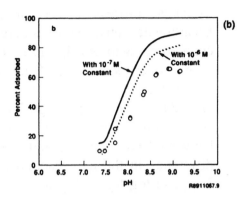

FIGURE 6. *Comparison of Zinc Adsorption on 10 g/L of Calcite with Half-reaction Exchange Model Predictions.* a) 10^{-7} M Zn with 10^{-7} M constants; b) 10^{-6} M Zn with 10^{-7} and 10^{-6} M constants.

3.2   Metal Anion Exchange Model

Anion (A^{2-}) sorption on calcite can be described using a half-reaction approach based on Equation (2) and analogous to that used for metallic cations. The reactions are:

$$CO_3^{2-} + X' \;\rightleftharpoons\; X'-CO_3 \;; \qquad K_{CO_3}$$

$$HCO_3^- + X' \;\rightleftharpoons\; X'-HCO_3 \;; \qquad K_{HCO_3}$$

and

$$A^{2-} + X' \;\rightleftharpoons\; X'-A \;; \qquad K_A$$

where X' is an anion-specific surface site. The half-reaction approach for anions differs from one for metallic cations in that two surface-saturating species (CO_3^{2-} and HCO_3^-) must be explicitly considered to successfully describe anion sorption data [29]. Thus, the surface exchange process is, at minimum, a ternary exchange reaction. Bicarbonate must be considered as a surface-reactive species and competitor with A^{2-} because it is the dominant carbonate solution species over most of the pH range of environmental interest (Figure 2). It is recognized, however, that a recent study suggests that only Ca^{2+} and CO_3^{2-} exist as exchangeable species on the surface of calcite [18]. As performed for the metallic cations, the surface-exchange constants between CO_3^{2-}, HCO_3^-, and the anion (A^{2-}) are calculated from the difference in the half-reaction constants, i.e. $\log K_{ex1} = \log K_A - \log K_{CO3}$ and $\log K_{ex2} = \log K_A - \log K_{HCO3}$.

An example of the half-reaction exchange approach described above is provided for SeO_3 sorption on calcite in the absence and presence of PO_4. The sorption data and modeling approach for this example have been presented in detail in Cowan et al. [29]. The set of half-reactions for CO_3 and SeO_3 species that gave the best approximation to the sorption data of SeO_3 as a single sorbate in Figure 3 using the chemical equilibrium program FITEQL is given in Table 3.

TABLE 3.   *Half-Exchange Reactions for Anionic Metal Exchange Model on the Calcite Surface and Fitted Constants [29].*

Reaction	$\log {}^*K^{(a)}$	$\log K^{(b)}$
$CO_3^{2-} + X' \rightleftarrows X'\text{-}CO_3$	12.5	12.5
$H^+ + CO_3^{2-} + X' \rightleftarrows X'\text{-}HCO_3$	27.06	16.76
$SeO_3^{2-} + X' \rightleftarrows X'\text{-}SeO_3$	16.70	16.70
$H^+ + SeO_3^{2-} + X' \rightleftarrows X'\text{-}HSeO_3$ 24.96	16.92	
$Ca^{2+} + H^+ + PO_4^{3-} + X' \rightleftarrows X'\text{-}CaHPO_4$		32.34
17.26		
$PO_4^{3-} + X' \rightleftarrows X'\text{-}PO_4$	20.79	20.79

(a)   The exchange constant for the half-exchange reaction in the first column with the anion exchange site, X'.

(b)   The exchange constant with the effect of the complexation of the anion removed.

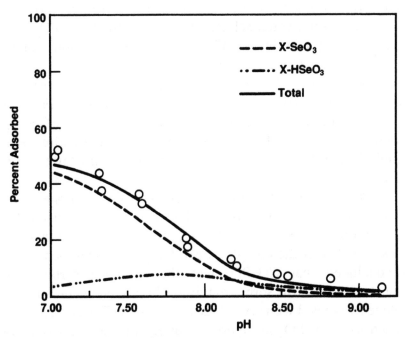

FIGURE 7. *Modeling (10^{-7} M) Selenite Adsorption on Calcite (100 g/L).*

The fit of these half-reactions to the data is shown in Figure 7. For these calculations, the total number of anion-specific surface sites, X'_t, for calcite in $CaCO_3(aq)$ solutions (2.06×10^{-6} mol/g) was estimated from isotopic exchange of $H^{14}CO_3^-$ [29]. Because it was hypothesized that SeO_3 sorption was an exchange reaction with surface-associated carbonate, the value of the CO_3^{2-} exchange constant was chosen to ensure that all the surface sites would be occupied by either CO_3^{2-} /HCO_3^- or SeO_3^{2-}/$HSeO_3^-$ and that the concentration of "free" or unoccupied sites would be very small.

Removal of the effects of CO_3 or SeO_3 ionization from the exchange constants (Table 3) indicated that the exchange of HCO_3^- was stronger than the exchange of CO_3^{2-}, and the exchange of SeO_3^{2-} was slightly stronger than the exchange of $HSeO_3^-$. The difference between the exchange constants for CO_3^{2-} and HCO_3^- was significant because assuming equal selectivity between the two surface CO_3 species resulted in inaccurate predictions of the measured sorption (not shown). However, assuming equal selectivity between the two SeO_3 constants did not result in significant differences in the predictions, suggesting that SeO_3 may exist on the surface as a single species. The HCO_3^- surface species ($X-HCO_3^-$) was calculated to predominate on the $CaCO_3(s)$ surface over the entire pH range (not shown). For both 10^{-7} and 10^{-6} M initial SeO_3 concentrations, the SeO_3^{2-} surface species accounted for most of the SeO_3 adsorption over the entire pH range, although the $HSeO_3^-$ surface species contributed at the lower pH values (Figure 7).

Like SeO_3, PO_4 sorption on calcite as a single sorbate could also be described as an exchange process between surface-associated HCO_3^- /CO_3^{2-} (Figure 8a, [29]). Unlike SeO_3, however, the dominant surface species of PO_4 was the complex $CaHPO_4$. House and Donaldson [30] also concluded that $CaHPO_4$ was the dominant sorbing species. Phosphate is known to sorb strongly on highly energetic surface sites on calcite associated with crystal growth [44-46]. The surface-exchange approach used for PO_4 in Figure 8 requires no specific assumption or hypothesis regarding the degree of hydration or bonding in the surface complexes ($X'-PO_4$, $X'-CaHPO_4$).

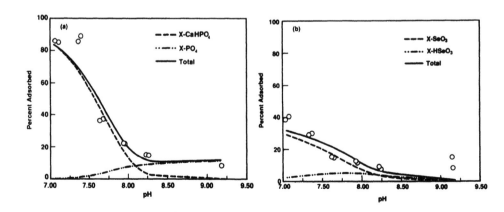

FIGURE 8.   *Modeling (10^{-7} M) Selenite Adsorption on Calcite (100 g/L) in Presence of Phosphate (10^{-4} M).* a) 10^{-4} M PO_4; b) 10^{-7} M SeO_3.

Binary solute sorption experiments with mixtures of PO_4 and SeO_3 showed that PO_4 reduces SeO_3 sorption on calcite [29]. The competitive sorption of the two anions is consistent with a mass action effect where both PO_4 and SeO_3 are binding to the same group of sites. The fitted exchange constants for PO_4, SeO_3, and CO_3 species on $CaCO_3(s)$ derived from single sorbate-sorption data (Table 3) were able to predict the resulting SeO_3 sorption in the presence of PO_4 quite well (Figure 8b). This good agreement confirmed that exchange reactions involving the three components could explain the observed decrease in SeO_3 sorption in the presence of PO_4. Similarly, SO_4 was also found to reduce SeO_3 sorption, and its effects could also be predicted with single sorbate-exchange constants [29]. The sorption of PO_4 did not result in 1:1 reduction in the sorption of SeO_3 because surface binding sites were in excess. The sorption of SeO_3 was reduced by only 1.5 X 10^{-10} mol/g, while the amount of sorbed PO_4 was 8.9 X 10^{-7} mol/g.

3.3   Exchange Versus Solid Phase Formation

The use of surface-exchange reactions to successfully describe the sorption of cations and anions on calcite does not presuppose a specific molecular nature of the surface complexes. Exchange reactions can be employed to describe replacement reactions between hydrated ions on layer silicates or between components of a solid solution [47], such as $CaCO_3(s)/CaSeO_3(s)$ or $CaCO_3(s)/ZnCO_3(s)$. As argued by Sposito [48], the fact that the ion activity product (IAP) for a solution is below the solubility product of the pure phase and that the sorption conforms to an adsorption isotherm is insufficient evidence to eliminate coprecipitation. In fact, both adsorption and co-precipitation can result in an IAP below the solubility product for the pure solid phase. The ability of two substances to form solid solutions is determined primarily by geometrical factors [49]. If two substances have similar structures and if the radii differ by no more than 15% of the smaller, then a wide range of solid solutions can occur [49]. Physically, SeO_3^{2-} and CO_3^{2-} are similar in structure and size, and thus could be readily exchanging and forming a solid solution on the surface. Even if substances differ in structure, as do PO_4^{3-} and CO_3^{2-}, solid solutions can be formed, but their composition will be more limited [49]. Thus, although the sorption of these ions can be modeled by exchange reactions, this is not evidence for the presence of hydrated surface complexes on the $CaCO_3(s)$ surface.

4. IMPORTANCE OF CALCITE AS A SORBENT IN AQUIFER MATERIALS

Natural soils and aquifer materials are composed of a mixture of mineral and organic material spanning the clay, silt, sand, and gravel size range (Ainsworth and Zachara [50] and references therein). Clay- and silt-sized materials typically dominate the sorptivity of an aquifer material for metallic anions and cations because most of the surface area of the porous media is associated with the smaller size fraction material [14,51]. Crystalline and amorphous oxides of Fe and Mn and layer silicates, such as smectites, illites, vermiculites, and kaolinites, have been consistently implicated as the clay or silt-sized material that dominates metal ion sorption in soil and aquifer sediments [48,49,52-57]. The relative importance of oxides versus layer silicates within a given aquifer material depends on many complex variables including sorbent site concentration [58], presence of strongly sorbed indigenous ions [14], pH [48], electrolyte composition and ionic strength (for example, Ca versus Na, [59-61]), as well as the crystalline nature and chemical composition of the layer silicate and oxide sorbents [51,62].

While this section has shown that specimen calcite sorbs both cations and anions, the importance of calcite as a sorbent relative to oxides and layer silicates remains in question. Most studies with natural materials containing calcite have implicated it as an important sorbent [6,63-66], while still others have suggested that calcite is not a major contributor to sorption [67-69]. To evaluate the conditions under which calcite may be an important sorbent for metallic cations, model calculations were made for Cd sorption on a hypothetical aquifer material containing amorphous iron oxide, smectite clay, and calcite. The calculation was performed using: 1) the calcite surface-exchange model discussed earlier in this chapter, 2) a non-electrostatic surface-exchange model for amorphous iron oxyhydroxide [70] and 3) an ion-exchange model for the basal plane sites on smectite [51,71] three models describe the sorption process in a conceptually similar manner using exchange half-reactions (see Table 4).

TABLE 4.   *Adsorption and Exchange Reactions with Constants for Modeling Adsorption of Cd on Aquifer Material.*

Reaction	*log K*
Calcite	
$Cd^{2+} + CaX = X-Cd + Ca^{2+}$	$-3.02^{(a)}$
$Fe_2O_3 \cdot H_2O$	
$SOH + H^+ \leftrightarrow SOH_2^+$	$4.7^{(b)}$
$SOH \leftrightarrow SO^- + H^+$	$-10.3^{(b)}$
$Cd^{2+} + SOH \leftrightarrow SO^--Cd^{2+} + H^+$	$-3.0^{(b)}$
$Ca^{2+} + SOH \leftrightarrow SO^--Ca^{2+} + H^+$	$-7.1^{(b)}$
$Na^+ + SOH \leftrightarrow SO^--Na^+ + H^+$	$-9.1^{(b)}$
Smectite	
$Cd^{2+} = CaZ_2 = CdZ_2 + Ca^{2+}$	$-0.5^{(c)}$
$2CdOH^+ + CaZ_2 = 2CdOHZ + Ca^{2+}$	$7.08^{(c)}$
$2Na^+ + CaZ_2 = 2NaZ + Cd^{2+}$	$-3.62^{(c)}$

(a) [28]
(b) [70]
(c) [51]

The sorbent site concentrations for the hypothetical aquifer material were a surface area of 2.2 m²/g of calcite, a cation exchange capacity of 1.0 meq/100g as smectite, and a NH_2OH-HCl extractable amorphous iron oxide content of 0.01% and 0.001%. Calcite was assumed to exhibit a site concentration of 8.31×10^{-6} mol/m² [33], and the entire extractable amorphous iron oxyhydroxide surface was assumed to be available for Cd complexation and to exhibit a site density as reported by [72]. The sorbent concentrations were not actual measurements on a single aquifer material, but were based on analyses of many relevant sub-surface materials presented in Holford and Mattingly [15], Borrero *et al.* [69], Davis *et al.* [24], and Loux *et al.* [57]. The aquifer material was assumed to have a bulk density of 1.5 g/cm^3 and a particle density of 2.65 g/cm^3. The test calculation is very sensitive to the assumed sorbent concentrations.

Cadmium was used for the test calculation because a consistent set of constants was available for its surface exchange on all three sorbents from high Ca electrolyte solutions where the competitive effects of Ca had been explicitly addressed. Cd is a very strongly sorbed ion on calcite [24,28], and its use in the test calculation will, therefore, demonstrate a maximum effect for the importance of calcite as a sorbent. While the surface exchange reaction for Cd on calcite given in Table 4 very accurately describes Cd sorption over a range in pH, it must be recognized that the surface-exchange complex apparently dehydrates rapidly, as Cd is poorly desorbable [28]. The equilibrium calculation presented here must, therefore, be viewed within context of this irreversibility. The model calculations were performed with an initial Cd concentration of 10^{-5} M in equilibrium $CaCO_3$(aq) solutions over the pH range of approximately 7.0 to 9.5, with ionic strength of approximately 0.1 M (Table 5).

Cadmium was calculated to be strongly sorbed in this hypothetical calcareous aquifer material (Figure 9). Under the assumed conditions of the calculation, only calcite and amorphous Fe are important sorbents. Amorphous Fe is a dominant sorbent for Cd in the simulation in spite of the competition with Ca, which has been shown to suppress Cd sorption [70]. The high Na and Ca concentrations in the equilibrium $CaCO_3$ solutions suppress Cd sorption on smectite through ion-exchange mass action; similar results have been reported experimentally [51,71]. While both amorphous Fe and calcite sorb Cd (Figure 9), calcite out-competes the amorphous Fe for Cd at higher pH where Ca concentrations are low. The pH region where calcite predominates as a sorbent is related to its total site concentration (governed by its exposed surface area) relative to the total site concentration of amorphous Fe in the aquifer material. As shown in

TABLE 5.   *Composition of CaCO₃(aq) Solutions.*

pH	Ca	Na	DIC	ClO₄	K	Mg	Si	Sr
				(mol/L)				
7.25	7.11E-2	5.87E-6	2.67E-4	1.29E-1	6.4E-5	6.58E-4	1.17E-4	4.0E-6
7.47	2.19E-2	6.35E-2	3.71E-4	9.59E-2	3.3E-5	2.01E-4	9.6E-5	1.4E-6
7.71	7.14E-3	8.57E-2	6.05E-4	9.53E-2	3.1E-5	7.0E-5	3.6E-5	4.3E-7
7.98	1.75E-3	9.44E-2	1.15E-3	8.61E-2	3.8E-5	9.9E-6	2.0E-5	1.4E-7
8.40	3.14E-4	9.34E-2	2.68E-3	8.58E-2	2.2E-5	BD	2.6E-6	6.0E-8
8.60	8.21E-5	9.66E-2	5.39E-3	8.21E-2	2.6E-5	BD	2.0E-6	BD
8.88	3.27E-5	9.92E-2	1.03E-2	7.95E-2	3.3E-5	5.8E-6	3.6E-6	6.0E-8
9.22	1.21E-5	1.06E-1	2.25E-2	7.36E-2	2.6E-5	BD	3.2E-6	6.0E-8

(a)   BD = below detection.

Figure 9a, amorphous Fe predominates as a sorbent over a greater range in pH when it is present in equal total site concentration to calcite. Calcite, in turn, is the dominant sorbent when its site concentration exceeds that of amorphous Fe (Figure 9b).

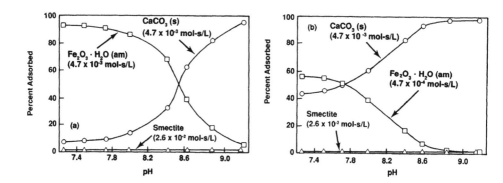

FIGURE 9.   *Calculated Adsorption of Cadmium (10^{-5} M) in Sorbent Mixtures Containing Calcite (4.7 X 10^{-3} mol-s/L) and Smectite (2.6 X 10^{-2} mol-s/L).* a) Fe_2O_3 (am) 4.7 X 10^{-3} mol-s/L; b) Fe_2O_3 (am) 4.7 X 10^{-4} mol-s/L.

The multisorbent model calculation underscores the need for accurate site concentration measurements of different sorbents in a aquifer material to understand and predict metal ion sorption. Such measurements, which are currently problematic and often at times ambiguous, are the focus of ongoing research. For example, while the total moles of amorphous Fe may be estimated by chemical extraction with NH_2OH-HCl [73], translating moles of extractable Fe into a site concentration is problematic because the effective surface area, accessibility, and surface site saturation of the amorphous Fe in the aquifer material are not known. Similarly, the site concentration of $CaCO_3$ in an aquifer material may be estimated by isotopic exchange with ^{45}Ca or $H^{14}CO_3$ [15,35], but ambiguities arise because corrections must be applied for the exchange of these same tracers on other mineral surfaces that may be present with greater surface areas (i.e., layer silicates and Fe, Mn, and Al oxides).

5.   CONCLUSIONS

Both metallic cations and anions sorb on calcite. Their sorption behavior over ranges in pH and CO_2 gas pressure is consistent with an initial surface-exchange process, where cations exchange with surface Ca and anions exchange with surface carbonate. A general surface-exchange model containing exchange half-reactions was developed to describe the effects of Ca and CO_3 concentration, pH, and calcite surface area on metal ion sorption. The model is conceptually consistent with the surface reaction approach applied to oxide and layer silicate minerals, and can be readily incorporated into geochemical models such as MINTEQ [91]. The model was able to describe the influence of pH on both anion and cation sorption, and provided good predictions of the effects of variable $CO_2(g)$ pressure on Zn sorption and of PO_4 on SeO_3 sorption. The model does not currently consider reactions such as metal dehydration, precipitation, or solid-solution formation that have been observed to follow the surface-exchange reaction for certain solutes (e.g., Cd and Mn) on calcite. These latter reactions, however, must be considered to accurately describe metal ion concentrations and migration in calcareous groundwater.

Model calculations in a ternary sorbent mixture containing calcite, amorphous Fe, and smectite clay documented that calcite can be an important sorbent in aquifer materials for metals that strongly associate with the calcite surface (i.e., Cd and Zn). Groundwater pH and the site concentration of calcite, relative to other strongly sorbing phases such as Fe and Mn oxides, are the most important factors governing whether metal sorption to calcite is significant. Clearly, sorption reactions of both cations and anions to calcite will be important and potentially dominant in limestone or dolomitic aquifers, or in groundwater systems depleted of oxides. The model calculations presented here were based on sorption constants measured for specimen mineral phases. More accurate calculations should use sorption constants of natural carbonates, oxides, and layer silicates, as their sorptivity may differ significantly from specimen solids as a result of substitutional/compositional impurities and crystallinity differences. With specific regard to the carbonates, substitutions of Mg, Fe, and Mn, which are common in nature, affect calcite surface area and lattice dimension, which could significantly influence metallic cation and anion sorption.

Measurement of site concentrations of individual sorbents, or sorbent classes, in aquifer materials is needed to accurately describe metal sorption in both calcareous and non-calcareous groundwaters. The determination of sorbent mass or molar content alone is not adequate because the relationship of these parameters to site concentration has not been established. Even if the sorbent surface area is determined, the sorption magnitude may be inadequately predicted because the surfaces of oxide and calcite particles in subsoil or aquifer material may be modified by the sorption of natural organic compounds (humic substances, lower molecular weight organic acids) or strongly binding inorganic ions, such as Si, Al, or Fe, that may block or modify surface sites. Research to improve site concentration measurements on natural sub-surface materials is a high priority.

Acknowledgments: This research was funded by the U.S. Department of Energy Sub-surface Science Program within the Office of Energy Research under Contract DE-AC06-76RLO 1830 and the Electric Power Research Institute, Inc., under Contract RP2485-03, "Chemical Attenuation Studies," with Battelle Pacific Northwest Laboratories.

REFERENCES

1.   Plummer, L.N. "Defining Reactions and Mass Transfer in Part of the Floridian Aquifer," *Water Resour. Res.* 13:801-812 (1977).

2.   Thorstenson, D.C., D.W. Fisher and M.G. Croft.   "The Geochemistry of the Fox Hills-Basal Hell Creek Aquifer in Southwestern North Dakota and Northwestern South Dakota," *Water Resour. Res.* 15:1479-1498 (1979).

3.   Wood, W.W. and W.H. Low.   "Solute Geochemistry of the Snake River Plain Regional Aquifer System, Idaho and Eastern Oregon," *U.S. Geological Survey Professional Paper* 1408-D (1988).

4.   Deutsch, W.J., E.A. Jenne and K.M. Krupka.   "Solubility Equilibria in Basalt Aquifers: The Columbia Plateau, Eastern Washington, U.S.A.," *Chem. Geol.* 36:15-34 (1982).

5.   Toran, L.   "Sulfate Contamination in Groundwater from a Carbonate-Hosted Mine," *J. Contam. Hydrol.* 2:1-29 (1987).

6.   Fuller, C.C. and J.A. Davis.   "Processes and Kinetics of Cd^{2+} Sorption by Calcareous Aquifer Sand," *Geochim. Cosmochim. Acta* 51:1491-1502 (1987).

7.   Morin, K.A., J.A. Cherry, H.K. Dave, T.P. Lin and A.J. Vivurka.   "Migration of Acidic Groundwater Seepage from Uranium-Tailings Impoundments. 2. Geochemical Behavior of Radionuclides in Groundwater," *J. Contam. Hydrol.* 2:305-322 (1988).

8.   Barker, J.F., J.E. Barbash and M. Labonte.   "Groundwater Contamination at a Landfill Site on Fractured Carbonate and Shale," *J. Contam. Hydrol.* 3:1-27 (1988).

9.   Garland, T.R., J.M. Zachara and R.E. Wildung. "A Case Study of the Effects of Oil Shale Operations on Surface and Groundwater Quality: II. Major Inorganic Ions," *J. Environ. Qual.* 17:660-666 (1988).

10. Langmuir, D. "The Geochemistry of Some Carbonate Groundwaters in Central Pennsylvania," *Geochim. Cosmochim. Acta* 35:1023-1045 (1971).

11. Butler, J.N. *Carbon Dioxide Equilibria and Their Applications.* (Reading, MA, Addison-Wesley Publishing Company, 1982).

12. Yanful, E.K., H.W. Nesbitt and R.M. Quigley. "Heavy Metal Migration at a Landfill Site, Sarnia, Ontario, Canada. 1. Thermodynamic Assessment and Chemical Interpretations," *Appl. Geochem.* 3:523-534 (1988).

13. Yanful, E.K., R.M. Quigley and H.W. Nesbitt. "Heavy Metal Migration at a Landfill Site, Sarnia, Ontario, Canada. 2. Metal Partitioning and Geotechnical Implications," *Appl. Geochem.* 3:623-631 (1988).

14. Zachara, J.M., J.A. Kittrick, L.S. Dake and J.B. Harsh. "Solubility and Surface Spectroscopy of Zinc Precipitates on Calcite," *Geochim. Cosmochim. Acta* 53:9-19 (1989).

15. Holford, I.C.R. and C.E.G. Mattingly. "Surface Area of Calcium Carbonate in Soils," *Geoderma* 13:247-255 (1975).

16. Parks, G.A. "Adsorption in the Marine Environment," in *Chemical Oceanography,* J.P. Riley and G. Skirrow, Eds. (Vol. 1, New York: Academic Press, 1975), pp. 241-308.

17. Foxall, T., G. Peterson, H.M. Rendall and A.L. Smith. "Charge Determination at the Calcium Salt/Aqueous Solution Interface," *J. Chem. Soc. Faraday Trans.* 75:1034-1039 (1979).

18. Thompson, D.W. and P.G. Pownall. "Surface Electrical Properties of Calcite," *J. Colloid Interface Sci.* 131:74-82 (1989).

19.   Somasundaran, P. and G.E. Agar. "The Zero Point of Charge of Calcite," *J. Colloid Interface Sci.* 24:433-440 (1967).

20.   Siffert, B. and P. Fimbel. "Parameters Effecting the Sign and Magnitude of the Electrokinetic Potential of Calcite," *Colloids Surf.* 11:377-389 (1984).

21.   Kornicker, W.A., J.W. Morse and R.N. Damasceno. "The Chemistry of Co^{2+} Interaction with Calcite and Aragonite Surfaces," *Chem. Geology* 53:229-236 (1985).

22.   Zachara, J.M., J.A. Kittrick and J.B. Harsh. "The Mechanism of Zn^{2+} Adsorption on Calcite," *Geochim. Cosmochim. Acta* 52:2281-2291 (1988).

23.   Zachara, J.M. "A Solution Chemistry and Electron Spectroscopic Study of Zinc Adsorption and Precipitation on Calcite," Ph.D. Dissertation, Washington State University (1988).

24.   Davis, J.A., C.C. Fuller and A.D. Cook. "A Model for Trace Metal Sorption Processes at the Calcite Surface: Adsorption of Cd^{2+} and Subsequent Solid Solution Formation," *Geochim. Cosmochim. Acta* 51:1477-1490 (1987).

25.   Lyklema, J. "Discrimination Between Physical and Chemical Adsorption of Ions on Oxides," *Colloids Surf.* 37:197-204 (1989).

26.   Lorens, R.B. "Sr, Cd, Mn, and Co Distribution Coefficients in Calcite as a Function of Calcite Precipitation Rate," *Geochim. Cosmochim. Acta* 45:553-561 (1981).

27.   McBride, M.B. "Chemisorption of Cd^{2+} on Calcite Surfaces," *Soil Sci. Soc. Amer. J.* 44:26-28 (1980).

28.   Zachara, J.M., C.E. Cowan and C.T. Resch. "Adsorption of
       Divalent Metallic Cations on Calcite," *Geochim. Cosmochim.
       Acta* 55:1549-1562 (1991)

29.   Cowan, C.E., J.M. Zachara and C.T. Resch. "Solution Ion
       Effects on the Surface Exchange of Selenite on Calcite,"
       Geochim. Cosmochim. Acta 54:2223-2234 (1991).

30.   House, W.A. and L. Donaldson.   "Adsorption and
       Coprecipitation of Phosphate on Calcite," *J. Colloid Interface
       Sci.* 112:309-234 (1986).

31.   Moller, P. and G. Werr. "Influence of Anions on Ca^{2+} - Mg^{2+}
       Surface Exchange Process on Calcite in Artificial Seawater,"
       Radiochim. Acta 18:144-147 (1972).

32.   Moller, P. "Determination of the Composition of Surface
       Layers of Calcite in Solutions Containing Mg^{2+}," *J. Inorg. Nucl.
       Chem.* 35:395-401 (1973).

33.   Moller, P. and C.S. Sastri. "Estimation of the Number of
       Surface Layers of Calcite Involved in Ca-^{45}Ca Isotopic
       Exchange with Solution," *Z. Phys. Chem.* Neue Folge 84:80-87
       (1974).

34.   Mucci, A. and J.W. Morse. "Auger Spectroscopy Determination
       of the Surface-Most Adsorbed Layer Composition on Aragonite,
       Calcite, Dolomite and Magnesite in Synthetic Seawater," *Amer.
       J. Sci.* 285:306-317 (1985).

35.   Inks, C.G. and R.B. Hahn. "Determination of Surface Area of
       Calcium Carbonate by Isotopic Exchange," *Anal. Chem.* 39:625-
       628 (1967).

36.   Kukura, M., L.C. Bell, A.M. Posner, and J.P. Quirk. "Radioisotope Determination of the Surface Concentrations of Calcium and Phosphorous on Hydroxyapatite in Aqueous Solution," *J. Phys. Chem.* 76:900-904 (1972).

37.   Lieser, K.H., P.H. Gütlich and I. Rosenbaum. "Die Tempuraturäbhangigkeit des Heterogenen Isotopenaustausches an der Oberfläche von Ionenkristallen," *Radiochim. Acta* 5:38-42 (1966).

38.   Lin, J., S. Rayhavan and D.W. Fuerstenau. "The Adsorption of Fluoride Ions by Hydroxyapatite from Aqueous Solution," *Colloids Surf.* 3:357-370 (1981).

39.   Jonasson, R.G., G.M. Bancroft and L.A. Boatner. "Surface Reactions of Synthetic, End Member Analogues of Monazite, Xenotime, and Rhabiophane, and Evolution of Natural Waters," *Geochim. Cosmochim. Acta* 52:767-770 (1988).

40.   Pate, F.D., J.T. Hutton and K Morrish. "Ionic Exchange Between Soil Solutions and Bone: Toward a Predictive Model," *Appl. Geochem.* 4:303-316 (1989).

41.   Comans, R.N.J. and J.J. Middelburg. "Sorption of Trace Metals on Calcite: Applicability of the Surface Precipitation Model," *Geochim. Cosmochim. Acta* 51:2587-2591 (1987).

42.   Westall, J. "FITEQL, a Computer Program for Determination of Equilibrium Constants from Experimental Data. Version 1.2," *Report 82-01, Department of Chemistry,* Oregon State University, Corvallis, Oregon (1982).

43.   Westall, J. "FITEQL, a Computer Program for Determination of Equilibrium Constants from Experimental Data. Version 2.0," *Report 82-02, Department of Chemistry,* Oregon State University, Corvallis, Oregon (1982).

44.   Reddy, M.M.   "Crystallization of Calcium Carbonate in the Presence of Trace Concentrations of Phosphorous Containing Anions. I. Inhibition of Phosphate and Glycerophosphate Ions at pH 8.8 and 25°C," *J. Crystal Growth* 41:287-295 (1977).

45.   Walter, L.M. and E.A. Burton. "The Effect of Orthophosphate on Carbonate Mineral Dissolution Rates in Seawater," *Chem. Geol.* 56:313-323 (1986).

46.   Busenberg, E. and L.H. Plummer.   "Thermodynamics of Magnesian Calcite Solid-Solutions at 25° and 1 atm Total Pressure," *Geochim. Cosmochim. Acta* 53:1189-1208 (1989).

47.   Sposito, G. *The Thermodynamics of Soil Solutions.* (Oxford, Oxford University Press, 1981).

48.   Sposito, G.   *The Surface Chemistry of Soils.*   (Oxford, Clarendon Press, 1984).

49.   Evans, R.C. *An Introduction to Crystal Chemistry.* 2nd Ed. (Cambridge, Cambridge University Press, 1966).

50.   Ainsworth, C.C. and J.M. Zachara. "Selection of Subsoils, Saturated Zone Materials, and Related Sorbents for Subsurface Research," DOE/ER-0390 U.S. Department of Energy, Office of Health and Environmental Research, Washington, D.C. (1988).

51.   Zachara, J.M., C. E. Cowan and S. C. Smith. "Edge Site Contributions to Cd Sorption on Specimen and Soil-derived Smectites in Na^+ and Ca^{2+} Electrolytes," *Soil Sci. Soc. Am. J.* (Submitted, 1990b).

52.   Means, J.L., D.A. Crerar, M. Borcsik and J.O. Duguid. "Adsorption of Co and Selected Actinides by Mn and Fe Oxides in Soils and Sediments," *Geochim. Cosmochim. Acta* 42:1763-1773 (1978).

53.   Bruggenwert, M.G.M. and A. Kamphorst.   "Survey of Experimental Information on Cation Exchange in Soil Systems," in *Soil Chemistry B. Physics-Chemical Models*, G. H. Bolt, Ed. (New York, Elsevier, 1979) pp. 141-203.

54.   Jackson, R.E. and K.J. Inch.   "Partitioning of Strontium-90 Among Aqueous and Mineral Species in a Contaminated Aquifer," *Environ. Sci. Technol.* 17:231-237 (1983).

55.   Stollenwerk, K.G. and D.B. Grove. "Adsorption and Desorption of Hexavalent Chromium in an Alluvial Aquifer near Telluride, Colorado," *J. Environ. Qual.* 14:150-155 (1985).

56.   Kent, D.G., V.S. Tripathi, N.B. Ball, J.O. Leckie and M.D. Siegel.   "Surface-Complexation Modeling of Radionuclide Adsorption in Subsurface Environments," NUREG/CR-4807, U.S. Nuclear Regulatory Commission, Washington, D.C. (1988).

57.   Loux, N.T., D.S. Brown, C.R. Chafin, J.D. Allison and S.M. Hassan.   "Modeling Geochemical Processes Attenuating Inorganic Contaminant Transport in the Subsurface Region: Adsorption on Amorphous Ion Oxide," *Proceedings of the First EPRI-EPA Research Conference.* U.S. Environmental Protection Agency and the Electric Power Research Institute, Palo Alto, CA (1989).

58.   Luoma, S.N. and J.A. Davis.   "Requirements for Modeling Trace Metal Partitioning in Oxidized Estuarine Sediments," *Marine Chem.* 12:159-181 (1983).

59.   Bowman, R.S., M.E. Essington and G.A. O'Connor.   "Soil Sorption of Nickel: Influence of Solution Composition," *Soil Sci. Soc. Am. J.* 46:860-865 (1981).

60.   Elrashidi, M.A. and G.A. O'Conner. "Influence of Solution Composition on Sorption of Zinc by Soils," *Soil Sci. Soc. Am. J.* 46:1153-1158 (1982).

61.   Hayes, K.F. and J.O. Leckie. "Modeling Ionic Strength Effects on Cation Adsorption at Hydrous Oxide/Solution Interfaces," *J. Colloid Interface Sci.* 115:564-572 (1987).

62.   Ainsworth, C.C., D.C. Girvin, J.M. Zachara and S.C. Smith. "Chromate Adsorption on Goethite: Effects of Aluminum Substitution," *Soil Sci. Soc. Am. J.* 53:411-418 (1989).

63.   Dudley, L.M., J.E. McLean, R.C. Sims and J.J. Jurinak. "Sorption of Copper and Cadmium from the Water Soluble Fraction of an Acid Mine Waste by Two Calcareous Soils," *Soil Sci.* 45:207-214 (1988).

64.   Goldberg S. and R.A. Glaubig. "Anion Sorption on a Calcareous, Montmorillonitic Soil-Selenium," *Soil Sci. Soc. Am. J.* 52:954-958 (1988).

65.   Goldberg S. and R.A. Glaubig. "Anion Sorption on a Calcareous, Montmorillonitic Soil-Arsenic," *Soil Sci. Soc. Am. J.* 52:1297-1300 (1988).

66.   Papadopoulos, P. and D.L. Rowell. "The Reactions of Cadmium with Calcium Carbonate Surfaces," *J. Soil Sci.* 39:23-36 (1988).

67.   Ryan, J., D. Curtin and M.A. Cheema. "Significance of Iron Oxides and Calcium Carbonate Particle Size in Phosphate Sorption by Calcareous Soils," *Soil Sci. Soc. Am. J.* 48:74-76 (1984).

68.   Neal, R.H., G. Sposito, K.M. Holtzclaw and S.J. Traina. "Selenite Adsorption on Alluvial Soils: I. Soil Composition and pH Effects," *Soil Sci. Soc. Am. J.* 51:1161-1165 (1987).

69.   Borrero, C., F. Pena and J. Torrent. "Phosphate Sorption by Calcium Carbonate in Some Soils of the Mediterranean Part of Spain," *Geoderma* 42:261-269 (1988).

70.   Cowan, C.E., J.M. Zachara and C.T. Resch. "Cadmium Adsorption on Iron Oxides in the Presence of Alkaline Earth Elements," *Environ. Sci. Tech.* 25:437-446 (1991).

71.   Zachara, J.M., C.E. Cowan and C.T. Resch. "Cadmium Exchange Sorption on Specimen and Natural Smectite in Ca and Mixed Electrolyte Solutions," *Soil Sci. Soc. Am. J.* (Submitted, 1990).

72.   Davis, J.A. and J.O. Leckie. "Surface Ionization and Complexation at the Oxide/Water Interface. II. Surface Properties of Amorphous Iron Oxyhydroxide and Adsorption of Metal Ions," *J. Colloid Interface Sci.* 67(1):90-107 (1978).

73.   Chao, T.T. and L. Zhou. "Extraction Techniques for Selective Dissolution of Iron Oxides from Soils and Sediments," *Soil Sci. Soc. Am. J.* 47:225-232 (1983).

74.   Bancroft, G.M., J.R. Brown and W.S. Fyfe. "Quantitative X-ray Photoelectron Spectroscopy (ESCA): Studies of Ba^{2+} Sorption on Calcite," *Chem. Geol.* 19:131-144 (1977).

75.   Kitano, Y., N. Kanamori and S. Yoshioka. "Adsorption of Zinc and Copper Ions on Calcite and Aragonite and Its Influence on the Transformation of Aragonite to Calcite," *Geochem. J.* 10:175-179 (1976).

76.   Franklin, M.L. and J.W. Morse. "The Interaction of Copper with the Surface of Calcite," *Ocean Sci. Eng.* 7:147-174 (1982).

77.   Mucci, A., J.W. Morse and M.S. Kaminsky. "Auger
       Spectroscopy Analysis of Magnesian Calcite Overgrowths
       Precipitated from Seawater and Solutions of Similar
       Composition," *Am. J. Sci.* 285:289-305 (1985).

78.   Sastri, C.S. and P. Moller. "Study of the Influence of Mg^{2+} Ions
       on Ca-^{45}Ca Isotopic Exchange on the Surface Layers of Calcite
       Single Crystals," *Chem. Phys. Lett.* 26:116-120 (1974).

79.   McBride, M.B. "Chemisorption and Precipitation of Mn^{2+} at
       $CaCO_3$ Surfaces," *Soil Sci. Soc. Am. J.* 43:693-698 (1979).

80.   Franklin, M.L. and J.W. Morse. "The Interaction of Manganese
       (II) with the Surface of Calcite in Dilute Solutions and
       Seawater," *Marine Chem.* 12:241-254 (1983).

81.   Keeney-Kennicutt, W.L. and J.W. Morse. "The Interaction of
       $Np(V)O^{2+}$ with Common Mineral Surfaces in Dilute Aqueous
       Solutions and Seawater," *Marine Chem.* 15:133-150 (1984).

82.   Keeney-Kennicutt, W.L. and J.W. Morse. "The Redox
       Chemistry of $Pu(V)O^{2+}$ Interaction with Common Mineral
       Surfaces in Dilute Aqueous Solutions and Seawater," *Geochim.
       Cosmochim. Acta* 49:2477-2588 (1985).

83.   Nelson, D.K., K.A. Orlandini and W.R. Penrose. "Oxidation
       States of Plutonium in Carbonate-Rich Natural Waters," *J.
       Environ. Radioactivity* 9:189-198 (1989).

84.   Jurinak, J.J. and N. Bauer. "Thermodynamics of Zinc
       Adsorption on Calcite, Dolomite, and Magnesite Type
       Minerals," *Soil Sci. Soc. Am. Proc.* 20:466-471 (1956).

85.   Cole, C.V., S.R. Olsen and C.O. Scott. "The Nature of
       Phosphate Adsorption by Calcium Carbonate," *Soil Sci. Soc.
       Amer.* 17:352-356 (1953).

86.   Dekanel, J. and J.W. Morse. "The Chemistry of Orthophosphate Uptake from Seawater onto Calcite and Aragonite," *Geochim. Cosmochim. Acta* 42:1335-1340 (1978).

87.   Truesdell, A.H. and B.F. Jones. "WATEQ, a Computer Program for Calculating Chemical Equilibria of Natural Waters," *J. Res. U.S. Geol. Survey* 2:233-248 (1974).

88.   Ball J.W., D.K. Nordstrom and E.A. Jenne. "Additional and Revised Thermodynamic Data and Computer Code for WATEQ2 -- A Computerized Chemical Model for Trace and Major Element Speciation and Mineral Equilibria of Natural Waters," WRI 78-116, U.S. Geological Survey, Menlo Park, CA (1980).

89.   Zirano, A. and S. Yamamoto. "A pH Dependent Model for the Chemical Speciation of Copper, Zinc, Cadmium, and Lead in Seawater," *Limnol. Oceanogr.* 17:661-671 (1972).

90.   Bilinski, H., R. Huston and W. Stumm. "Determination of the Stability Constants of Some Hydroxo Carbonato Complexes of Pb(II), Cu(II), Cd(II), and Zn(II) in Dilute Solutions by Anodic Stripping Voltammetry and Differential Pulse Polarography," *Anal. Chim. Acta* 84:157-164 (1976).

91.   Felmy, A.R., D.C. Girvin and E.A. Jenne. "MINTEQ, a Computer Model for Calculating Aqueous Geochemical Equilibria," NTIS-PB-84157148, National Technical Information Service, Springfield, VA (1984).

3

SUB-SURFACE REDOX CHEMISTRY: A COMPARISON OF EQUILIBRIUM AND REACTION-BASED APPROACHES

William Fish
Department of Environmental Science and Engineering
Oregon Graduate Institute of Science and Technology
Beaverton, OR 97006

1.   INTRODUCTION

Reactions involving the transfer of electrons from one reactant to another are an important category of material transformation in the environment. Oxidation, the loss of electrons, couples with reduction, the gain of electrons. The resultant changes in oxidation states generate products that are dramatically different from the reactants in their solubilities, toxicities, reactivities, and mobilities. Furthermore, changes in the oxidation states of naturally occurring elements can cause the dissolution of solid minerals or the precipitation of new mineral phases. These phase changes alter the sorptive properties of aquifer materials and thereby alter the mobilities of contaminants and other groundwater constituents [1]. An understanding of redox reactions is thus essential for evaluating the transport, fate, and hazard potential of many natural and anthropogenic materials.

Broadly speaking, there are two opposite perceptions of the redox chemistry of the sub-surface, at least among environmental scientists who are not closely involved with research on this subject.

0-87371-277-3/93/$0.00 + $.50

The optimistic view holds that redox chemistry is no different than any other environmental chemistry and that it simply entails the analysis of samples and the computation of the appropriate mathematical formulations. The pessimistic view is that it is very difficult to make any measurements of chemical parameters that are meaningful to redox chemistry and that it is essentially only possible to model a trivial redox system.

Although the pessimists are not too far off the mark, the truth is that natural redox chemistry is not an impossible subject for scientific elucidation. A careful reading of the historical and current literature in this area reveals that, even with imperfect measurements and models, certain chemical trends or patterns are clearly discernible in sub-surface environments and that these patterns are roughly predictable on the basis of simple, deterministic chemical principles. The challenges lie in improving the accuracy and resolution of these predictions so that they can serve as useful tools for determining the present and future chemical states of an aquifer.

In aerated waters, the H_2O/O_2 redox couple dominates all others due to the effectively infinite reservoir of atmospheric oxygen. Most groundwaters have relatively little contact, often none at all, with the atmosphere. Consequently, oxygen may be depleted or totally absent. The H_2O/O_2 redox couple, therefore, is often not dominant in the sub-surface and many electron transfer reactions are possible.

In an unperturbed aquifer, the presence of natural organic matter from the near-surface recharge zone can induce a thermodynamic gradient which drives a series of redox reactions. As will be demonstrated later, such gradients occur on large spatial scales because the rates of reactions can be relatively slow. Nonetheless, redox reactions are almost invariably dramatic in the chemical changes they produce, and even a weak chemical gradient can produce sharp discontinuities in the distribution of redox-active species. In general, redox reactions are an essential feature of the geochemical evolution of natural groundwaters. The dissolution of certain metal oxides and the precipitation of new phases is typically observed. The phase changes also induce secondary effects, such as alterations of the adsorptive properties of aquifer materials.

The consequences of deliberate or accidental additions of redox-active materials to the sub-surface are of special significance to regulatory agencies, such as the U.S. Environmental Protection Agency. These additions are of two types. The first type is the addition of aerobically biodegradable organic compounds such as hydrocarbons or landfill leachates that stimulate microbial respiration. Aerobic respiration strips O_2 from the system and triggers subsequent redox reactions, both biologically mediated (such as sulfate reduction) or primarily abiotic (such as the reductive dissolution of MnO_2).

In these circumstances, redox reactions have many practical implications for contaminant hydrology. Degradation of organic wastes is usually fastest under aerobic conditions. After O_2 is consumed, biodegradation must proceed via alternative pathways such as nitrate reduction (denitrification). On the other hand, halogenated compounds, which are slow to break down under aerobic conditions, degrade more readily under strongly reducing conditions via reductive dehalogenation [2]. Finally, the inorganic products of redox reactions may create new water quality problems at a contamination site. For example, reductive dissolution of natural Fe and Mn oxides or the reduction of sulfate to H_2S in an organic waste plume can produce offensive levels of Fe, Mn, and sulfides in the groundwater. These nuisance constituents may persist in the aquifer well beyond the original organic contaminant [3].

The second type of induced redox perturbation is the addition of an inorganic compound that directly engages in highly energetic reactions. An example would be the release of a strong oxidant, such as chromate or permanganate. These oxidants will drive a sequence of reactions that are not necessarily biologically mediated. As a class, these reactions have been far less studied than have microbial redox reactions, but they can be significant at chromium waste sites or other industrial disposal sites.

Regardless of the type of material driving the redox chemistry, anthropogenic contamination generally causes concentrations of active compounds that are much greater than would occur in a natural aquifer. Furthermore, the redox-active compounds are usually confined to relatively limited areas. Thus, the sharp and intense redox gradients that form at contamination sites are essentially different in their effect than the diffuse gradients of natural systems. At a contaminated site, the physical and temporal scales of redox chemistry relevant to sampling, monitoring, and modeling become compressed compared to the unperturbed aquifer. Therefore, approaches that are suited to the study of natural aquifers may be entirely unsuited to a zone of contamination.

This chapter provides an overview of the current state of knowledge of redox chemistry in the sub-surface, with special emphasis on the chemical rather than the microbiological aspects of the problem. The appropriate roles of thermodynamic, mechanistic, and kinetic methods are evaluated with a focus on practical methods of assessing the behavior of metals and other contaminants under dynamic redox conditions.

2.   THE Sub-surface REDOX LITERATURE

2.1   Redox Potential as a Geochemical Indicator

The study of redox chemistry in geochemical environments is not a recent phenomenon. However, as with many of the field sciences, the early studies were hampered by a lack of suitable measurement techniques. The earliest application of the standard electrode potential (Eh) to natural geochemical systems appears to be the work of Gillespie [4] who, in 1920, examined the reduction potentials of waterlogged soils using a platinum electrode. Other studies of electrode potentials in soils were conducted from the 1920s onward, but the measurements were often made on samples altered in some way and measured in the laboratory. In a review of the literature from this period, Baas Becking *et al.* [5] concluded that only a few studies made field measurements of potential by directly inserting an electrode into the soil [6,7,8]. Even where direct measurements were made, no uniform method was used to standardize electrode response, so all such data must be viewed as approximate.

Modern notions of the equilibrium chemistry of electron transfer reactions in natural environments can be traced to the seminal work of ZoBell [9] who studied the redox potential of marine sediments. ZoBell outlined the utility of Pt electrode measurements as a qualitative tool for assessing the equilibrium redox state of an environment. He also improved the practical methods of obtaining reliable field measurements of Eh, including the development of the eponymous ZoBell's solution, which provides a known and stable redox couple for standardizing the electrode. (ZoBell's solution is a 0.1 M KCl solution containing equimolar amounts of $K_4Fe(CN)_6$ and $K_3Fe(CN)_6$) [9].

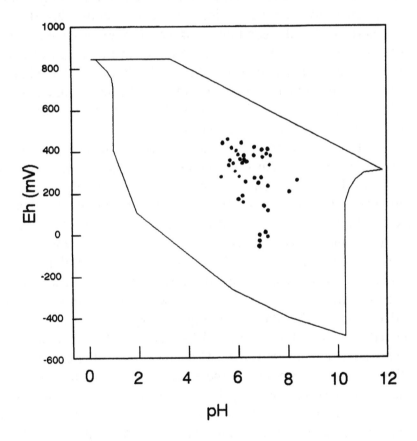

FIGURE 1. **_Eh-pH Characteristics of Natural Waters._** Outline shows the region delineated by Baas Becking *et al.* [5] as containing all available data for Eh and pH pairs for 6200 natural water samples. Small circles indicate the then-available (1960) data for groundwaters. Redrawn from Baas Becking *et al.* [5].

Although many researchers recognized the value of identifying the apparent redox state of individual natural systems, Baas Becking [5] and co-workers were the first to place environmental redox potential measurements into a unified framework. In their 1960 compendium, these authors placed over 6,200 pairs of Eh and pH measurements into a single Eh-pH diagram. Although the reliability of most of these data would now be regarded as questionable, the authors were able to relate specific aquatic environments, such as peat bogs, waterlogged soils, and groundwater, to general regions of the overall Eh-pH diagram. Characteristic Eh-pH conditions of specific environments could be stated, for the first time, on a general scientific basis rather than on narrow or anecdotal evidence. Although Krumbein and Garrels [10] earlier had developed a similar geochemical categorization scheme on the basis of Eh-pH pairs, the work of Baas Becking was the first to use actual field data rather than inductive reasoning.

The aggregated data, when plotted, indicated that all known geochemical environments fell within a broad but circumscribed region of the Eh-pH diagram, and that some zones of possible aqueous Eh-pH pairs were rarely, or never, observed in nature (Figure 1). Thus it was possible to rule out the natural occurrence of certain theoretical redox couples. One of the important conclusions of this work, now rather taken for granted, is that the Eh-pH limits of biological systems coincided almost exactly with the observed limits on natural aquatic systems. This implied that there are few, if any, environments on earth rendered sterile by the prevailing Eh-pH conditions; i.e., life has adapted to every known Eh-pH condition. The converse, that all aquatic environments have Eh-pH characteristics dictated by biological activity, is not supported by these data, but the authors were able to assert that most aquatic Eh-pH characteristics on earth are determined chiefly by four major biochemical processes: photosynthesis, respiration, oxidation/reduction of iron, and oxidation/reduction of sulfur.

The interpretations of Eh-pH regions by Baas Becking were considerably strengthened by comparisons with the Eh-pH domains of natural and laboratory cultures of various types of microorganisms. For example, the milieu of sulfate reducing bacteria could be assigned to a definite region of the Eh-pH diagram. Besides defining the redox domains of different classes of organisms, the comparison of biological and geochemical data allowed some assessment of which natural redox reactions appeared to be dominated by biological mediation and which reactions might be abiotic.

The general success of this work, despite the poor data, can be attributed to the fact that Eh and pH data are effectively logarithmic phenomena in which a 50% error translates into a fairly modest uncertainty on the Eh-pH diagram. Also, the great volume of data points lent an overall statistical reliability to the work that would not have been obtained from only a few measurements.

For the purposes of this chapter it is interesting to note that Baas Becking *et al.* [5] in their 1960 paper commented specifically on the striking paucity of Eh data for groundwaters. Out of more than 6,200 data points in the study, only 76 corresponded to groundwaters (Figure 1). The authors speculated that groundwater might have Eh and pH characteristics similar to waterlogged soils, apparently in an attempt to generalize from the sparse ground-water data. In fact, before 1960, essentially no systematic study of the redox characteristics of groundwater had ever been conducted. The data available to Becking were based on samples taken from flowing springs and boreholes or else bailed from a few isolated wells. Thus the ground-water Eh measurements reported by Becking were probably artificially high due to oxygen contamination. The first meaningful investigations of redox chemistry in groundwater were conducted in the early 1960s by William Back and Ivan Barnes [11,12,13].

2.2   Trends in the Redox Chemistry of Natural Aquifers

In the early 1960s Back and Barnes [11,12,13] carried out extensive geochemical studies of an aquifer underlying Anne Arundel County, Maryland, in the Atlantic Coastal Plain. These investigations are significant, in the history of groundwater research, in that they are the first in which oxidation potentials were measured on site in groundwater flowing through a sealed cell from a pumped well, a procedure that is now a standard practice. Sampling at some forty wells revealed distinct geochemical trends along the flow lines of the aquifer. Measured Eh values declined from an oxic potential of +700 mV near the recharge zone to a reducing potential of -40 mV about 15 km downgradient. Dissolved Fe was shown to correlate well with Eh and the measured Eh in the anoxic region could be predicted reasonably well by an equilibrium calculation assuming a dominant Fe(II)/Fe(III) couple and an Fe(III) solubility in between that of amorphous Fe(OH)$_3$ and hematite.

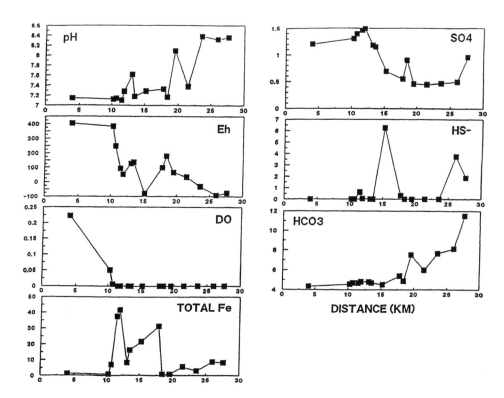

FIGURE 2. *Changes in the Geochemical Composition in the Lincolnshire Limestone Aquifer (England) as a Function of Distance Along the Direction of Flow.* Eh values are given in millivolts. All other concentrations are given in mg/L. (Drawn from data published by Edmunds [14]).

The notion that natural groundwaters evolve through a series of thermodynamically predictable redox states was demonstrated more thoroughly by Edmunds [14] who examined eighteen wells lying along a 42 km flow path in the Lincolnshire Limestone of eastern England. The favorable geometry of this formation, as it dips from an upland recharge zone through a zone confined beneath a clay layer, allowed Edmunds to observe a very clear picture of redox succession (Figure 2). In the outcrop near the recharge zone, the groundwater is oxygenated, contains modest amounts of dissolved organic carbon and no reduced metal or sulfur species. As the formation dips beneath the clay cap, oxygen vanishes, the Eh drops by some 200 mV, the dissolved Fe content increases by an order of magnitude, and the pH drifts upward. Further downgradient, sulfate concentrations decline as sulfide species appear. Dissolved Fe is depleted in sections of this flow region, apparently due to precipitation of Fe-sulfide minerals. Nitrate declines precipitously along the first 20 km of the formation, although most other major ions are relatively invariant. All of these observations were consistent with thermodynamic predictions of the sequence of reactions that should occur as dissolved organic matter is successively oxidized by oxygen, nitrate, ferric iron, and sulfate. Each stage of the succession appeared to be in equilibrium with respect to the major redox-active species.

A similar succession of quasi-equilibrium oxidation states has been observed in a fluvial sands aquifer at the Chalk River Nuclear Laboratories site in northwest Ontario. Champ and coworkers [15,16] in a series of papers demonstrated that a thermodynamic model of the aquifer, assuming the system is closed with respect to available oxidants, satisfactorily accounted for the successive reduction of oxygen, ferric oxides, and sulfate by the dissolved organic carbon.

These three sites, although geologically disparate, show consistent, and readily discernible trends in chemical evolution that agree well with a thermodynamic description. Thus, the utility of the thermodynamic approach in interpreting redox conditions in the sub-surface should not be minimized. The framework for interpreting these trends was provided by Stumm and Morgan [17] and Thorstenson

Fish

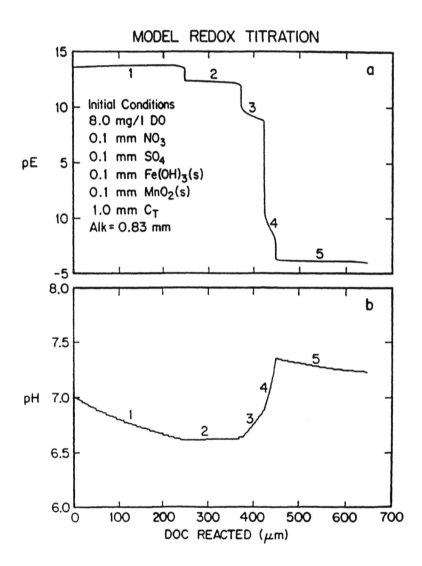

MODEL REDOX TITRATION

FIGURE 3. *Redox Titration Curve of a Model Groundwater System: a) pE Response and b) pH Response.* Numbered segments correspond to sequential reduction of 1) O_2(aq), 2) NO_3^-(aq), 3) MnO_2(s), 4) $Fe(OH)_3$(s), and 5) SO_4^{2-}(aq). From Scott and Morgan [19].

[18] who developed rigorous thermodynamic formulations for predicting the chemical evolution of oxic waters mixed with excess reductants. A central feature of these formulations is that redox reactions are well buffered or "poised" at each step in the sequence. At each step, the calculated (Nernstian) Eh and redox speciation are fixed by a dominant couple involving the oxidant that is most favored energetically, and that all other (less energetic) couples are inactive during that step. The Nernstian Eh of the system is determined by the active couple and is almost invariant until essentially all of that oxidant is reduced. The chemical stability of the system within each step can be quantified by a redox buffer intensity, β_E, as described by Stumm and Morgan [17] and Thorstenson [18]. Redox buffer intensity is defined in a manner analogous to a pH buffer intensity. When an oxidant is depleted, the system undergoes a rather sharp change in Eh as a new redox couple becomes dominant. Overall, the system can be regarded as a type of titration in which the excess reductant (organic matter) is titrated sequentially by a discrete series of oxidants in the following order: O_2, nitrate, Mn(VI), Fe(III), sulfate, and CO_2 (Figure 3). This theoretical framework has been recently revisited in the specific context of groundwater evolution by Scott and Morgan [19].

2.3   Thermodynamic Limitations to Redox Chemistry Approaches

Although the thermodynamic approach described above is clearly useful for evaluating the evolutionary composition of a variety of different aquifers, the success of thermodynamic analysis in these systems is largely due to the well defined, directional flow regimes, the gentle gradients of organic carbon, and the fact that the aquifers could be reasonably modeled as closed with respect to oxidants. Equilibrium models have inherent shortcomings that become most pronounced when flow patterns are complex, when the chemical changes in an aquifer occur over short spatial and temporal scales, and when the system has multiple inputs of oxidants and reductants.

The assumption of true thermodynamic equilibrium among all possible reactants is probably never strictly valid in natural waters. Many electron-transfer reactions, such as nitrate and sulfate reduction, are limited by high activation energies and must be catalyzed by microbial enzymes. Furthermore, it is easy to write a thermodynamic description for a reaction, such as the direct reduction of nitrate to ammonia, which does not occur in a single step in nature. Such a simple formulation, while formally correct, may not properly describe the chemistry of a complex, biologically mediated system.

Although the problem of kinetic limitation of redox equilibria had long been recognized, Lindberg and Runnells in their 1984 paper [20] were the first to provide comprehensive evidence of the general disequilibrium among natural redox couples in groundwater. In that work, analyses of 611 groundwaters were used to compute equilibrium Eh values that could be compared to the Eh values measured with an electrode. Almost no agreement was observed between the electrode measurements and the Eh computed by the Nernst equation using chemical measurements (Figure 4). The authors argued that Eh is of no value as a master variable for predicting the chemical state of a system. Although that study was somewhat limited by the poor quality of the database available to the authors [21], their main conclusion is almost certainly correct. Electrode measurements of Eh can be a valid determination of the state of certain, highly reversible redox couples, most notably the Fe(II)/Fe(III) couple [22], but it is incorrect to extrapolate from electrode measurements to the state of other redox couples not active at the electrode surface.

The problem of general disequilibrium of redox couples is not peculiar to electrode measurements and, in fact, limits the usefulness of **any** single indicator of redox state. For example, Cherry et al. [23] suggested the measurement of As(III)/As(V) speciation as an indicator of redox conditions in groundwater. This couple equilibrates rapidly enough to respond to environmental conditions, but slowly enough so that it is easy to obtain samples and make accurate determinations of speciation. While the technique has merit as a qualitative indicator of redox state, it is functionally no different than an electrode in that the As(III)/As(V) couple can only interact with other suitably reactive couples and it cannot indicate the redox state of any couple that is not in equilibrium with it. For example, the nitrate/nitrite couple will probably not react with As(III)/As(V).

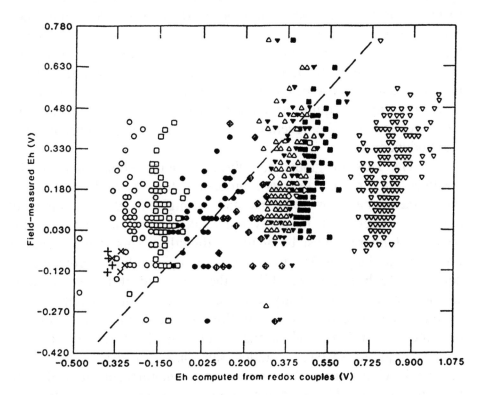

FIGURE 4.   *Diagram Showing Computed Nernstian Eh Values Versus Field-measured (Pt Electrode) Eh Values for Ten Redox Couples in 611 Groundwater Samples.* The dashed line indicates the expected locus of points if all computed values matched the measured values and all couples were in internal equilibrium. The symbols for the couples are: ($\lozenge$) Fe^{2+}/Fe^{3+}, (∇) $O_2(aq)/H_2O$, (O) HS^-/SO_4^{2-}, ($\square$) $HS^-/S_{rhombic}$, (■) NO_2^-/NO_3^-, (▾) NH_4^+/NO_3^-, ($\triangle$) NH_4^+/NO_2^-, (+) $CH_4(aq)/HCO_3^-$, (×) NH_4^+/N_2, (•) $Fe^{2+}/Fe(OH)_3(s)$. From Lindberg and Runnells [20].

The conclusions of Lindberg and Runnells [20] are by no means a repudiation of the use of thermodynamics in the redox chemistry of groundwater. Nor, as discussed above, do they contradict the thermodynamic interpretation of groundwater evolution. In fact, a groundwater system such as the Lincolnshire Limestone is never in "global" equilibrium. At each stage in the ongoing evolution, the system approaches a "local" equilibrium for only the dominant redox couple. A selective formulation which assumes equilibrium only among the active species in that zone will account for the observed chemistry. Redox evolution of natural groundwaters often can be satisfactorily modeled as a sequence of quasi-equilibrium states. Even if the system is not at complete equilibrium, the order of the reaction sequence and the direction of the overall process are still predictable from thermodynamics.

The lack of global equilibrium among redox-active species becomes a more serious problem for groundwaters with multidimensional flow patterns and strongly localized sources of reactants. Lack of a master variable like Eh means that all redox couples of interest must be analyzed directly. This is not a serious limitation in large-scale, gradually evolving aquifers. However, in zones of high redox variability, the cost and difficulty of fully characterizing the chemistry can be severe. For example, in an uncontaminated shallow aquifer in Illinois, Barcelona *et al.* [24] found that local hydrology and strong vertical gradients of redox-active species required spatially detailed sampling in order to adequately resolve the chemistry of the system. At a contaminated site in Illinois, the same authors [24] found even more pronounced redox gradients in both the horizontal and vertical directions, requiring that sampling well-points be placed at spatial frequencies of a few meters or less. Similar observations have been made in other aquifers contaminated by landfill leachate [25], organic wastes [26], creosote [27,28], and gasoline [29].

At the last site mentioned, a shallow, unconsolidated aquifer in Yakima, Washington, was contaminated with an unknown quantity of gasoline that had leaked from an underground storage tank over approximately eight years. The transect data, shown in Figure 5, indicate that the plume of gasoline contamination was depleted in

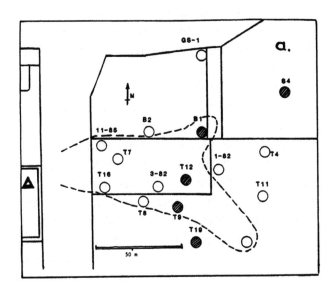

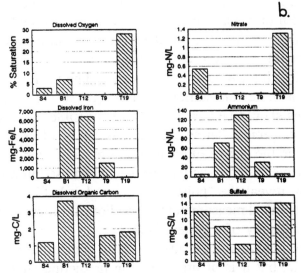

FIGURE 5. *Gradients of Redox-active Constituents Across a Gasoline-contaminated Aquifer.* a) map showing the main body of the gasoline plume, as defined by the isopleth of the 200 ppb concentration of xylenes (dashed line), and the location of the five sampling wells used to construct the transect in Figure 5b (shaded circles); b) concentration of redox-active constituents along the cross-sectional transect of wells identified in Figure 5a.

oxygen, nitrate, and sulfate, and rich in dissolved Fe and ammonium compared to surrounding, uncontaminated waters. As at most such sites, acutely contaminated with organic compounds, the general pattern of sequential consumption of oxygen, nitrate, iron, and sulfate can be detected in the spatial distribution of these constituents. For example, nitrate and Fe reduction occur only near the zone depleted of oxygen, while sulfate reduction is even less widespread and is limited only to the central core of the plume. Thus, thermodynamics is a useful framework in which to interpret the observations, even though the contaminated zone of the aquifer is clearly far from complete equilibrium. The zones of nitrate, Fe, and sulfate reduction overlap, and the chemical characteristics of the system, could be identified only by direct measurement at many closely spaced sampling points.

In summary, thermodynamic calculations play an important, but not exclusive, role in characterizing redox chemistry in the sub-surface. Redox couples are rarely if ever in complete equilibrium, so the measured Eh of a water is not a useful master variable for calculating the speciation of all redox couples. Electrode measurements of Eh can be a practical indicator of certain couples of interest, particularly Fe(II)/Fe(III), but Eh measurements should be interpreted with extreme caution for oxic or sulfidic waters [24,30,31]. Thermodynamic calculations should be based, whenever possible, on direct analyses of the reactive species. These calculations can be of use in predicting the direction and the degree of evolution of reducing conditions. Complete measurements of redox-active species, when used in a thermodynamic model, can reveal the "poisedness" or redox buffering of the system. Detailed chemical measurements also allow a computation of the overall oxidative capacity of a system [19].

Thermodynamics is always useful in determining what reactions are possible, but it is of little value in predicting which reactions will actually occur, or when they will occur. The utility of thermodynamic calculations is especially restricted in highly dynamic, contaminated systems and other regions of high chemical variability. In such situations, the acquisition of the chemical data needed to develop useful thermodynamic information may be prohibitively expensive or otherwise unavailable. Under these conditions, which are typical of most contamination sites, a more refined focus, using a "reaction-based" approach may provide more useful information than could be obtained by a general thermodynamic characterization alone.

3.   REACTION-BASED APPROACH TO REDOX CHEMISTRY IN GROUNDWATER

Although equilibrium calculations are an essential step in the evaluation of the redox condition of an aquifer, the method is clearly unable to provide detailed characterization of redox reactions at a contamination site. More detailed information may be obtained by adopting a reaction-based strategy of the sort that has been used successfully to elucidate hydrolysis, photolysis, and other reactions of organic compounds.

In the reaction-based approach: 1) a chemical component of interest is specified, 2) the most important reactions the compound can undergo in the environment are determined,   3) the environmental variables that control the reactions are identified, and 4) the rates and stoichiometries of the reactions are measured under conditions expected in the environment. This strategy yields practical information about the rates at which a parent compound is transformed. It can serve as the basis for understanding the transport properties of reactants and products, and it allows chemical time scales to be related to transport time scales.

The initial step of choosing the reactants of importance requires an assessment of the chemical constituents known, or suspected to be present in the system, coupled with knowledge (or that elusive quality: "chemical intuition") about the reactions likely to occur. *A priori* knowledge of all reactions is not essential as long as the major reactions can be predicted and a means of rechecking and correcting the system is in place. Thus, if additional reactions are detected once experiments are conducted, the study can be expanded or modified to include these additional reactions.   The thermodynamic states of all suspected reactions should be calculated to insure that the reactions are energetically feasible under the postulated conditions.   Once   the system of reactants is defined for study, experiments should be conducted to determine the main parameters that govern the rates of reactions. Environmental conditions that typically are expected to influence reactions are the pH, the substituents and isomeric form of organic reactants, the temperature, the ionic strength of the solution, and the presence or absence of solid or (non-aqueous) liquid phases that serve as reactants or as catalytic surfaces. Of these conditions, pH is almost always important because the proton is usually a reactant in redox reactions. Temperature influences all chemical reactions, but temperature corrections for kinetics are generally a straightforward procedure. The ionic strength of many groundwaters is low enough so that it has only a minor effect on reaction kinetics and can consequently be safely neglected in preliminary studies.

Intrinsic reaction parameters that typically must be evaluated are stoichiometry and the order of the reactions. Stoichiometry is critical to the correct formulation of any rate law, and is obviously important in keeping track of mass balances during a reaction. The exact order of a reaction may not be critical to a preliminary or approximate formulation of a reaction. Often it is sufficient to obtain a "characteristic order" for an overall reaction. The characteristic order is an approximate relationship between the concentrations of reactants and the rates that are observed.

Finally, all of the information gathered from the literature and from new experimentation should be put in the context of the field situation. Ideally this is an iterative process. Available field data, waste disposal manifests, and so forth are gathered to insure that all important reactants are considered. The suspected reactants are tested in the laboratory under conditions expected to prevail in the field. Then, a second sampling campaign in the field is instituted to gather new data to verify the hypotheses that are logically inferred from laboratory results. If necessary, the study returns to the laboratory to obtain detailed data under the conditions now known with greater certainty to occur in the field. This iterative process, while time consuming (in some circumstances prohibitively so) will likely be required because rare is the field campaign that obtains all the "right" data on the first try, and rare is the laboratory study that correctly reproduces the field conditions. In principle, however, this approach should yield an understanding of redox reactions that is well worth the effort.

Other than studies of microbial processes, reaction-based approaches have been little applied to redox reactions relevant to the sub-surface. Detailed studies of abiotic electron-transfer reactions in a geochemical context have focused primarily on the reductive dissolution of metal oxides by natural and contaminant reductants (e.g., the work of Stone [32,33,34]). The reaction-based approach has not been widely applied to redox reactions among waste components and, to the author's knowledge, no full-scale, integrated field and laboratory investigation of the type proposed above has been yet conducted for these reactions. However, the utility of reaction-based investigations for understanding the redox behavior of waste mixtures is illustrated by recent experiments in our laboratory on redox reactions between Cr(VI) and methylphenol (cresol) isomers.

Hydroxylated compounds such as phenols are quite reactive with Cr(VI) [35,36]. Redox reactions between alcohols and Cr(VI) have been the most extensively researched, but phenolic compounds are of greater environmental concern. Cresols (methylphenol isomers) are a major constituent of creosote wood preservatives and are found as soil and sub-surface contaminants at wood-treating plants, and at lumber or utility-pole storage yards. Chromated cupric arsenate (CCA) is also a common wood preservative. Combinations of preservatives are used at most wood treating sites and historically were disposed of in common pits or drainage areas, leading to mixtures of chromate and cresols.

As stated above, a reaction-based study of a redox process should begin by assessing all available information about the relevant reactions or reactions similar to these. Oxidative decomposition of cresols and other substituted phenols may proceed by any of several pathways [33]. Phenoxy radicals (ArO·) may be formed, followed by oxidative coupling which leads to polymeric products [37,38,39]. At low pH, oxidation is more likely to proceed via phenoxenium ions (ArO$^+$) that can hydrolyze and oxidize further to benzoquinone or else which can electrophilically attack unaltered phenols to form oxidized phenol couples that subsequently polymerize [37].

Westheimer *et al.* [35,36] investigated the oxidation of isopropanol by Cr(VI) and found evidence for a pre-oxidation step involving the formation of a metastable chromate ester. Subsequent studies documented the formation of chromate esters in primary and secondary alcohols [40] and in tertiary alcohols [41,42]. Chromate esters of primary and secondary alcohols break down relatively rapidly into Cr(V) and Cr(IV) intermediates which, in turn, quickly decompose to Cr(III) and an oxidized organic molecule, typically an aldehyde, ketone, or carboxylic acid [40,43]. Many oxidation pathways have been proposed for alcohol-based chromate esters, but the exact mechanisms are not known [40,43]. Little information was available for the specific case of chromate reacting with cresols, but the information sketched out above was used to guide the design of experiments with cresols and chromate. For example, the importance of the proton in the reaction, the importance of isomeric effects, and the possibility of multiple reaction pathways were all anticipated.

Experiments were conducted using saturated aqueous solutions of the ortho, meta, and para isomers of cresol. Cresol solutions were buffered to pH 1.0, 2.0, 3.0, and 5.0. Reactions were initiated by adding Cr(VI) to a final concentration of 1 mM, 5 mM, or 10 mM. Reactions were carried out in amber-glass 40-mL EPA-type sample vials with screw caps containing Teflon-lined silicone rubber septa. Blanks were prepared by adding Cr(VI) to buffer solutions without cresol. All reactions and corresponding blanks were performed in triplicate. The progress of reactions was followed by withdrawing small (6-10 μL) aliquots and analyzing for Cr(VI) by the diphenylcarbazide method. At the same time, the pH of the solution was measured directly with a Ross-type semi-micro combination electrode (Orion).

Rates of reactions of chromate with the model organic compounds were extremely fast at pH 1.0 but became much slower as pH values approached neutrality. This strong pH dependence reflects the expected consumption of protons in the reaction and was clearly illustrated in the reduction of Cr(VI) by *p*-cresol. At pH 1.0, a saturated *p*-cresol solution completely reduced a 10 mM Cr(VI) solution within 60 seconds. This rapid rate agrees with rates found in earlier studies of Cr(VI) reactions with alcohols under acid conditions.

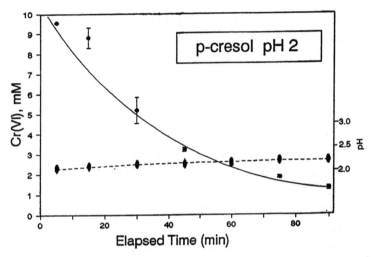

FIGURE 6. *Reduction of Cr(VI) by p-Cresol as a Function of Time.*
Initial [Cr(VI)] = 10 mM, initial pH 2.0; as in all experiments cresol
was a saturated aqueous solution (~0.2 M). Error bars mean one
standard deviation about the mean value.

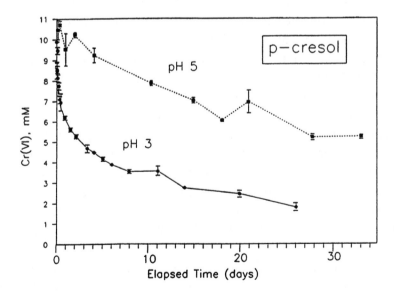

FIGURE 7. *Reduction of Cr(VI) by p-Cresol as a Function of Time.*
Initial [Cr(VI)] = 10 mM, initial pH 3.0 (•), pH 5.0 (■)

For reactions with an initial pH of 2.0 the reduction was slower: after 25 min, 50% of the Cr(VI) was reduced. At 60 min, 90% of the Cr(VI) was reduced and the solution pH had drifted up to pH 2.11, indicating the consumption of protons (Figure 6). In reactions with an initial pH of 3.0, reduction of 50% of the Cr(VI) required approximately 24 hours and, for reactions starting at pH 5.0, a 50% reduction of Cr(VI) required 25 days (Figure 7).

It is important to note that, although the solutions were buffered, the pH was not constant in any of the experimental reactions. As the reduction of Cr(VI) progressed, the consumption of protons drove up the pH by an amount that depended on the extent of reaction and the buffer capacity of the solution. Because the rate of reduction was so sensitive to pH, the diminishing proton activity in solution over the course of the reaction caused the overall rate to progressively slow down. Reduction of Cr(VI), therefore, could not be accurately modeled by a simple first-order dependence on Cr(VI) concentration. A second-order model in which the reduction rate depended on both Cr(VI) and pH provided a reasonable fit of the kinetic data [44].

The organic reaction products were not identified, but the reactions produced various types of brownish, insoluble residues. The amount and characteristics of the insoluble organic matter were variable and depended in part on the initial pH of the reaction. The variability of the products showed the expected dependence of reaction pathway on initial pH and other reaction conditions.

In general, focusing on a limited set of electron-transfer reactions, as was done in this example, can provide a wealth of practical information about the rate of reaction and the environmental factors that govern that rate. A kinetic or mechanistic study does not replace a thermodynamic evaluation of redox reactions in a contaminated aquifer, rather, it greatly supplements that which can be learned from thermodynamics. The laboratory investigations of the type described here ultimately should be compared with field data to ascertain if the conditions of the laboratory accurately describe conditions in the field. However, well conceived laboratory experiments are a highly cost-effective way of learning about the general behavior of specific redox reactions before launching an expensive and time-consuming campaign of field sampling. The need for preliminary experiments in the laboratory is at least equally important for studies of biologically mediated redox reactions.

4.   SUMMARY

Oxidation-reduction reactions alter the chemistry of groundwater in many ways. In uncontaminated aquifers, natural organic matter is oxidized by a sequence of electron-accepting compounds, starting with dissolved oxygen and progressing through nitrate, Mn oxides, Fe oxides, sulfate, and ultimately CO_2. Thus, redox reactions cause minerals and adsorptive phases to dissolve or precipitate. In aquifers contaminated with organic compounds, a similar sequence of redox reactions accompanies the degradation of contaminants, although the reactions may be far more compressed in space and time when compared to the natural redox evolution. Redox reactions dramatically affect the rate and pathway of degradation, the mobility, and the toxicity of contaminants.

The sequence of redox reactions is predicted by thermodynamic considerations and has been observed in many types of aquifers. The success of thermodynamics in explaining the reaction sequence has prompted the development of thermodynamic equilibrium models of the redox state of aquifers in which an electrode potential, or some equivalent chemical indicator of electron potential, can serve as a master variable for defining the state of all redox couples. However, there is ample evidence that many redox-active species in groundwater are not in equilibrium, so there is no single variable that can characterize the "redox state" of an aquifer. A potential measured by an electrode indicates the state only of those couples that are reactive enough to produce a sufficient current at the electrode surface. In natural waters, the Pt electrode can accurately indicate the state of the Fe(II)/Fe(III) couple but its response to other couples is difficult to interpret and it does not respond at all to many common couples. The speciation of an indicator couple, such as As(III)/As(V), is similarly limited in that it measures the state of only those couples in equilibrium with the indicator. Therefore, thermodynamics is only one of the tools which must be employed to obtain a detailed picture of redox chemistry in the sub-surface, particularly at a chemically complex contamination site.

A reaction-based paradigm of redox chemistry appears to be a fruitful means of elucidating the effects of specific electron transfer reactions on the behavior of sub-surface contaminants. In this approach, redox-active constituents are identified and the specific reactions they undergo are individually characterized. The effects of environmental conditions, such as temperature, ionic strength, and concentration range, on the rates of mechanisms of reactions can be quantified, leading to refined models of contaminant behavior. Thermodynamic calculations still play an important role in these studies by defining the energetically feasible reactions and the theoretical limits of reaction processes. An ideal study of the redox behavior of contaminants would entail iteratively alternating between well controlled laboratory studies of reactions and field validation of the rates and products of reactions measured in the laboratory.

REFERENCES

1.   Förstner, U. and A. Carstens. "In-situ Experiments on Changes of Solid Heavy Metal Phases in Aerobic and Anaerobic Groundwater Aquifers," *Environ. Technol. Lett.* 10:823-832. (1989).

2.   Vogel, T.M., C.S. Criddle, P.L. McCarty. "Transformations of Halogenated Aliphatic Compounds," *Environ. Sci. Technol.* 21:722-736 (1987).

3.   Palmer, C.D. Personal communication (1989).

4.   Gillespie, L.J. "Reduction Potential of Bacterial Cultures and of Waterlogged Soils," *Soil Sci.* 9:199-216 (1920).

5.   Baas Becking, L.G.M., I.R. Kaplan, and D. Moore. "Limits of the Natural Environment in Terms of pH and Oxidation-Reduction Potentials," *J. Geol.* 68:243-284 (1960).

6.   Pearsall, W.H. "The Soil Complex in Relation to Plant Communities," *J. Ecol.* 26:180-193.

7.   Pearsall, W.H. and C.H. Mortimer. "Oxidation Reduction Potential in Waterlogged Soils, Natural Waters, and Muds," *J. Ecol.* 27:483-501 (1939).

8.   Starkey, R.L. and K.M. Wight. "Anaerobic Corrosion of Iron in Soil," Tech. Rept. Amer. Gas Assoc., New York (1945).

9.   ZoBell, C.E. "Studies on Redox Potential of Marine Sediments," *Am. Assoc. Petrol. Geol. Bull.* 30:477-513 (1946)

10.   Krumbein, W.C. and R.M. Garrels. "Origin and Classification of Sediments in Terms of pH and Oxidation-Reduction Potential," *J. Geol.* 60:1-33 (1952).

11.   Back, W. and I. Barnes. "Relation of Electrochemical Potentials and Iron Content to Ground-Water Flow Patterns," *U.S. Geological Survey Prof. Paper* 498-C (1965).

12.   Back, W. and I. Barnes. "Equipment for Field Measurement of Electrochemical Potentials," *U.S. Geological Survey Prof. Paper* 424-C (1961).

13.   Barnes, I., and W. Back. "Geochemistry of Iron-Rich Groundwaters of Southern Maryland," *J. Geol.* 72:435-447 (1964).

14.   Edmunds, W.M. "Trace Element Variations Across an Oxidation-Reduction Barrier in a Limestone Aquifer," in *Proceedings of Symposium on Hydrogeochemistry and Biogeochemistry, Vol. I - Hydrogeochemistry* (Washington, DC: Clarke Co., 1973), pp. 500-526.

15.   Champ, D.R., J. Gulens, and R.E. Jackson. "Oxidation-Reduction Sequences in Ground-water Flow Systems," *Can. J. Earth. Sci.* 16:12-23 (1979).

16.   Jackson R.E. and R.J. Patterson. "Interpretation of pH and Eh Trends in a Fluvial Sand Aquifer System," *Water. Resour. Res.* 18:1255-1268 (1982).

17.   Stumm, W. and J.J. Morgan. *Aquatic Chemistry* (New York: Wiley Interscience, 1970), pp. 300-382.

18.   Thorstenson, D.C. "Equilibrium Distribution of Small Organic Molecules in Natural Waters," *Geochim. Cosmochim Acta* 34:745-770 (1970).

19.   Scott, M.J. and J.J. Morgan. "Energetics and Conservative Properties of Redox Systems," in *Chemical Modeling in Aqueous Systems II*, D.C. Melchior and R.L. Bassett, Eds. (Washington, DC: American Chemical Society, 1990), p. 368.

20.   Lindberg, R.D. and D.D. Runnells. "Ground Water Redox Reactions: An Analysis of Equilibrium State Applied to Eh Measurements and Geochemical Modeling," *Science* 225:925-927 (1984).

21.   Jenne, E. Personal communication (1989).

22.   Grundl, T.J. and D.L Macalady. "Electrode Measurement of Redox Potential in Anaerobic Ferric/Ferrous Chloride Systems," *J. Contam. Hydrol.* 5:97-117 (1989).

23. Cherry, J.A., A.U. Shaikh, D.E. Tallman, and R.V. Nicholson. "Arsenic Species as an Indicator of Redox Conditions in Groundwater," *J. Hydrol.* 43:373-392 (1979).

24. Barcelona, M.J., T.R. Holm, M.R. Schock, and G.K. George. "Spatial and Temporal Gradients in Aquifer Oxidation-Reduction Conditions," *Wat. Resour. Res.* 25:991-1003 (1989).

25. Baedecker, M.J and W. Back. "Modern Marine Sediments as a Natural Analog to the Chemically Stressed Environment of a Landfill," *J. Hydrol.* 43:393-414 (1979).

26. Jackson R.E., R.J. Patterson, B.W. Graham, J. Bahr, D. Belanger, J. Lockwood, and M. Priddle. "Contaminant Hydrology of Toxic Organic Chemicals at a Disposal Site, Gloucester, Ontario. 1. Chemical Concepts and Site Assessment," *Paper 23, Environ. Can. Nat. Hydrol. Res. Inst.*, Inland Waters Directorate, Ottawa (1985).

27. Bedient, P.B., A.C. Rodgers, T.C. Bouvette, M.B. Tomson, and T.H. Wang. "Ground-water Quality at a Creosote Waste Site", *Ground Water* 22:318-329 (1986).

28. Cozzarelli, I.M., M.J. Baedecker, and J.A. Hopple. "Effects of Creosote Products on the Aqueous Geochemistry of Unstable Constituents in a Surficial Aquifer," in *Proceedings of the Third Technical Meeting: U.S. Geol. Survey Program on Toxic Waste -- Ground-Water Contamination, Pensacola, Florida,* B.J. Franks, Ed. (Washington, DC: U.S. Geol. Survey Open-File Report 87-109, 1987) pp. A15-A16.

29. Fish, W. and J. Ebbert, 1987. Subsurface Transport of Gasoline-Derived Lead at a Filling-Station Contamination Site in Yakima, Washington, *Proceedings of the NWWA Focus Conference on Northwestern Ground-Water Issues,* Portland, OR, May 5-7, 1987, p. 557-567.

30. Stumm, W. "Interpretation and Measurement of Redox Intensity in Natural Waters," *Schweiz. Z. Hydrol.* 46:291-296 (1984).

31. Boulegue, J. and G. Michard. "Sulfur Speciation and Redox Processes in Reducing Environments," in *Chemical Modeling in Aqueous Systems,* E.A. Jenne, Ed. (Washington, DC: American Chemical Society, 1979), p. 25.

32. Stone, A.T. "Adsorption of Organic Reductants and Subsequent Electron Transfer on Metal Oxide Surfaces," in *Geochemical Processes at Mineral Surfaces, ACS Symp. Series 323,* J.A. Davis and K.F. Hayes, Eds. (Washington, DC: American Chemical Society, 1986).

33. Stone, A.T. "Reductive Dissolution of Manganese (III/IV) Oxides by Substituted Phenols," *Environ. Sci. Technol.* 21:979-988 (1987).

34. Stone, A.T. "Microbial Metabolites and the Reductive Dissolution of Manganese Oxides: Oxalate and Pyruvate," *Geochim. Cosmochim. Acta* 51:919-925 (1987).

35. Westheimer, F.H. and N. Nicolaides. "Chromic Acid Oxidation of Isopropyl Alcohol," *J. Am. Chem. Soc.* 71:25-31 (1949).

36. Westheimer, F.H. and A. Novick. "The Kinetics of the Oxidation of Isopropanol by H_2CrO_4," *J. Chem. Phys.* 11:506-512 (1943).

37. Waters, W.A. "Mechanism of One-electron Oxidation of Phenols. Fresh Interpretation of Oxidative Coupling Reaction of Plant Phenols," *J. Chem. Soc.* B 10:2026-2029 (1971).

38. Bailey, S.I., I.M. Ritchie, and F.R. Hewgill. "The Construction and Use of Potential-pH Diagrams in Organic-Reduction Reactions," *J. Chem. Soc., Perkins Trans.* 2:645-652 (1983).

39. Tanaka, H., I. Sakata, and R. Senju. "Oxidative Coupling Reaction of Phenols with Dichromate," *Bull. Chem Soc.* Japan, 43:212-215 (1970).

40. Sengupta, K.K., T. Samanta, and S.N. Basu. "Kinetics and Mechanism of Oxidation of Ethanol, Isopropanol, and Benzyl Alcohol by Chromium(VI) in Perchloric Acid Medium," *Tetrahedron* 42:681-685 (1986).

41. Richer, J.C. and J.M. Hachey. "Chromic Acid Esters Derived from Tertiary Alcohols," *Can. J. Chem.* 53:3087-3093 (1975).

42. Zeiss, H.H. and C.N. Mathews. "Chromate Esters. I. Solvolysis of Di-(2,4-dimethyl-4-hexyl) Chromate," *J. Am. Chem. Soc.* 78:1694-1698 (1956).

43. Wiberg, K.B. and H. Schäfer. "Chromic Acid Oxidation of Isopropyl Alcohol. The Pre-oxidation Steps," *J. Am. Chem. Soc.* 91:923-932 (1969).

44. Fish W. and M.S. Elovitz. "Kinetics of Electron Transfer Reactions Between Cr(VI) and Methylphenol Isomers. *Environ. Sci. Technol.* (Submitted).

4

LANTHANIDE ION PROBE SPECTROSCOPY FOR METAL ION SPECIATION

Wisnu Susetyo and Lionel A. Carreira
Department of Chemistry
University of Georgia
Athens, GA 30602

Leo V. Azarraga
U.S. Environmental Protection Agency
Athens, GA 30613

David M. Grimm
Technology Application, Inc.
U.S. Environmental Protection Agency
Athens, GA 30613

1.   INTRODUCTION

1.1   Humic Substances

The majority of the organic materials in waters and soils are humic substances, which are polyfunctional, acidic, macromolecular compounds, that are produced by the biodegradation of biomass. It is well known that humic substances are structurally heterogeneous and cannot be represented by a single structure. For the purpose of studying metal-humic interactions, humic substances can be considered as a mixture of phenol-carboxylate polyelectrolytes [1]. Due to the many structural variations in a humic molecule and between one humic molecule and another, these functional groups constitute a vast collection of proton- or metal-binding sites which have a relatively broad distribution of complex stability constants, K_i's.

1.2   Gaussian Model

Consider an interacting system consisting of metal ions, M, and metal-binding sites of a humic sample, L. Let subscript i denote a group of metal binding sites in humic macromolecules that have the same complex stability constant to a metal M, i.e. K_{Mi}. Then, for a binding site of type i, the complexation equilibrium reaction for a 1:1 metal-ligand complex is

$$M + L_i \rightleftarrows ML_i$$

and the stability constant K_{Mi} for this complex is:

$$K_{Mi} = \frac{[ML_i]}{[M]\,[L_i]} \tag{1}$$

where $[ML_i]$ represents the concentration of the metal species that is bound to the humic binding sites, $[M]$ represents the concentration of free metal ion, and $[L_i]$ is the concentration of the binding sites of type i that are not bound to the metal ion M. If C_i is the total concentration of the binding sites of type i, then it is easy to show from the mass balance of L_i and Equation (1) that:

$$[ML_i] = \frac{C_i\,K_{Mi}\,[M]}{1 + K_{Mi}\,[M]} \tag{2}$$

A model that treats the ligands in humic substances as a continuous distribution of binding sites in which individual ligand concentrations C_i are normally distributed with respect to the logarithm of the K_i values for proton or metal binding was proposed by Perdue [2] and used as the basis of this work:

$$C_i = \frac{C_L}{\sigma \sqrt{2\pi}} \exp\left(-\frac{(\mu - \log K_i)^2}{2\sigma^2}\right) d \log K \qquad (3)$$

where C_i/C_L is the mole fraction of binding sites of type i. Notice that Equation (3) is parallel to the famous Gaussian error function. The total concentration of the ligands (C_L), the mean of $\log K_i$ value (μ), and the standard deviation (σ) for the $\log K_i$ distribution of the binding sites can be estimated by fitting experimental data to the calculated data using a Jacobian nonlinear least square technique [3].

Figure 1 illustrates the model and its important parameters. By combining Equations (2) and (3) and summing over all i's, the total concentration of the bound species can be calculated as follows:

$$\Sigma[ML_i]_{calc} = \frac{C_L}{\sigma\sqrt{2\pi}} \int \frac{K_{Mi}[M]}{1 + K_{Mi}[M]} \exp\left(-\frac{(\log K_{Mi} - \mu_M)^2}{2\sigma^2}\right) d \log K_M \qquad (4)$$

In calculating $\Sigma[ML_i]_{calc}$ using Equation (4), μ_M and σ can be provided as guessed values. C_L can be provided as a known value from acid-base titration data of the humic sample or can also be provided as a guessed value. The integral cannot be solved analytically and numerical integration must be performed. The integral is thus replaced by summation and the limits of the integration are set to $\mu \pm 4\sigma$. The Gaussian distribution curve is vertically sliced into an arbitrarily high number of slices (of uniform width) such as one hundred or higher. This arbitrary number is denoted as i_{max}. The area of slice number i now represents C_i, the concentration of binding site of type i. The uniform width of the slices is obviously equal to $(8\sigma)/i_{max}$ which is equal to $d \log K_M$, such that each individual K_{Mi} value can now be known in terms of μ_M, σ, and i_{max}. All of the parameter and variable values, with the exception of [M] on the right hand side of Equation (4) are now known or guessed. This value of [M] needs also to be given as an arbitrarily guessed value. Once all of the parameter values are known or guessed, $\Sigma[ML_i]_{calc}$ can be calculated from evaluating the right hand side of the Equation (4). The $\Sigma[ML_i]_{calc}$ value and the guessed value of [M] are then compared with the known value of C_M, the total concentration of the metal ion M. If they are the same, the mass balance condition for metal is achieved and the value of $\Sigma[ML_i]_{calc}$ is accepted, otherwise [M] is systematically reguessed and the iteration process continued until convergence is achieved.

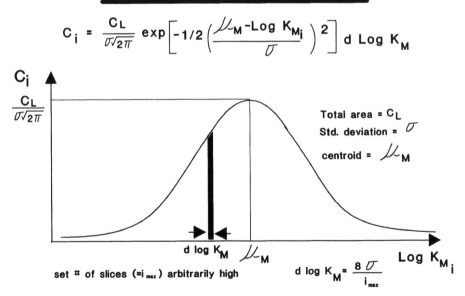

FIGURE 1.　*Gaussian Distribution Model and Its Important Parameters.*

1.3   Lanthanide Ion Probe Spectroscopy

Humic samples were titrated with Eu^{3+}, which acts as the fluorescent probe ion. The spectral titration data were generated by a lanthanide ion probe spectrofluorometry technique introduced by Horrocks and Sudnick [4]. Two of the emission lines of Eu^{3+} used in this study are located at 592 and 616 nm. The 616 nm emission line is produced by a hypersensitive transition [5], i.e. $^5D_0 \longrightarrow {}^7F_2$, which is normally a forbidden transition, but interaction with different ligand environments, other than water, often makes the transition allowed and enhances the intensity of its emission. The more probe ions (i.e. Eu^{3+}) bound to the humic binding sites, the larger the area of the band at 616 nm. In contrast, the 592 nm emission band originates from a nonhypersensitive transition, $^5D_0 \longrightarrow {}^7F_0$, and its intensity depends on the Eu^{3+} concentration and is not affected by different ligand environments. Figure 2 a-c illustrates how the relative intensities of the two emission lines change with % bound of the probe ions (i.e Eu^{3+}). Figure 2a shows the emission profile of a Eu^{3+} solution at pH 3.5 in water excited at a wavelength of 394.3 nm. The two peaks are located at 592 and 616 nm. Notice that the peak at 592 nm is much taller than the peak at 616 nm. Figure 2b shows the emission profile of a Eu^{3+} solution in distilled water where a small amount of humic material had been added. Note that the peak at 592 nm is still taller than the peak at 616 nm, but not by much. Figure 2c shows the emission profile when some more humic material is added. The peak at 592 nm is now lower than the peak at 616 nm. Based on these observations, the ratio of the intensities of the two bands, defined as $\mathbf{R} = I_{592}/I_{616}$, has the following property: lower $\mathbf{R}$ values relate to higher concentrations of bound europium species, $\Sigma[ML_i]$, and vice versa. With proper calibration, this intensity ratio can be used to quantitatively determine the amount of bound and free europium species for a given total amount of Eu^{3+}, C_M.

Susetyo, Carreira, Azarraga and Grimm

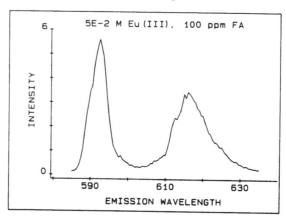

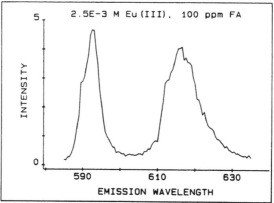

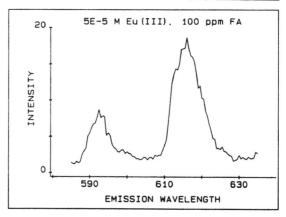

FIGURE 2.   *Emission Spectrum of Eu^{3+} and 100 ppm Suwannee River Fulvic Acid (FA).* The excitation wavelength was 394.3 nm and the pH of the solution was 3.5. a) 5.0 X 10^{-2} M Eu^{3+}; b) 2.5 X 10^{-3} M Eu^{3+}; c) 5.0 X 10^{-5} M Eu^{3+}.

It has been shown [6] that:

$$R = \frac{C_M \, X_B \, X_S}{[M] \, X_S + X_B \, \Sigma ML_i} \qquad (5)$$

where C_M is the total concentration of Eu^{3+} added, $[M]$ is the total concentration of the species of Eu^{3+} which is not bound to the humic binding sites, $\Sigma[ML_i]$ is the total concentration of the species of Eu^{3+} which is bound to the humic binding sites, X_B is the upper limit of **R**, i.e. when all or most of the Eu^{3+} ions are free, and X_S the lower limit of **R**, i.e. when all or most of the Eu^{3+} ions are bound.

SPECTRAL TITRATION PLOT

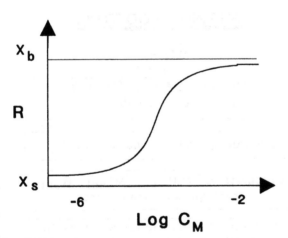

Typical values :

X_b = 2.15 C_L = 1.0 x 10^{-5} M is hold constant

X_s = 0.30 C_M is varied between 10^{-6} to 10^{-2} M

FIGURE 3.   *Representation of a Typical Spectral Titration Plot, R Versus log C_M.* (X_B = upper limit of R, X_S = lower limit of R.)

Once an **R** value of a sample is measured, the concentration of free and bound species of the probe ions can be determined:

$$[M]_{observed} = \frac{C_M\, X_B\, (R - X_S)}{R\, (X_B - X_S)} \tag{6}$$

$$\Sigma[ML_i]_{observed} = C_M - [M] \tag{7}$$

Notice that C_M, X_B, X_S, and **R** are observables. The spectral titration plot is presented as **R** versus log C_M, and has a symmetric sigmoidal form. This plot and its important parameters, which is illustrated in Figure 3, can be used as a basis to study metal-humic interactions. To find the values of the model parameters, i.e. μ_M, σ, and C_L, the experimental titration plot is fitted to a calculated plot. The simplified flowchart of the fitting algorithm is illustrated by Figure 4.

FITTING ALGORITHM

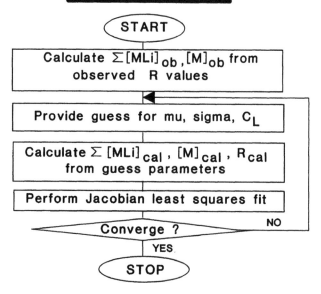

FIGURE 4.　*Fitting Algorithm to Extract the Values of Model Parameters (μ_M, σ, C_L, etc.) from an Experimental Titration Plot.*

1.4   Model Development

The model is developed further by including the inherent competition effects of protons (H) [7]. While proton binding can be studied in the absence of competitive metal binding in metal-free solutions, it is obvious that metal complexation must inevitably be studied in competition with proton binding [2]. Consider an interacting system consisting of metal ions (M), protons (H), and metal-binding sites (L_i) of a humic sample:

$$M + L_i \rightleftarrows ML_i$$

$$H + L_i \rightleftarrows HL_i$$

For this system, it has been shown that [7]:

$$[ML_i] = \frac{C_i \, K_{Mi} \, [M]}{1 + K_{Mi}[M] + K_{Hi}[H]} \tag{8}$$

The only difference between Equations (8) and (2) is the term K_{Hi} [H] in the denominator. [H] can be estimated from pH measurement of the sample. The only way to get K_{Hi} is by making an assumption. Note that for each type of metal ion X (proton included), there exists a Gaussian distribution curve, C_i versus log K_{Xi} (see Figure 1). It is assumed that for one type of humic sample the width of those Gaussian distribution curves (i.e. σ) are the same, and independent of X. Figure 5 illustrates the situation for a system that consists of one type of humic sample, metal ions M and protons H. The two distribution curves have exactly the same form, height, width and area. They are just shifted one from the other along the log K_{Xi} axis indicating that M and H bind to the humic sites with a different stability constant. As a logical consequence of this assumption, the ratio of the two formation constants K_{Mi}/K_{Hi} are constant for all i's.

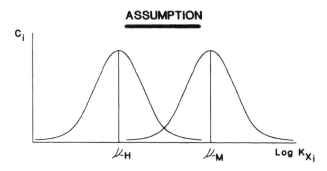

ASSUMPTION

M and H have their own Gaussian distribution
Only centroids (i.e. mu) depend on the type of cations
The width of the distributions (i.e. sigma) and
the total areas (i.e. C_L) are independent of the type of cations

FIGURE 5. *Assumptions that Have to be Made in Order to Get K_{Hi} Values.* Only μ is dependent on the type of cation, σ and C_L are characteristics of the humic substance and independent of the type of cation.

Let P be the constant such that:

$$P = \frac{K_{Mi}}{K_{Hi}} \tag{9}$$

If this constant P can be estimated, then all of the K_{Hi} values can be calculated since the corresponding K_{Mi} values are known. In recalling that μ_M and μ_H are equal to the mean values of log K_{Mi} and log K_{Hi} respectively, P can be found as:

$$P = \frac{10^{\mu_M}}{10^{\mu_H}} \tag{10}$$

Because μ_M is either known or provided as a guessed value, the only term needed is μ_H. Fortunately, data on proton binding are available. For example, the value of $\mu_H = 3.8$ for Suwannee River fulvic acids was determined from acid-base titration data of E.M. Perdue [8]. The K_{Hi} value for each type of binding site, i, corresponding to each K_{Mi} can then be calculated from Equation (9) using the value of P found from applying Equation (10). Once all the parameter values are known, the total concentration of the bound metal species, $\Sigma[ML_i]_{calc}$, can then be calculated by combining Equations (8) and (3) and summing for all i's.

1.5 Why Eu^{3+}

Actually, Eu^{3+} is not usually the metal ion of interest. Eu^{3+} is used here as a fluorescent probe ion. There are several good reasons why Eu^{3+} was chosen as the probe ion. First, a good fluorescent probe must bind to the oxygen-containing ligands such as carboxylate, phenol, etc. in the humic samples, and the free and bound species of the probe ions must be able to be differentiated. As a strong Lewis acid, Eu^{3+} binds strongly to oxygen or any other electron pair donor. The hypersensitivity of one of the emission lines of Eu^{3+} provides a means to differentiate its bound and free species. Because the probe ions are very often not the only fluorescent species in the sample, the fluorescence of the probe ions must be differentiated from the background fluorescence. The fluorescence from Eu^{3+} has a very long lifetime compared to that of the humic samples, so the two fluorescence signals can be discriminated by a standard time-resolved technique [9]. Furthermore, Eu^{3+} has very narrow emission lines in an easily observable spectral region which makes detection of the signal easier. To study the interaction of metals other than the probe ion with the humic binding sites, the probe ion must have a large stability constant for complexing the humic binding sites. The reason for this requirement will be clear after the following discussion of the three cation-multiligand system. The model demands a large amount of the other metal to displace the probe ions from the humic binding sites.

1.6 Three Cation-Multiligand System

The current objective of this project is to functionally characterize interaction between any metal ion N and any humic sample, using a single probe metal ion M, in a three-cation multiligand system, i.e. M, H, and N. A three-cation multiligand interaction system is described by putting an additional metal ion N into the model. Consider the three complexation equilibria which are involved:

$$M + L_i \rightleftarrows ML_i$$
$$H + L_i \rightleftarrows HL_i$$
$$N + L_i \rightleftarrows NL_i$$

For this three cation-multiligand system, it can be shown [10] that:

$$[ML_i] = \frac{C_i \, K_{Mi} \, [M]}{1 + K_{Mi}[M] + K_{Ni}[N] + K_{Hi}[H]} \qquad (11)$$

Using the same assumption described previously to get K_{Hi}, the individual K_{Ni} values can be known. Note that this can only be done after the μ_M value is first established from a two cation (i.e. M and H) experiment. The concentration of the free species of the other metal, [N], in general cannot be determined directly (except for a few of them that can be determined from the ion specific electrode technique). To circumvent this problem, the total concentration of the ion N, C_N, is deliberately made two or three orders of magnitude greater than the total concentration of the binding sites, C_L, so that [N] can be accurately estimated by C_N, which is a known value. This explains why the probe ions must have a large binding constant. Otherwise, after the addition of large excess of C_N (i.e. $C_N \gg C_L$), no more probe ions will be left bound to the humic binding sites and the measurement would become useless. Once all of the parameter values in Equation (11) are known, the total concentration of the bound probe metal species, $\Sigma[ML_i]_{calc}$, can be calculated by combining Equations (11) and (3) and summing over all i's.

1.7   Ionic Strength

For normal titrations of humic samples with Eu^{3+}, ionic strength effects are not important because of the low concentrations of the humic samples (20 ppm) and Eu^{3+} (micro- to millimolar range). However, titrations involving metal ion N (i.e. metal ions other than the probe ion) need to take ionic strength into account. This is due to the limitation of the experimental methodology which demands that C_N be made relatively high (hence also the ionic strength of the system) so that [N] can be estimated accurately enough by C_N. In the previous discussion, the stability constants K_{Mi}, K_{Ni}, and K_{Hi} are not really true thermodynamic constants as they were used, instead they are merely concentration quotients. The relations between the concentration quotients K_{Mi}, K_{Ni}, and K_{Hi} and their respective true thermodynamic constants K^T_{Mi}, K^T_{Ni}, and K^T_{Hi} are:

$$K^T_{Mi} = \frac{\{ML_i\}}{\{M\}\{L_i\}} = \frac{[ML_i]}{[M][L_i]} \times \frac{\gamma_{MLi}}{\gamma_M \cdot \gamma_{Li}} = K_{Mi}\,\Gamma_{Mi} \qquad (12)$$

$$K^T_{Ni} = \frac{\{NL_i\}}{\{N\}\{L_i\}} = \frac{[NL_i]}{[N][L_i]} \times \frac{\gamma_{NLi}}{\gamma_N \cdot \gamma_{Li}} = K_{Ni}\,\Gamma_{Ni} \qquad (13)$$

$$K^T_{Hi} = \frac{\{HL_i\}}{\{H\}\{L_i\}} = \frac{[HL_i]}{[H][L_i]} \times \frac{\gamma_{HLi}}{\gamma_H \cdot \gamma_{Li}} = K_{Hi}\,\Gamma_{Hi} \qquad (14)$$

where braces { } denote activities and square brackets [ ] denote concentrations, and g denotes an activity coefficient. Notice that because, at a given temperature, K^T_{Mi} will remain constant, the concentration quotient K_{Mi} and the activity coefficient ratio Γ_{Mi} will vary in opposite directions as ionic strength changes. The activity coefficient ratio Γ_{Mi} equals one at zero ionic strength and increases with increasing ionic strength. Therefore K_{Mi} values are equal to K^T_{Mi} at zero ionic strength and tend to decrease with increasing ionic strength. In a given solution of metal and ligand, the total concentration of the bound species, $\Sigma[ML_i]$, is thus expected to decrease upon addition of a background electrolyte [11]. This implies that the intensity ratio, **R**, is expected to increase upon addition of a background electrolyte. The ionic strength effect can be incorporated into the model by substituting the concentration quotients K_{Mi}, K_{Ni}, and K_{Hi} by their respective thermodynamic constants and activity coefficient ratios given by Equations (12)-(14):

$$[ML_i] = \frac{C_i [M] \dfrac{K_{Mi}^T}{\Gamma_{Mi}}}{1 + \dfrac{K_{Mi}^T [M]}{\Gamma_{Mi}} + \dfrac{K_{Ni} [N]}{\Gamma_{Ni}} + \dfrac{K_{Hi} [H]}{\Gamma_{Hi}}} \qquad (15)$$

The activity coefficient ratios can be estimated by estimating each individual activity coefficient involved using the standard Davies Equation. For ionic strength I, the activity coefficients are:

$$\log \gamma_M = -0.509 \, Z_M^2 \{ \sqrt{I} / (1 + \sqrt{I}) - 0.2I \} \qquad (16a)$$

$$\log \gamma_H = -0.509 \{ \sqrt{I} / (1 + \sqrt{I}) - 0.2I \} \qquad (16b)$$

$$\log \gamma_N = -0.509 \, Z_N^2 \{ \sqrt{I} / (1 + \sqrt{I}) - 0.2I \} \qquad (16c)$$

$$\log \gamma_L = -0.509 \, Z_L^2 \{ \sqrt{I} / (1 + \sqrt{I}) - 0.2I \} \qquad (16d)$$

$$\log \gamma_{ML} = -0.509 \, (Z_L - Z_M)^2 \{ \sqrt{I} / (1 + \sqrt{I}) - 0.2I \} \qquad (16e)$$

$$\log \gamma_{HL} = -0.509 \, (Z_L - 1)^2 \{ \sqrt{I} / (1 + \sqrt{I}) - 0.2I \} \qquad (16f)$$

$$\log \gamma_{NL} = -0.509 \, (Z_L - Z_N)^2 \{ \sqrt{I} / (1 + \sqrt{I}) - 0.2I \} \qquad (16g)$$

where Z_X (X = M or N) denotes the charge of the metal ion X and Z_L denotes the effective humic (organic) anion charge.

1.8   Simulation and Experimental Study

Once the model was established, computer programs were written to carry out simulation studies. A series of computer simulation studies were then performed to predict and describe the behavior of various spectral titration plots generated from different complex multiligand equilibria. The effects of varying important parameters such as μ, σ, C_L, C_N, pH and ionic strength are discussed below. This is followed by presentation of the corresponding experimental data which support the prediction from the model.

2.   EXPERIMENTAL

2.1   Experimental Setup

A block diagram of the experimental setup is illustrated in Figure 6. In order to induce fluorescence from the Eu^{3+} probe ions, a Lambda Physik EMG 102 XeCl excimer laser and a FL3002 dye laser were used in tandem to supply a high intensity tunable light source. The average output of the XeCl excimer laser was around 1.6 W (10 Hz repetition rate) at 308 nm. To pump the resonant transition of Eu^{3+} ion at 394.3 nm, an ultraviolet dye QUI (Exiton), was used. A small portion of the beam generated by the dye laser was split off via a beam splitter and monitored by a photodiode detector. This signal was sampled by one of the two boxcar integrators (Model SR 250 Stanford Research Systems) used in this experiment and served as a reference signal to monitor any changes in the power of the excitation pump beam as a function of time or wavelength.

The rest of the beam was directed by several mirrors and focused with a lens vertically through a cuvette which contained the sample. The fluorescence was collected 90° off axis with respect to the incident beam to minimize the amount of stray radiation entering the monochromator. The fluorescent light was then collimated and focused onto the entrance slit of a one meter double focusing Jarell Ash monochromator.

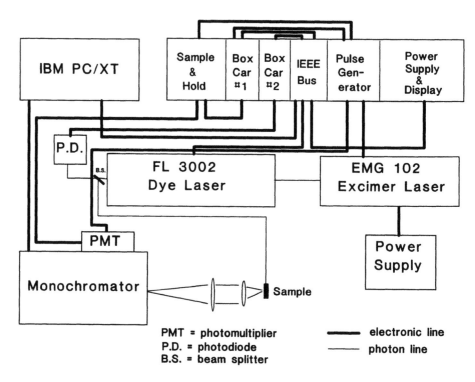

FIGURE 6.   *Block Diagram of Experimental Setup and Electronic*
Interface.

At the exit slit of the monochromator, a gated photomultiplier
unit (an RCA C31034A tube + a Products-For-Research gated socket)
was used to detect the dispersed fluorescence. The output current of the
anode of the photomultiplier tube was passed to a current-to-voltage
converter and subsequently sampled by a sample-and-hold amplifier.
This sample-and-hold amplifier has the ability to sample the voltage for
a predetermined amount of time defined by the trigger pulse width and
hold that voltage until the device is triggered again. The conditioned
output signal of the sample-and-hold amplifier, which was a DC
voltage, was then sampled by another boxcar integrator.   For
measurement of low concentrations of Eu^{3+}, the background signal
becomes important and needs to be corrected.   An automatic

background corrector was designed and installed in the sample and hold module. The output of the two boxcar integrators, which represent the reference and fluorescence signals, was stored by an SR 245 Stanford Research Systems interface, equipped with a 12-bit analog to digital converter. At the completion of the scan, the interface was polled and the entire set of data points was read back to the computer. All measurements were taken as the ratio of the fluorescence and the reference signals.

2.2 Time-Resolved Technique

The transition states between the $^5D_0$ and $4f^n$ configurations in Eu^{3+} are formally electric dipole forbidden. Consequently, the absorption and emission of energy between these states is weak. A very small molar extinction coefficient, approximately 10 $M^{-1}cm^{-1}$, results in low transition probabilities and long fluorescence lifetimes. Lifetimes of several milliseconds are experimentally observable. In contrast, the humic materials have molar extinction coefficients which can range in the tens of thousands $M^{-1}cm^{-1}$ and fluorescence lifetimes that are typically in the nanosecond range. This produces a tremendous amount of background fluorescence in the measurement of the very weak signal of Eu^{3+} fluorescence. Clearly, the utilization of techniques that can discriminate the specific Eu^{3+} fluorescence from background fluorescence deriving from the humic material in the sample solution is of critical importance in this experiment. The disparity in fluorescence lifetimes means that the Eu^{3+} fluorescence can be temporally resolved from the background fluorescence of the humic material by means of a fairly simple time-resolved gated detector. In order to implement the time-resolved technique, a programmable pulse delay generator (Model SR 535 Stanford Research System) was used for precise and accurate control of each timed event in the experiment. This pulse generator was triggered by the excimer laser. After a 200 microsecond delay, which was needed to wait for the background fluorescence to die away, a pulse with a width of 200 microseconds was generated to turn the photomultiplier on. After the photomultiplier was turned on for 110 microseconds, a second pulse with the width of 10 microseconds was generated to trigger the boxcar integrator and sample conditioning circuits. Although the boxcar and the sample-and-hold amplifier were triggered simultaneously, the actual sampling by the boxcar was delayed 200 microseconds via an internal delay to allow the sample conditioning circuits sufficient time to settle. The other boxcar integrator which sampled the reference signal from the photodiode detector was triggered by the time zero pulse (i.e. initial firing of the excimer laser). The timing sequence of the experiment is illustrated in Figure 7.

Timing Sequence

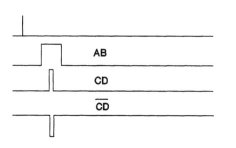

fire excimer laser ($=T_0$) and trigger box car 1

wait 200 us, turn PMT on (for 200 us)

wait 110 us, trigger sample & hold amplifier (for 10 us)

at the same time, trigger box car 2 and make a read

FIGURE 7.   Timing Sequence to Implement Time-resolved Technique.

2.3   Sample Preparation

The humic materials used in this study were dissolved organic materials (DOM) from Suwannee River provided by E.M. Perdue. Because, in one titration plot C_L, C_N, and pH need to be held constant, the humic materials were discretely titrated with Eu^{3+} while holding the volume of sample constant. This is done simply by 1:1:1 (by volume) mixing of different concentrations of Eu^{3+} solution with solutions of the humic materials and the other metal ion, N. The final concentrations of M were varied between 1.0×10^{-6} to 1.0×10^{-2} M. The final concentration of the humic material was kept constant at 20 ppm, except in the study of the effect of C_L. The final concentration of N was kept constant for each titration, e.g. 2.0×10^{-4} M for Pb^{2+}. The pH of each solution was adjusted using concentrated solutions of HCl and NaOH. No buffer solutions were used to avoid unwanted binding between Eu^{3+} and materials other than the humics.

2.4   Optimum Experimental Conditions

Optimum titration conditions were found at low pH and low humic concentration. Low pH was chosen to minimize the speciation of metals and maximize the sensitivity to metal N competition. Low humic concentration was chosen to minimize ionic strength effects and also to minimize precipitation of the humic sample by the metal ions.

3.   RESULTS AND DISCUSSION

3.1   Model Parameter Determination

The effect on the spectral titration plot of varying μ_M while holding the other parameters constant at pH of 6.0 is illustrated in Figure 8. Notice that at this high pH value, while it is easy to discriminate low μ_M values (i.e. between 2.4 and 4.4), it is rather difficult to differentiate high μ_M values (i.e. 6.4 and 4.4). It should be mentioned here that the precision in measuring **R** is around 0.1. Since the probe ion Eu^{3+} interacts with most of humic samples with μ_M values between 4.4 and 6.4, it is important to find a condition where the titration plots are sensitive to the change of μ_M. It turns out that if the pH is lowered to 3.5, as illustrated in Figure 9, the titration plots become quite sensitive to the change of μ_M value in the region between 4.4 to 6.4.

Susetyo, Carreira, Azarraga and Grimm

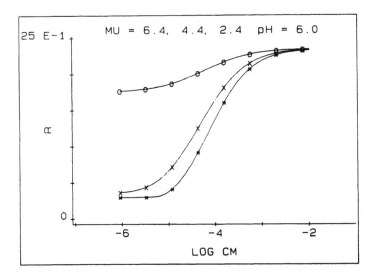

FIGURE 8.   *Effect of Varying μ_M While Holding the Other Parameters Constant at pH = 6.0.*   (O) μ_M = 2.4, (X) μ_M = 4.4, (*)μ_M = 6.4 (simulation).

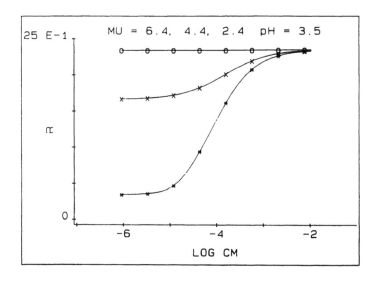

FIGURE 9.   *Effect of Varying μ_M While Holding the Other Parameters Constant at pH = 3.5.*   (O) μ_M = 2.4, (X) μ_M = 4.4, (*)μ_M = 6.4 (simulation).

Similar simulations are also made in the effect of varying σ, while holding the other parameters constant. Figure 10 illustrates this effect at pH 6.0. Notice that the titration plots are not sensitive to the change of σ. However, even though difficult, it is experimentally still possible to discriminate σ values between 1.0 and 2.0. This is an important region since most humic substances have σ values in this range. At pH of 3.5, the titration plots become much more insensitive to a change of σ as illustrated by Figure 11. (Notice that it is misleading to judge the sensitivity of the titration plot to the change of σ only by comparing the physical distances between titration curves in the Figures 10 and 11 because they are at different scales). Since the precision in measuring **R** is around 0.1, according to the simulation results presented in Figure 11, it is impossible to experimentally distinguish different σ values between 0.0 and 2.0 at low pH.

From the above simulation results (Figures 8 - 11), a good strategy for accurately determining μ_M and σ values was developed. First, make a complete titration at pH 3.5 and guess σ with any reasonable value (between 1.0 and 2.0) and extract μ_M value from the data. After that, make another complete titration at pH 6.0, use the accurate μ_M value determined from previous titration and extract σ. Titrations of 20 ppm Suwannee River DOM were done at pH 3.5 and 6.0 and the following parameters were determined: $\mu_M = 6.4$, $\sigma = 1.72$, and $C_L = 1.0 \times 10^{-5}$ M. These values were determined using a μ_H value of 3.8 determined from acid-base titrations data of the Suwannee River DOM done by E.M Perdue. This μ_H value can also be determined accurately by the Lanthanide Ion Probe Spectroscopy (LIPS) technique.

Figure 12 illustrates the effect of varying pH, holding the other parameters constant. Notice that the titration plots are not very sensitive to a change in pH in the pH range of 3.0 to 6.0, but quite sensitive in the pH range of 2.0 to 3.0. A strategy to determine μ_H accurately is to do a titration at pH between 4.0 and 5.0 and guess μ_H with any reasonable guess and determine the other model parameters. Under these conditions, a large error in guessing μ_H does not significantly affect the values of the parameters produced in the fitting process. After that, titration at low pH, say between 2.5 and 2.0, is done and the data are fitted to get a more accurate μ_H value.

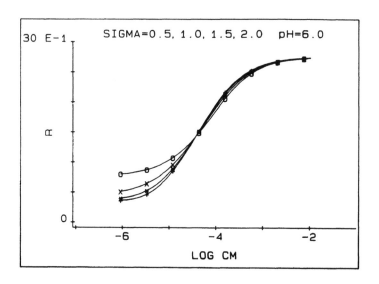

FIGURE 10. *Effect of Varying σ While Holding the Other Parameters Constant at pH = 6.0.* (O) σ = 0.5, (X) σ = 1.0, (*) σ = 1.5, (+) σ = 2.0 (simulation).

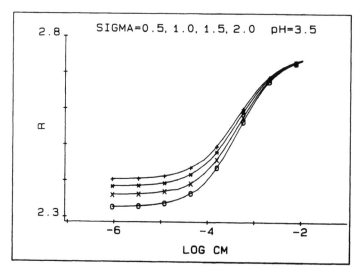

FIGURE 11. *Effect of Varying σ While Holding the Other Parameters Constant at pH = 3.5.* (O) σ = 0.5, (X) σ = 1.0, (*) σ = 1.5, (+) σ = 2.0 (simulation).

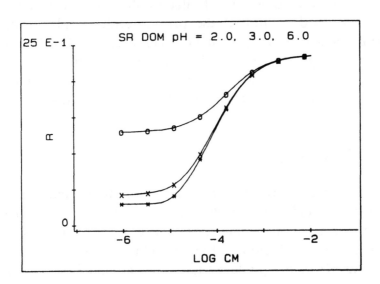

FIGURE 12. *Effect of Varying pH on the Spectral Titration Plot.*
(O) pH = 2.0, (X) pH = 3.0, (*) pH = 6.0 (simulation).

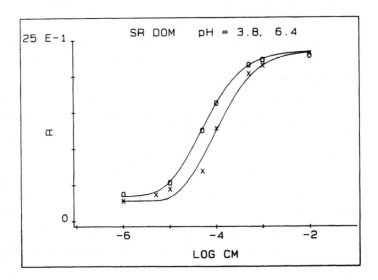

FIGURE 13. *Effect of Varying pH on the Spectral Titration Plot.*
(O) pH = 3.8, (X) pH = 6.4 (experimental).

It is clear from Figure 12 that titrations at low pH must be done with very good control of the pH for each sample, otherwise the titration plots will not be a smooth curve at all. On the other hand, titrations at high pH, say pH> 4.0, can be done without very accurately controlling pH adjustment of each sample. Figure 13 illustrates the experimental results of changing pH from 3.8 to 6.4 to the titration plots.

3.2   Effect of μ_H

It is well known that the majority of humic binding sites consist of carboxylic and phenolic sites. There is always a question as to how these two classes of binding sites behave in the interaction with metal ions in different pH regions. Experimental titrations that were made for phenol at pH of 6.4 produced μ_M values around 8.6. This value was produced using a μ_H value of 9.89. The previously mentioned results show that, for Suwannee River DOM, μ_M is equal to 6.4. This implies that the most abundant, or the most probable, binding sites of the DOM have stability constant of $10^{6.4}$ in complexing the probe ions, Eu^{3+}. Since the μ_H of the DOM is equal to 3.8, it is probably fair to assume that the most abundant sites are carboxylic and not phenolic. A simulation study using these parameter values was conducted to investigate the behavior of these two classes of binding sites in interaction with the probe ions at pH 6.4 and 3.5 and C_L of 2.5 X 10^{-5} M. Figure 14 illustrates the titration plots at pH 6.4. From the simulation study presented previously (see Figures 8 and 9) one would expect that the two curves would be shifted either horizontally or vertically such that for a given C_M value, the one with a higher μ_M would have a lower **R**. Recall that a higher μ_M implies a greater stability constant, hence higher $\Sigma[ML_i]$ and higher emission intensity of the 616 nm band and, therefore, a lower **R**. The two experimental curves in Figure 14 are indeed shifted vertically one from the other, however, the curve related to phenol (which has the higher μ_M) is consistently above the carboxylic curve throughout the titration.

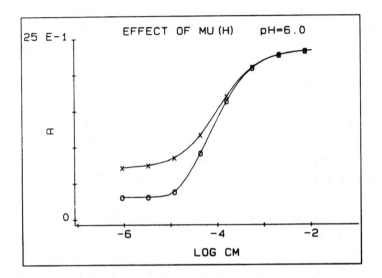

FIGURE 14. *Effect of μ_H on the Spectral Titration Plot at pH = 6.0.*
(O) Carboxylic sites $\mu_M = 6.4$ and $\mu_H = 3.8$ (X) Phenolic sites
$\mu_M = 8.6$ and $\mu_H = 9.9$ (simulation).

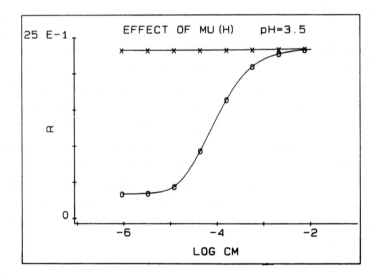

FIGURE 15. *Effect of μ_H on the Spectral Titration Plot at pH = 3.5.*
(O) Carboxylic sites $\mu_M = 6.4$ and $\mu_H = 3.8$ (X) Phenolic sites μ_M
$= 8.6$ and $\mu_H = 9.9$ (simulation).

The following explanation for this 'discrepancy' is given: Since phenolic sites have a very strong affinity to proton binding (μ_H of 9.89), even at a pH as high as 6.4, the competition effect due to the inherent competition of protons has forced the phenol curve to level off before reaching the lower limit value, X_S. At the same pH value, the same competition has not significantly affected the carboxylic curve which still reaches the X_S value, because the proton affinity of the carboxylic sites is by six orders of magnitude smaller, than that of phenol. If this simulation were done at very high pH, say 14 or so, such that the competition effect from protons could be neglected for both phenol and benzoic acid systems, then the two curves would be shifted as predicted by Figure 8 or 9. Notice that in creating Figures 8 and 9, μ_M was varied while μ_H was held constant, and this is certainly not the case with carboxylic and phenolic sites which have very different μ_H values. Figure 15 shows the results of a simulation at pH of 3.5. At this low pH, the inherent proton competition effect becomes much more important in the phenolic sites and pushes the phenol curve to its upper limit (X_B), consistent with the above explanation. Even though the μ_M of phenol is three orders of magnitude greater than that of the carboxylic sites, the latter are able to compete efficiently with the phenolic sites and consistently form more europium complexes throughout the titration. The phenolic sites are very sensitive to the pH due to their strong proton affinity. The carboxylic sites, on the other hand, are not as sensitive to a change in pH because of their relatively weak affinity toward proton binding. In Figure 15 the titration curve for the carboxylic sites still reaches close to its lower limit X_S even at pH 3.5. This implies that the competition effect that renders the phenolic site 'inactive' at this low pH does not significantly affect the interaction between the carboxylic sites and the probe ions. The importance of the above observation is that at low pH, metal-humic interaction is mostly with the carboxylic sites, while the phenolic sites are deactivated by protonation.

3.3 Experimental Validation of the Model

The effect of varying C_L, the concentration of the binding sites, was measured at two different pH's. Figure 16 illustrates the effect of varying C_L at pH 6.0 where the inherent proton competition is not important. Three titrations were done using three different C_L values corresponding to 20, 40 and 80 ppm Suwannee River DOM. It is expected that the three curves will be shifted horizontally and/or vertically such that at a given C_M value, the curve that represents higher C_L will have lower **R** value. Recall that higher C_L produces higher $\Sigma[ML_i]$, higher 616 nm band, and hence lower **R**. The curve related to 80 ppm DOM is then expected to be at the rightmost position, followed by the 40 ppm and 20 ppm curves, and, indeed, this is the case. Figure 16 also indicates that the higher the C_L the less sensitive the spectral titration plot is to a change of C_L. Notice that even though the distance between one curve and the other is fairly small, it is still distinguishable by the lanthanide ion probe technique. Figure 17 illustrates the effect of varying C_L at pH = 3.5, where the inherent proton competition becomes more important. Three different C_L values corresponding to 10, 20, and 40 ppm Suwannee River DOM were used in the titration. The three titration curves are shifted horizontally following the predicted pattern discussed above. In addition to that, slight vertical shifts between the three curves are also observed, indicating that the proton competition effect is large. It should be noticed that at low pH, the spectral titration plot is more sensitive to the change of C_L as demonstrated by the larger distance between the three curves compared to its analog in Figure 16. The inherent proton competition at low pH makes the effective C_L significantly lower than the real C_L value and thus makes the titration plot more sensitive to the change of C_L.

The effect of ionic strength on the spectral titration plot was studied by varying the concentration of background electrolyte, NaCl. The choice of NaCl as the background electrolyte was based on the fact that Na^+ ion is very weakly bound, if at all, to the humic metal binding sites, and Cl^- is already in the system as the counter ion that comes from dissolving $EuCl_3$ or from pH adjustment using HCl.

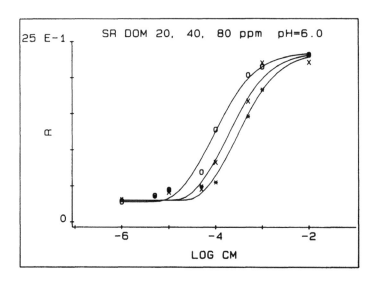

FIGURE 16. *Effect of Varying C_L on the Spectral Titration Plot at pH = 6.0.* (O) 20 ppm DOM, (X) 40 ppm DOM, (*) 80 ppm Suwannee River Dissolved Organic Matter (experimental).

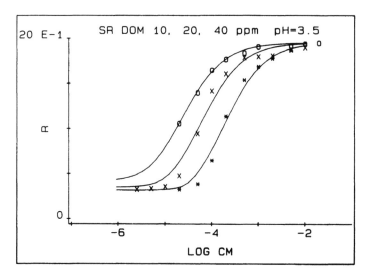

FIGURE 17. *Effect of Varying C_L on the Spectral Titration Plot at pH = 3.5.* (O) 10 ppm DOM, (X) 20 ppm DOM, (*) 40 ppm Suwannee River Dissolved Organic Matter (experimental).

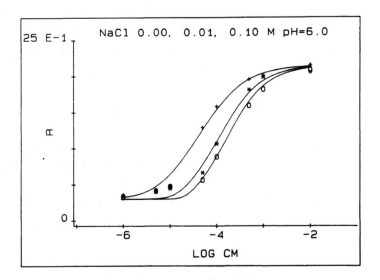

FIGURE 18. *Effect of Varying Ionic Strength on the Titration Plot at* *pH = 6.0.* (O) NaCl = 0.00 M, (*) NaCl = 0.01 M, (+) NaCl = 0.10 M (experimental).

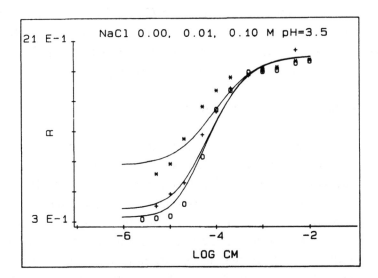

FIGURE 19. *Effect of Varying Ionic Strength on the Titration Plot at* *pH = 3.5.* (O) NaCl = 0.00 M, (*) NaCl = 0.01 M, (+) NaCl = 0.10 M (experimental).

Figure 18 illustrates the effect of varying ionic strength on the spectral titration plot at pH of 6.4. Three titration plots shown here were done using three different ionic strengths corresponding to 0.0, 0.01, and 0.1 M NaCl. From the previous discussion of the model, it is expected that addition of background electrolyte will decrease the total concentration of bound probe ion. It should be emphasized here that this decrease is **not** due to the competition effect of Na^+ (similar to proton competition) to occupy the same set of humic metal binding sites, but purely due to thermodynamic activity. Figure 18 shows that the prediction from the model is accurate and supported by the experimental data. The three curves shift horizontally so that the one associated with the higher ionic strength always has the higher **R** value (i.e. lower $\Sigma[ML_i]$). Figure 19 illustrates the effect of varying ionic strength at pH = 3.5. At this low pH, the titration plots are pushed upward not only by the ionic strength but also by the proton competition effect. These combined effects make the titration plot at low pH very sensitive to the variation of ionic strength. As a result, the titration plots are shifted vertically one from the other by relatively large distances, especially in the low C_M region. Because all of the model parameters except Z_L (i.e. the average humic anion charge) are known, the data in Figures 18 and 19 can be used to find it. A value of $Z_L = 2.8$ was produced by the fitting routine.

3.4   Significance of the Model

All of the experimental data that are related to one humic sample and one metal species can be fitted to the model with a single set of parameter values. For Suwannee River Dissolved Organic Matter, those values are: $\mu_M = 6.4$, $\sigma = 1.7$, $\mu_H = 3.87$, and $Z_L = 2.8$; where Z_L is the effective charge of a humic molecule. The fact that these parameters do not change even though the chemical parameters of the samples (such as pH, ionic strength, concentration of binding sites, concentration of metal ion other than probe ions) are varied indicates that the parameters found are true thermodynamic constants and not just conditional constants. Other existing models dealing with metal-humic interactions either produce conditional constants, which are constant only if pH and ionic strength are constant or cannot be used if more than one type of cation is present that can be bound by humic material [7].

3.5 Quick Screening Technique

Once the model parameter values such as μ_M, μ_H, σ, C_L, and Z_L are well established, a one-point-titration can be done to do quick screening of many metal ions to determine their μ_N values. As in the full titration procedure, this quick screening technique must be done at low pH (around 3.5) to keep the speciation of the metal ions as simple as possible. To maximize the effects of competition at the low pH, C_M of 1.0×10^{-5} M was chosen and C_N was chosen such that it gave an intensity ratio (**R**) around 1.0. All of the parameter values μ_M, μ_H, σ, C_L, Z_L, pH, C_M, C_N were the input to a computer program which calculated the value of μ_N.

TABLE 1.   *List of μ_N Values for Several Metals.*

Ion	Conc. (M)		% in soln.	pH	μ_N
Al^{3+}	3	E-4	97	3.5	5.2
Ba^{2+}	1	E-2	100	3.5	3.1
Be^{2+}	6	E-3	100	3.8	3.5
Ca^{2+}	1	E-2	100	3.5	2.9
Cd^{2+}	1	E-2	54	3.2	3.3
Cr^{3+}	1	E-4	78	3.5	5.6
Cu^{2+}	1.3	E-4	99	3.5	4.9
$FeOH^{2+}$	6	E-5	47	3-4	5.5
Fe^{3+}	6	E-5	3	3-4	7.7
Mg^{2+}	1	E-2	100	3.5	< 2
Ni^{2+}	6	E-3	98	3.5	3.3
Pb^{2+}	4	E-4	97	3.6	5.2
Zn^{2+}	1	E-2	97	3.5	3.5

Table 1 shows the list of μ_N values of several metal ions of environmental and agricultural interest, such as Al^{3+}, Pb^{2+}, Ca^{2+}, Cu^{2+}, Cd^{2+}, etc., which were produced by this quick screening technique. The pH at which the titrations were done, the concentrations of the metal ions N (C_N) and their fractions (f) at the corresponding pH are also given in Table 1. These fractions were generated using the program MINTEQA2 [12].

3.6   Stability Constants of Individual Species

Sometimes significant speciation cannot be avoided even at low pH, such as in the case of Fe(III). Table 2 shows the fractions of the Fe(III) species at three different pH's: 3.0, 3.5, and 4.0. The quick screening technique described above was used at these pH's to determine the μ values for each individual species. Titration at each pH produced a $K_{effective}$ value (K is 10^μ). This effective value is related to the individual K values by a simple relation:

$$K_{effective} = K_{Fe} \, f_{Fe} + K_{FeOH} \, f_{FeOH} + K_{Fe(OH)_2} \, f_{Fe(OH)_2} \qquad (17)$$

Titration at three different pH's produced three different $K_{effective}$. Because the individual fractions at different pH's were known, the three Equations with three unknown (K_{Fe}, K_{FeOH}, and $K_{Fe(OH)2}$) could be solved simultaneously with the following results: log K_{Fe} = 7.7 (μ_{Fe}), log K_{FeOH} = 5.7 (μ_{FeOH}), and $K_{Fe(OH)2}$ = 0.

TABLE 2.   Fe(III) Speciation at Low pH (from MINTEQA2).

Species	Percentage of Species at		
	pH 3	pH 4	pH 5
Fe^{3+}	11.3	2.6	0
$FeOH^{2+}$	67.3	48.6	24.3
$Fe(OH)_2^+$	21.2	48.7	75.7

4.   CONCLUSIONS

The Lanthanide Ion Probe Spectrofluorometry (LIPS) technique used in this study was built around the hypersensitive property of one of the transitions of the fluorescent probe ion, Eu^{3+}, at 616 nm. This transition is normally a forbidden transition, but interaction with different ligand environments (such as the humic binding sites) often makes the transition more allowed and enhances the intensity of its fluorescence emission. The intensity ratio **R**, defined as I_{592}/I_{616}, was used as a measure of percent/bound of the probe ions. The emission fluorescence at 592 nm comes from a non-hypersensitive transition and is not affected by different ligand environments that are bound to the probe ions.

Various different binding sites in humic substances are assumed to produce an equal amount of intensity enhancement from the hypersensitive effect. This assumption can be justified because most of the active binding sites of the humic samples at the pH where this work was done are carboxylic sites and thus are expected to give the same amount of perturbation to the probe ions.

The LIPS technique, in conjunction with the Gaussian model, can be used to predict and explain the observations found in experiments designed to study metal speciation in the context of metal-humic interactions. The competition effects that come from protons and metal ions other than the probe ions as well as the thermodynamic effects originating from ionic strength can be satisfactorily described by the model. The simulation results are supported by the experimental data.

All of the experimental data under various conditions can be fitted to the model with just one single set of parameter values. For Suwannee River DOM, those values are: $\mu_M = 6.4$, $\sigma = 1.7$, $\mu_H = 3.87$, and $Z_L = 2.8$. This indicates that the parameters found are true thermodynamic constants and not just conditional constants or mere curve-fitting parameters.

A one-point titration technique was established to do a quick screening of various metal ions of interest. This method is based on the established parameter values generated from the full titration experiments. Furthermore, stability constants of individual species of a metal ion can be determined by doing measurements at several different pH's. This technique has been demonstrated in the case of iron (III).

In addition to multiligand system, the LIPS technique can be applied to any single ligand system (i.e. for $\sigma = 0$) or well defined ligand mixtures (i.e. several different well defined single ligands), and expands its area of application to determine complex stability constants between any metal ion (or proton) and any ligands.

REFERENCES

1.   Christman, R.F., and E.T. Gjessing, Eds. *Aquatic and Terrestrial Humic Materials,* (Ann Arbor, MI: Ann Arbor Science, 1983), p. 219.

2.   Perdue, E.M., and C.R. Lytle. "Distribution Model for Binding of Protons and Metal Ions by Humic Substances," *Environ. Sci. Technol.* 17:654-660 (1983).

3.   Kim, H., "Computer Programming in Physical Chemistry Laboratory, Least-Squares Analysis," *J. Chem. Ed.* 47:120-122 (1970)

4. Horrocks, W.DeW., and D.R. Sudnick. "Lanthanide Ion Probes of Structure in Biology. Laser Induced Luminescence Decay Constants Provide a Direct Measure of the Number of Metal-coordinated Water Molecules," *J. Am. Chem. Soc.* 101:334-340 (1979).

5. Sinha, S.P. Ed. *Systematics and the Properties of the Lanthanides*, (Dordrecht, Holland: D. Reidel Publishing Co., 1982), pp. 449-450.

6. Dobbs, J.C., W. Susetyo, F.E. Knight, M.A. Castles, L.A. Carreira, and L.V. Azarraga. "Characterization of Metal Binding Sites in Fulvic Acids by Lanthanide Ion Probe Spectroscopy," *Anal. Chem.* 61:483-488 (1989).

7. Dobbs, J.C., W. Susetyo, L.A. Carreira, and L.V. Azarraga. "Competitive Binding of Protons and Metal Ions in Humic Substances by Lanthanide Ion Probe Spectroscopy," *Anal. Chem.* 61:1519-1524 (1989).

8. Perdue, E.M., Personal Communication (1988).

9. Butt, W.R. Ed. *Practical Immunoassay-The State of The Art,* (New York: Marcel Dekker, 1984).

10. Susetyo, W., J.C. Dobbs , L.A. Carreira, D.M. Grimm, and L.V. Azarraga, "Development of a Statistical Model for Metal Humic Interactions," *Anal. Chem.* 62:1215-1221 (1990).

11. Perdue, E.M., "The Effects of Humic Substances on Metal Speciation," in *Aquatic Humic Substances: Influence on Fate and Treatment of Pollutants*, I. H. Suffet and P. MacCarthy, Eds., (Washington, DC: American Chemical Society, Advances in Chemistry Series No. 219, 1989), pp. 281-295.

12.   Brown, D. S., and J. D. Allison. *"MINTEQA1, Equilibrium Metal Speciation Model,"* Users Manual EPA/600/3/87/012 USEPA, Athens, Georgia (1987)

5

AN OVERVIEW OF MODELING TECHNIQUES FOR SOLUTE TRANSPORT IN GROUNDWATER

Peter S. Huyakorn, Jan B. Kool and T. Neil Blandford
HydroGeoLogic, Inc.
Herndon, VA 22070

1.   INTRODUCTION

Recent concerns over the environmental impact of land disposal of hazardous wastes have led to the rapidly increasing use of computer models to perform predictive assessment of groundwater contamination problems. Modeling has been used in support of field investigation, remediation effort, regulatory development and demonstration of regulatory compliance. In some cases, simulation results have also been used to provide technical support in groundwater litigation cases.

In this chapter, an overview of a number of modeling approaches currently used to predict the fate and transport of metals and other hazardous components in groundwater is presented.   The approaches considered are: (1) analytical and semi-analytical solutions that can be effectively combined with the Monte Carlo technique for uncertainty analysis,   (2) particle tracking methods for migration pathline and travel time analysis, and (3) numerical (finite element and Laplace Transformed Galerkin (LTG) finite-element) models that incorporate various site-specific conditions and are tailored for operation on personal computers. Specific examples highlighting the development and application of the various modeling techniques are presented.

0-87371-277-3/93/$0.00 + $.50

These examples have been designed to demonstrate the utility of models in support of risk assessment, site screening or performance assessment, and site-specific groundwater contamination investigations and design of remedial measures.

2.   PROBLEM IDENTIFICATION

The important features of a given groundwater contamination problem should be identified before attempting to perform modeling or related quantitative analysis. There are four important aspects that merit consideration.

The first aspect concerns the sub-surface media types. Types of media generally encountered are: porous versus fractured media, and aquifers versus aquitards.

The second aspect concerns the type of flow system. Two major system types, confined or unconfined, may be encountered. The third aspect concerns flow conditions which include: saturated versus variably saturated flow, steady versus transient flow, and uniform versus nonuniform flow.

The final aspect concerns transport processes believed to be important. For metals, these usually include advection, hydrodynamic dispersion, sorption, speciation, and dissolution/precipitation. In some cases, degradation resulting from chemical reactions or radioactive decay may be important. For fractured porous media, the effects of diffusion of contaminants into porous blocks (matrix diffusion) may also merit consideration.

Once the important features of the problem are recognized, selection of a proper modeling approach can proceed.

3.   DESCRIPTION OF A TYPICAL PROBLEM AND MODELING OBJECTIVES

Consider a typical situation involving an unconfined groundwater system overlain by a surficial source of contamination such as a land disposal unit (Figure 1). Leachate emanates from the bottom of the source and moves through the unsaturated zone toward the underlying water table. Upon entering the saturated zone, the contaminants move in conjunction with the groundwater flow; the contaminant migration is controlled by advection and hydrodynamic dispersion. For certain metals, sorption and speciation may also be controlling processes. The contaminants may be intercepted by receptor wells and/or discharged into streams or surface water bodies downstream from the source.

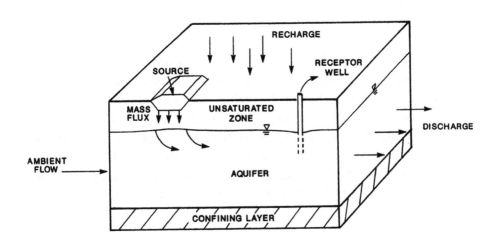

FIGURE 1.   Schematic Diagram of Commonly Encountered Groundwater Contamination Scenario.

A number of alternative modeling approaches are available to handle this type of problem. In general, selection of the appropriate approach is dependent upon the modeling objectives as well as the salient features of each approach. Typical objectives include:

1.   Pathline and travel time analysis. Modeling may be used to determine migration pathlines of contaminants and the time of travel from the source to a specified location of interest.

2.   Plume migration assessment. Modeling may be used to delineate the contaminant plume and assess its rate of migration toward discharge points or other specified locations.

3.   Concentration distributions and breakthrough curves. Modeling may be used to predict the spatial concentration distribution of dissolved contaminants in the plume and the breakthrough curves of concentration versus time for specified locations of interest.

4.   Exposure and risk assessment. Modeling may be used to provide information needed for subsequent exposure and risk assessments. Such information may be presented in the form of breakthrough curves, dose rates, and values of dilution attenuation factors (DAFs) at specified points of exposure. If the uncertainty of one or more model input parameters is deemed to be important, the modeling results may be presented in the form of cumulative probability functions of concentration or DAFs.

4.   SOLUTION TECHNIQUES

Having identified the modeling objectives, the alternative solution approaches that may be used to meet such objectives are described, with evaluation of the advantages and disadvantages of each approach.

The first approach considered is particle tracking [1,2,3]. The particle tracking method is suitable for the calculation of contaminant pathlines and travel times, but not concentrations. It is computationally efficient and simple to use. The method can account for advection and equilibrium sorption only; dispersion and other attenuation processes are neglected.

The second approach considered is solution of the advection-dispersive transport equation via analytical methods. Analytical modeling may provide temporal and spatial distributions of concentration, and breakthrough curves at downstream locations [4,5,6]. Analytical solutions are usually efficient and simple to use. They are, however, limited to homogeneous and isotropic aquifer systems with regular geometries and simple boundary and flow conditions. (Typically, the analytical solution approach assumes that flow in the aquifer is uniform, radial or a combination of uniform and radial flow fields).

The high computational efficiency of the analytical solution approach makes it conducive for linkage to Monte Carlo techniques for uncertainty analysis. Analysis of the uncertainty in model output, due to the uncertainty in model input parameters, may be desirable for exposure and risk assessments.

The third modeling approach is to use numerical solution techniques for the flow and transport equations. Among these techniques are finite difference and finite element methods [7,8], as well as recently developed boundary element methods [9]. Other techniques specifically developed for handling the advective-dispersive transport equation include the method of characteristics and the random walk method [10,11]. The numerical solution approach is versatile and capable of handling site-specific conditions (such as material heterogeneity and anisotropy, complex boundary conditions and chemical reactions, and transport in nonuniform and/or transient flow fields) that are not amenable to analytical solution. However, numerical solution techniques, when applied to the transport equation, may be prone to potential difficulties such as numerical dispersion and oscillations. Multi-dimensional numerical simulations (particularly 3-D) are computationally intensive. Numerical codes are generally more difficult to use than analytical codes. Because of computational requirements, the numerical solution approach is often not conducive to Monte Carlo uncertainty analysis.

The fourth transport modeling approach considered herein is the recently developed Laplace-Transform-Galerkin (LTG) method [12]. This technique provides better accuracy and higher efficiency for time dependent problems than finite difference or finite element methods because it uses the Laplace Transformation to incorporate the time domain, which eliminates the need for time stepping and reduces the occurrence of numerical oscillations. The LTG approach is, however, limited to linear transport equations and steady-state velocity fields.

5.   SPECIFIC SCENARIOS AND SELECTED MODELING APPROACHES

In this section, five specific scenarios involving various applications of the solution techniques presented in the foregoing section are described. Recognizing the major advantages and drawbacks of each of the four approaches, their use is blended as necessary to satisfy the modeling objectives of each scenario.

The first two scenarios involve the migration of leachate from a land disposal unit overlying an unconfined groundwater system. Scenario 1 involves a landfill, and scenario 2 involves a surface impoundment. A major distinction between the two scenarios is the fact that higher infiltration rates generated by the surface impoundment unit may cause significant disturbance of the ambient groundwater flow field, and mounding of the water table may occur. In both cases, it is desired to determine contaminant concentration values at specified downstream locations that correspond to entry points of receptor wells in the saturated zone. Modeling results are used to provide support for site screening analyses and subsequent exposure and risk assessments.

The third scenario concerns the design and operation of a pumping well recovery system in a confined aquifer with a known contaminant plume. In this case, contaminant pathline analysis and delineation of well capture zones are desirable.

The fourth scenario concerns transient transport of conservative and chemically active (or radioactive) metals in a shallow unconfined aquifer with relatively uniform ambient groundwater flow. In this case, the unsaturated zone may be neglected. The modeling objective is to define the spatial distribution of concentration within the plume.

The fifth scenario concerns contaminant transport from a surface source overlying an anisotropic, unconfined aquifer subject to well pumping. In this case, the problem is fully three-dimensional. The spatial distribution of concentration within the unsaturated and saturated zones is required. Of particular interest is the determination of the steady-state (maximum attainable) concentration value in the pumping well.

Listed in Table 1 are the modeling approaches selected for the five scenarios. A description of these approaches and discussion of the simulation results are provided in the following sections.

TABLE 1. *Specific Transport Scenarios and Modeling Approaches.*

Transport Scenario	*Selected Modeling Approach*
• Leachate migration from landfill	• Composite semi-analytical unsaturated zone model analytical saturated zone model
• Leachate migration from surface impoundment	• 2-D numerical finite element variably saturated model
• Well pumping/recovery system	• Particle tracking model
• Transient transport of conservative metals and chained decay	• 2-D finite element and LTG models
• Three-dimensional transport from landfill to a pumping well	• 3-D numerical finite element variably saturated model

5.1   Leachate Migration from Landfill

The first scenario considered is that of leachate migration from a landfill. The typical modeling scenario is illustrated in Figure 2a. Both the unsaturated and saturated zones are simulated. The unsaturated zone may play an important role in situations where degradation occurs and/or transient effects are important. In addition, three-dimensional dispersion in the saturated zone is considered to be important. A composite unsaturated-saturated zone analytical model has been presented for the U.S. EPA to provide efficient simulation capability for contaminant transport [13,14]. The assumptions and modeling methodology employed in this model are detailed in a technical background document [15]. The following is a summary of salient features of the model:

1.   It contains both unsaturated-zone and saturated-zone components. The unsaturated-zone component consists of one-dimensional, semi-analytical infiltration and solute transport modules. These modules are linked to the saturated-zone analytical transport module describing three-dimensional dispersion in a uniform, horizontal groundwater flow field.

2.   Both infiltration and groundwater flow are assumed to be at a steady state. The infiltration rate is regarded as small enough so that it does not perturb the ambient flow field.

3.   A vertical strip Gaussian source with a prescribed concentration boundary condition is used to simulate transport in the saturated zone. Coupling of the unsaturated-zone and saturated-zone transport modules is performed via the use of a mass balance requirement as depicted in Figure 2b. In this figure, m is the contaminant mass flux entering the water table and leaving the Gaussian source, I is the infiltration rate, A is the area of the land disposal unit, C^* is the entry concentration (assumed to correspond to the peak concentration of the Gaussian source), V_x is the ambient groundwater flow velocity, σ is the standard deviation of the Gaussian source, H is the average depth of vertical mixing in the control volume, and ℓ and W are the dimensions of the landfill.

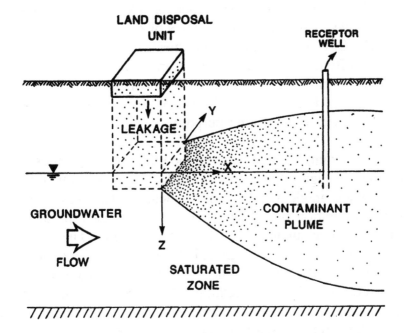

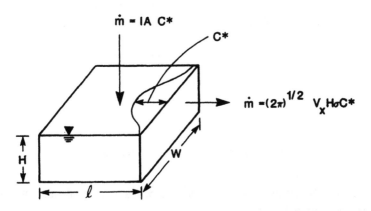

FIGURE 2. *The Modeling Scenario for Leachate Migration from a Landfill (a), and Mass Balance Requirement for Coupling the Unsaturated and Saturated Zone Transport Modules (b).*

4.   The composite transport model is linked to a Monte Carlo driver which allows evaluation of the effects of model input data uncertainty [16]. The CPU and memory requirements of the completely assembled code are sufficiently low that Monte Carlo simulations consisting of one thousand or more runs can be accommodated by AT-class personal computers. However, the model is not designed for site-specific use. It may be used for screening purposes.

An example demonstrating the utility of the composite model is illustrated in Figure 3. In this example, the model is first applied in a deterministic mode to the case depicted. This case concerns the leachate migration of a conservative metal species through a vertical cross section of an unconfined aquifer system. The modeled cross section is bounded on the right side by a ditch in which the water level is 19 m above the datum. The left vertical boundary of the modeled region corresponds to a location where the water table position is known to be 21 m above the datum. The rate of infiltration through the base of the land disposal unit and the leachate concentration is 1 m/yr and 1 ppm, respectively. The hydraulic properties of the aquifer are assumed to be homogeneous and isotropic. The hydraulic conductivity, K, and effective porosity, ϕ, are 750 m/yr and 0.25, respectively. The constitutive relations used to describe soil moisture characteristics in the vadose zone are as follows:

$$k_{rw} = ((S_w - 0.25)/0.75)^4 \tag{1}$$

and

$$S_w = 0.25 + 0.75 \, [1 + (0.2\psi)^2]^{-1} \tag{2}$$

where k_{rw} is the relative permeability, S_w is water saturation, and ψ is capillary pressure head.

The transport parameters of the aquifer are the longitudinal and transverse dispersivities, the values of which are given in Figure 3a for both the unsaturated and saturated zones.

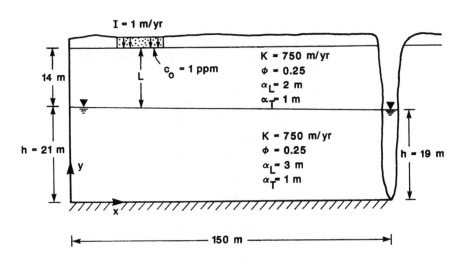

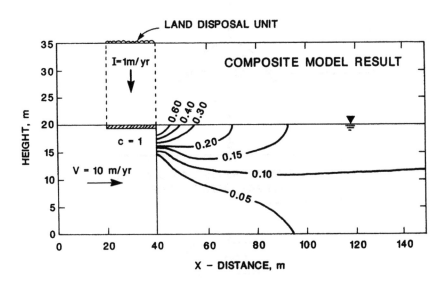

FIGURE 3. *Schematic Representation of Case Study Modeled Using EPACML Composite Model (a), and Model Results in Terms of Steady-state Concentration Contours (b).*

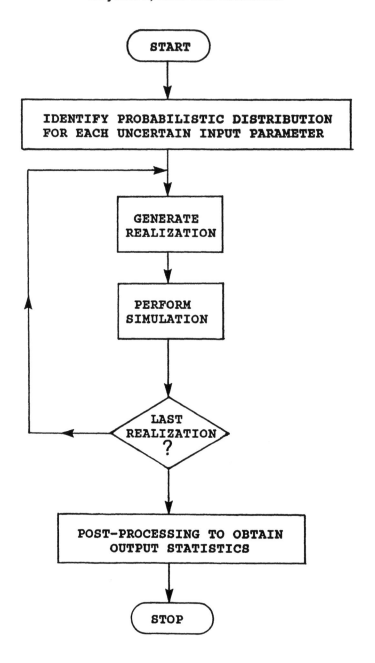

FIGURE 4.   Flow Chart Illustrating Implementation of the Monte Carlo Method for Uncertainty Analysis.

The composite model was used to compute transient concentration distributions, and steady-state concentration values at selected locations in the saturated zone. In the simulation, the thickness of the unsaturated-zone column, L, was set to 15 m, and the averaged saturated thickness of the aquifer was set to 20 m. Shown in Figure 3b is the steady-state concentration distribution predicted by the composite model. Note that the center line of the contaminant plume lies along the water table due to the inherent assumption of the analytical solution that groundwater flow transport is uniform and horizontal. The composite model result was shown by Huyakorn *et al.* [17] to compare reasonably well with a steady-state numerical solution obtained by running a two-dimensional, variably saturated, finite element code [18].

Another interesting application of the EPA's code is the use of the Monte Carlo method for data uncertainty analysis. The Monte Carlo analysis is performed in the manner described in Figure 4. The aquifer and soil hydraulic and transport parameters, as well as the location of a downstream receptor well, are treated as random variables with assigned probabilistic distributions. More than one thousand realizations of the random model input parameters are generated and used by the composite model to perform repetitive simulations. Model results are transmitted to the post-processor to obtain output statistics. Shown in Figure 5 is the cumulative distribution function (CDF) of normalized concentration (inverse of the dilution/attention factor). Results presented in this graph may be used to support exposure or risk assessments. From the CDF graph, the normalized concentration (or DAF) value corresponding to a prescribed confidence percentile level can be determined.

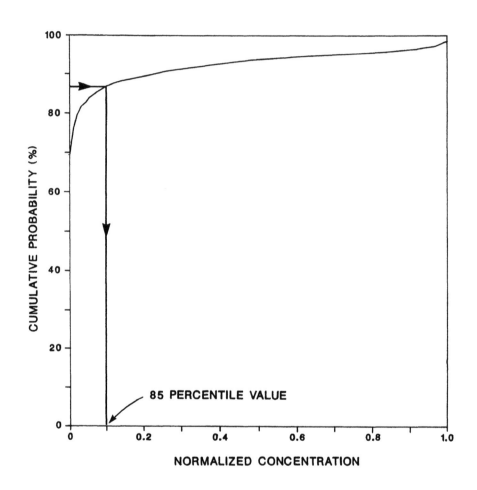

FIGURE 5.   *CDF of Normalized Concentration for EPACML*
Composite Model Example.

5.2   Leachate Migration from Surface Impoundment

In this example, the scenario shown in Figure 6a is considered. As illustrated, the leakage rate from the surface impoundment facility is sufficient to cause significant disturbance (mounding) to the ambient flow condition in the aquifer. This may lead to plume behavior that differs from that predicted by a landfill model. To account for nonuniformity of the groundwater velocity in this scenario, the modeling approach needs to account for groundwater mounding. One way of dealing with the surface impoundment case is to use a multi-dimensional numerical code that treats the unsaturated and saturated zones simultaneously. A second alternative is to adopt a modified composite modeling approach involving a combined use of one-dimensional unsaturated-zone and two-dimensional saturated-zone numerical models. These models are linked via the interface flux boundary condition at the water table directly below the surface impoundment (Figure 6b).

The two-dimensional, unsaturated-saturated numerical modeling approach is discussed here. The code used to simulate groundwater flow and solute transport beneath the surface impoundment is VAM2D [18]. This code is a two-dimensional finite element code with the following salient features:

1.   The Galerkin finite element method is used to solve the groundwater flow equation for variably saturated porous media. Nonlinearities are handled with Picard and Newton-Raphson algorithms designed to handle sharp saturation contrasts.

2.   The upstream-weighted finite element method is used to solve the advective-dispersive solute transport equation. The use of the upstream weighting technique reduces the occurrence of numerical oscillations and negative concentrations and thereby allows the otherwise strict Peclet number criterion for nodal spacing to be relaxed.

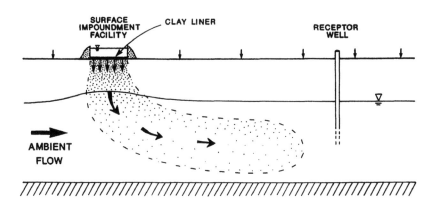

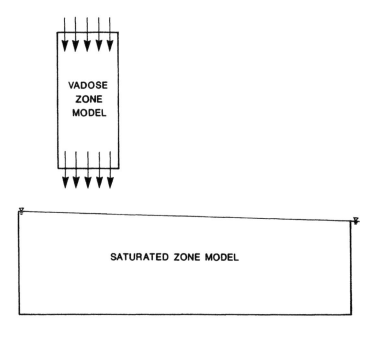

FIGURE 6. *Conceptual Physical Model for Groundwater Contamination Caused by a Leaky Surface Impoundment (a), and Corresponding Mathematical Model (b).*

3.   VAM2D is designed to address complex, site specific, conditions. The code permits modeling of heterogeneous aquifers with arbitrarily varying head or flux boundary conditions.

A typical scenario corresponding to sub-surface flow and transport from a surface impoundment is depicted in Figure 7. The figure shows a surface impoundment overlying an unconfined homogeneous aquifer. Regional groundwater flow in the aquifer is from left to right. A stream is located 110 m downstream of the surface impoundment. Values of the key hydraulic and transport parameters, as well as surface boundary conditions, are shown in this figure. The contaminant leaching from the impoundment is assumed to be non-reactive.

Predicted steady-state concentration contours in the unsaturated and saturated zones are shown in Figure 8. VAM2D indicates that there is relatively little horizontal spreading of the contaminant in the unsaturated zone, except for the capillary fringe immediately above the water table. This reflects the predominantly vertical downward water flow in the unsaturated zone. In contrast, there is substantial spreading in both the longitudinal and vertical direction in the saturated zone. The distinct downward dip of the contaminant plume in the saturated zone is due to the effect of the relatively high rate of water leakage from the surface impoundment. This leakage adds a downward component to the groundwater flow field and the plume migration path. The overall result is increased vertical mixing and deeper penetration of the contaminant plume into the aquifer than was the case for the landfill scenario shown in Figure 3.

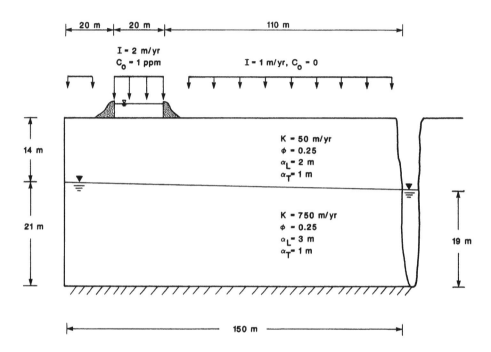

FIGURE 7. *Model Scenario for Leaky Surface Impoundment Overlying an Unconfined Aquifer.*

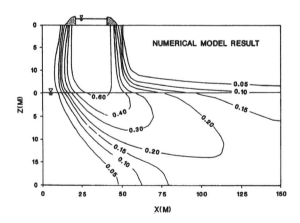

FIGURE 8. *Predicted Steady-state Concentration Contours Using VAM2D for Leaky Surface Impoundment Scenario.*

5.3 Well Pumping Recovery System

In this example, the remediation of existing groundwater contamination by means of a well recovery system is examined. Specifically, the situation where the extent of the contaminant plume is known and the effectiveness of a well recovery system at the site must be evaluated. The modeling objective is to perform pathline analysis and well capture-zone delineation.

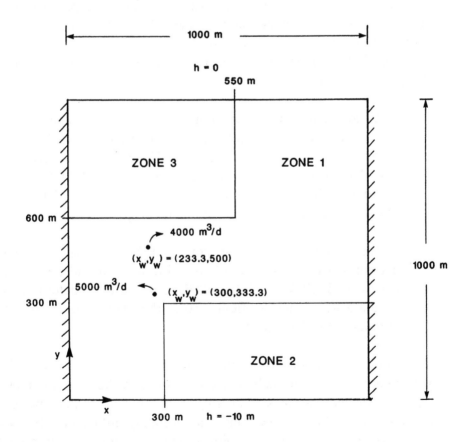

FIGURE 9. *Schematic Diagram of Heterogeneous Aquifer with Two Extraction Wells.*

The scenario is presented in Figure 9. The aquifer is assumed to be confined and consist of three zones of different hydraulic properties. The thickness of each zone, B, is 20 m. Zone 1 is isotropic with transmissivities in the x and y directions, T_{xx} and T_{yy}, equal to 2000 m^2/d. Zone 2 is anisotropic with T_{xx} = 1000 m^2/d and T_{yy} = 200 m^2/d. Zone 3 is also anisotropic with T_{xx} = 500 m^2/d and T_{yy} = 200 m^2/d. The effective porosities of the three zones are 0.25, 0.20 and 0.15, respectively. Boundaries of the aquifer parallel to the y-axis are assumed to be impermeable. The other two boundaries located at y = 0 and y = 1000 m, are constant head boundaries with hydraulic head values prescribed as -10 m and 0, respectively. Two pumping wells are used to remove the contaminant at the site. Their locations and pumping rates are depicted in Figure 9.

The following procedure was used to perform pathline migration and capture zone analysis:

1.   A steady-state simulation of groundwater flow in the aquifer was made using a finite-element numerical model. The computer code used was Saturated Zone Flow and Transport Two-Dimensional Finite Element Model, (SAFTMOD) [19,20]. By running SAFTMOD, hydraulic head values were determined at the nodal points of the finite element grid.

2.   Pathlines and the capture zones of both wells were then determined using the particle tracking method and a computer code called GPTRAC [3]. GPTRAC was used to determine the groundwater velocity field given the hydraulic head field computed by SAFTMOD.

Shown in Figure 10 is the result of the GPTRAC code. The 100-day capture zone of each well corresponds to the region enclosed by reverse-tracked pathlines emanating from the well. As can be seen, the shapes of both capture zones are strongly influenced by well interference effects and the variability and anisotropy of the aquifer materials. Evidently, much of the contaminant plume is intercepted by the well capture zones over the pumping period of one hundred days.

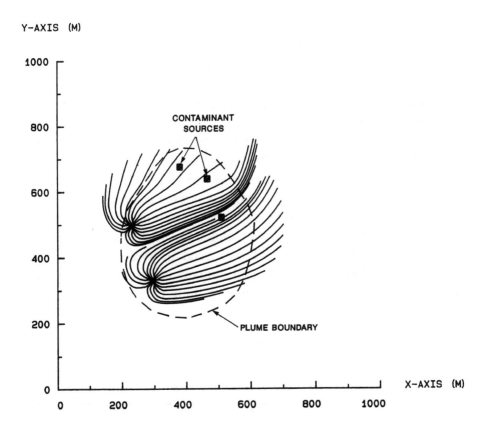

FIGURE 10. *One-hundred Day Capture Zones for Two Extraction Wells Overlain on Approximate Contaminant Plume Boundary.*

5.4   Transient Transport of Conservative and Decaying Species

Another scenario concerning leachate migration from a landfill is considered. In this example, we demonstrate model testing, validation, and application to transient, two-dimensional transport problems involving conservative metal species, and species subject to chained decay (e.g., radioactive decay) reactions. It is assumed that the effect of the unsaturated zone on contaminant concentrations is negligible. The modeling objectives are to predict plume migration and the areal distribution of concentration within the plume. Two models were used to achieve this purpose. The first model is the VAM2D model described previously. The second model is a Laplace Transform Galerkin two-dimensional transport model (LTG2DT) recently developed by Sudicky [12]. VAM2D was used to perform: (1) a field simulation of transport of hexavalent chromium at a site on Long Island, New York; and (2) a simulation of chained decay two-component transport. LTG2DT was used to evaluate a case where dispersivity values are small and numerical oscillations could be encountered with the standard Galerkin finite element technique.

TABLE 2.   Parameter Values Used in the Simulation of Two-Dimensional Transport in Uniform Groundwater Flow.

Parameter	Value
Darcy velocity, V	0.161 m/d
Porosity, ϕ	0.35
Longitudinal dispersivity, α_L	21.3 m
Transverse dispersivity, α_T	4.3 m
Aquifer saturated thickness, b	33.5 m
Contaminant mass flux, Q_{co} (per unit thickness of aquifer)	704 g/(m·d)
Bulk density, ρ_B	1.46 g/cm^3

The field simulation problem concerns groundwater contamination by hexavalent chromium due to the disposal of plating wastes and sewage on a site on Long Island. The site is located in the South Farmingdale Massapequa Area, Nassau County, New York. Data used in the modeling effort are presented in Table 2. The data are based on information reported by Perlmutter and Lieber [21], and Wilson and Miller [22]. The aquifer region was modeled using an areal rectangular finite element grid covering 2400 m in the longitudinal groundwater flow direction, and 300 m in the direction transverse to that of groundwater flow. The contaminant source was represented by one nodal point at the origin of the coordinate axes. The grid spacings in the longitudinal and transverse directions were selected as $\Delta x = 60$ m, and $\Delta y = 30$ m, respectively. These spacings were sufficient to provide good numerical accuracy for the assigned dispersivity values. The transient simulation was performed for 28 time steps with $\Delta t = 100$ days. Contours of predicted concentration at $t = 2800$ days are plotted in Figure 11. Superposed on the contours is a shaded region depicting the observed plume in mid-1949 as reported by Perlmutter and Lieber [21]. The 5 ppm contour compares reasonably well with the observed plume.

The application of the VAM2D code was extended to a second case involving two-component transport with chained decay reactions. Site conditions were identical to those in the previous case. It was assumed that metal species Component 1 transformed into a conservative species Component 2. The reaction was treated as a first-order decay process. The nonconservative species Component 1 was assumed to have values of decay and distribution coefficients equal to 1.9×10^{-4} d^{-1} and 0.308 cm^3/g, respectively. Its mass flux at the source was prescribed as 704 g/d per unit aquifer thickness. The second simulation was performed using the same discretization data as the first simulation.

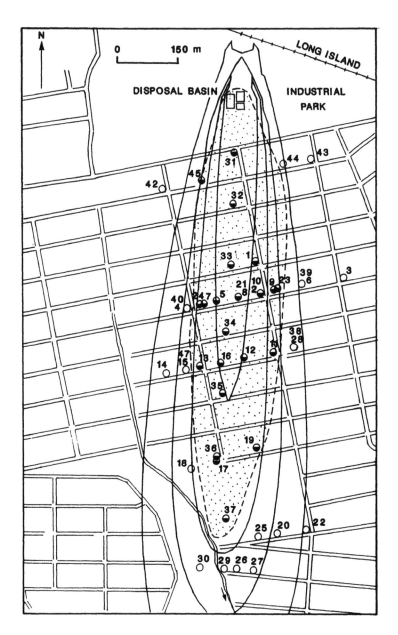

FIGURE 11.   *Predicted Concentration Contours for Hexavalent Chromium at t = 2800 Days Predicted Using VAM2D.*

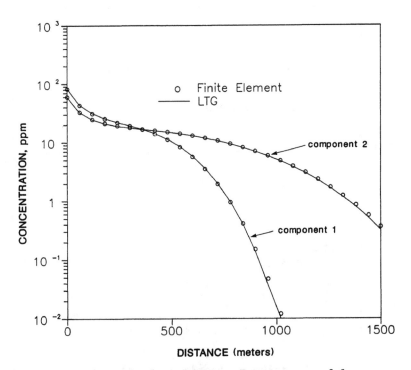

FIGURE 12. Concentration vs. Distance Downstream of the Contaminant Source for Components 1 and 2.

Figure 12 is the plot of concentration profiles along the plume center line for Components 1 and 2 for t = 2800 days. Results obtained from the VAM2D finite element simulation and LTG simulation are compared and shown to be in excellent agreement. The concentration of Component 1 decreases sharply with distance from the source, reflecting its transformation into Component 2. At distances greater than 1000 m, Component 1 has essentially disappeared. The concentration profile of Component 2 extends farther than that of Component 1. At distances greater than 360 m, concentration of Component 2 becomes greater than that of its parent. This behavior is as expected because Component 2 is a conservative species whereas its parent, Component 1, is subject to retardation and decay.

Huyakorn, Kool and Blandford

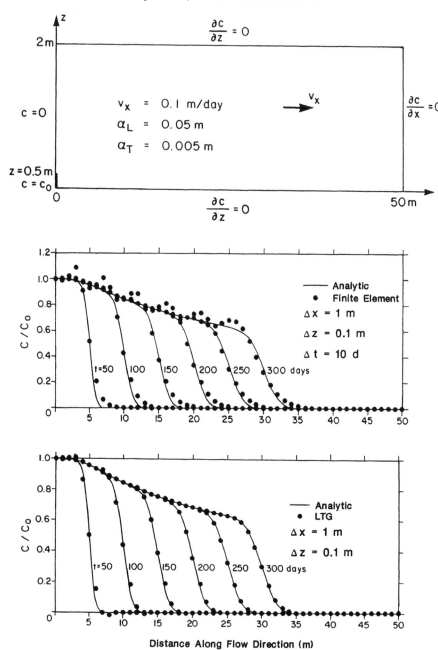

FIGURE 13. *Comparison of Model Solutions for a Simple Groundwater Flow and Contaminant Transport Scenario.*

The final example is included to highlight a situation where numerical difficulties are encountered with the standard numerical (finite element or finite difference) model. For this situation, it is demonstrated that an effective remedy to these difficulties is provided by the Laplace Transform Galerkin (LTG) technique in which the time domain of the transient problem is handled via Laplace Transformation. Shown in Figure 13 is a two-dimensional areal transport scenario involving simple boundary conditions and very small dispersivity values compared to the dimensions of the flow domain. It can be seen that the finite element numerical solution (obtained using the depicted values of grid spacings and time step size) exhibits pronounced oscillations and numerical dispersion. On the other hand, the result obtained using the LTG technique is highly accurate and in excellent agreement with the exact analytical solution of the same transport problem. Clearly, in this situation, the LTG technique provides an effective way to overcome the numerical difficulties experienced by the standard finite element technique.

5.5   Three-Dimensional Transport from a Landfill to a Pumping Well

Finally, a complex scenario involving three-dimensional flow and transport in both the unsaturated and saturated zones is considered. The problem of interest is depicted in Figure 14; it concerns leachate migration in an anisotropic, unconfined aquifer subject to well pumping. The modeling objectives are to perform a detailed evaluation of the extent of the contaminant plume, and to determine the concentration level of the contaminant upon entry to the pumping well.

The fully three-dimensional numerical modeling approach was selected to achieve these objectives. Note that this is the most complex modeling approach available. Prior to selecting this approach, several factors must be carefully considered. First of all, are sufficient data available to warrant the use of a three-dimensional numerical model? The data input requirements are substantial, as is the labor effort required to collect, analyze, and input such data.

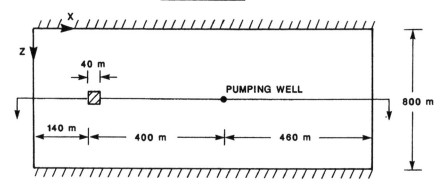

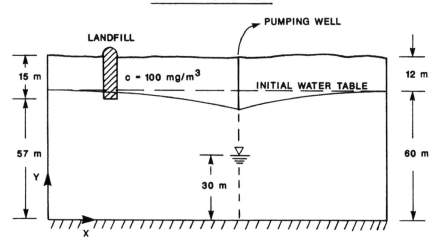

FIGURE 14. Plan-view and Cross-section Diagrams of an Unconfined Aquifer System Subject to Contamination from Leachate Emanating from a Landfill.

Secondly, are the computational facilities required to run the model available? The computational effort required to solve a three-dimensional numerical model may be drastically increased over that required for two-dimensional or one-dimensional models. Finally, is a three-dimensional code available that will solve for the desired variables robustly and efficiently? Selection of the appropriate code is an important step that may dictate the success or failure of a given study.

The selected model is VAM3D-CG (Variably Saturated Analysis Model in 3-Dimensions with Conjugate Gradient matrix solvers) [23]. VAM3D-CG is a recently developed, state-of-the-art three-dimensional code designed in a similar manner to the VAM2D code described earlier. VAM3D-CG utilizes robust iterative matrix solvers that allow simulation of problems with 10,000 to 20,000 nodals unknowns on 80386 class PC's. The matrix solvers use conjugate gradient and ORTHOMIN matrix solution techniques [24-26].

VAM3D-CG was used to perform steady-state flow and transient transport simulations for the scenario shown in Figure 12. For the flow simulation, saturated hydraulic conductivities in the x, y and z directions were assumed to 5, 2 and 0.5 m/d, respectively. The following soil moisture relations were also used:

$$k_{rw} = (S_w-0.05)^2/(0.95)^2 \qquad (3)$$

and

$$S_w = 0.05 + 0.95 \, [1 + (0.5\psi)^2]^{-1} \qquad (4)$$

where the symbols are as defined previously.

For the transient transport simulation, the steady-state velocity field and water saturations obtained from the steady flow analysis were used. Longitudinal and transverse dispersivities were taken as 2 m and 4 m, respectively. Aquifer porosity and solute retardation and decay coefficients were taken as 0.35, 2 and 10^{-5} d^{-1}, respectively. Initially, the flow system was assumed to be contaminant free. Concentration levels at the nodes representing the landfill were prescribed as 100 mg/m³. At other boundary nodes, a zero concentration gradient normal to the boundary was prescribed. The transient transport problem was solved using 40 time steps with a constant $\Delta t = 250$ days.

Simulated contours of concentration in the vertical plane passing through the contaminant source and the pumping well are plotted in Figure 15 for a time value of 4000 days. Also depicted in the figure is the predicted position of the water table. As can be seen, the contaminant plume extends through both the unsaturated and saturated zones. Also, at t = 4000 days, the plume intercepts the pumping well. The entry concentration of contaminant is slightly less than 2 ppm.

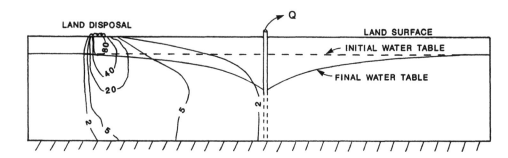

FIGURE 15. Cross-section View of Concentration Contours at
t = 4,000 Days.

6.   CONCLUDING REMARKS

In this chapter, an overview of alternative modeling approaches currently used to predict the fate and transport of metals and other hazardous components in groundwater has been presented. Appropriate example applications of analytical, semi-analytical, particle tracking and numerical methods were discussed. Each method has distinct advantages and disadvantages. The method most applicable to a specific problem is not only dependent upon the strengths and weaknesses of each method, but upon the objectives of the modeling study as well. In certain instances, two or more modeling approaches can be combined to meet the posed objectives in an efficient and satisfactory manner.

The preliminary steps involved in modeling the fate and transport of contaminants in groundwater may be outlined as follows: 1) identify the objectives of the modeling study, 2) identify the important site-specific flow and transport conditions, and 3) based upon the results of Steps 1 and 2, select the most appropriate, cost-effective modeling approaches.

REFERENCES

1. Javandel, I. and C. Tsang. "Capture-Zone Type Curves: A Tool for Aquifer Cleanup," *Ground Water* 24:616-625 (1986).

2. Shafer, J.M. "GWPATH: Interactive Ground-Water Flow Path Analysis," State of Illinois Dept. of Energy and Natural Resources, Bull. 69 (1987).

3. Blandford, T.N. and P.S. Huyakorn. "WHPA: An Integrated Semi-Analytical Model for the Delineation of Wellhead Protection Areas," HydroGeoLogic, Inc., Technical Report prepared for US EPA Office of Ground Water Protection (1989).

4. Yeh, G.T. "Analytical Transient One-, Two-, and Three-Dimensional Simulation of Waste Transport in the Aquifer System," Oak Ridge National Lab., Report ORNL-5602, 83 pp. (1981).

5. Huyakorn, P.S., M.J. Ungs, L.A. Mulkey, and E.A. Sudicky. "A Three-Dimensional Analytical Method for Predicting Leachate Migration," *Ground Water* 25:588-598 (1987).

6.   Galya, D.P. "A Horizontal Plane Source Model for Ground-Water Transport," *Ground Water* 25:733-739 (1987).

7.   Peaceman, D.W. *Fundamentals of Numerical Reservoir Simulation* (Amsterdam: Elsevier, 1977).

8.   Huyakorn, P.S. and G.F. Pinder. *Computational Methods in Subsurface Flow* (New York: Academic Press, 1983).

9.   Liggett, J.A. and P.L-F. Liu. *The Boundary Integral Equation Method for Porous Media Flow* (London: George Allen and Unwin, 1983).

10.   Konikow, L.F. and J.D. Bredehoeft. "Computer Model of Two-Dimensional Solute Transport and Dispersion in Ground Water," U.S. Geological Survey, Book 7, Chapter C2 (1978).

11.   Prickett, T.A., T.C. Naymik, and C.G. Lonnquist. "A Random-Walk Solute Transport Model for Selected Groundwater Quality Evaluations," Illinois State Water Survey, Champaign. (1981).

12.   Sudicky, E.A. "The Laplace Transform Galerkin Technique: A Time-Continuous Finite Element Theory and Application in Groundwater," *Water Resour. Res.,* 25:1833-1846 (1989).

13.   Federal Register. "Hazardous Waste Management System; Land Disposal Restrictions; Proposed Rule," Vol. 51, No. 9, pp. 1601-1766 (1986).

14.   Federal Register. "Hazardous Waste Management System, Identification and Listing of Hazardous Waste; New Data and Use of These Data Regarding the Establishment of Regulatory Levels for the Toxicity Characteristic; and Use of the Model for the Delisting Program," Vol. 53, No. 147, pp. 28892-28895 (1988).

15. U.S. Environmental Protection Agency. "Background Document for EPA's Composite Landfill Model (EPACML)," EPA Office of Solid Waste, Washington, D.C. Prepared by Woodward-Clyde Consultants, Oakland, California (1988).

16. McGrath, E.J. and D.C. Irving. "Techniques for Efficient Monte Carlo Simulation," Technical Report Prepared for the Department of the Navy, Office of Naval Research, Arlington, Virginia (1973).

17. Huyakorn, P.S., B.H. Lester, H.O. White, T.D. Wadsworth, and J.E. Buckley. "Analytical and Numerical Simulations of Leachate Migration in Unconfined Aquifers," in *Proceedings of Conference on Solving Ground Water Problems with Models* (National Water Well Assoc., 1987), pp. 727-748.

18. Huyakorn, P.S., J.B. Kool, and J.R. Robertson. "VAM2D - Variably Saturated Analysis Model in Two Dimensions," US NRC, Report NUREG/CR-5352 (1989).

19. Huyakorn, P.S. and J.E. Buckley. "SAFTMOD: Saturated Zone Flow and Transport Two-Dimensional Finite Element Model," HydroGeoLogic, Inc., Technical Report submitted to US EPA Environmental Research Lab., Georgia (1988).

20. Dean, J.D., P.S. Huyakorn, A.S. Donigian, K.A. Voos, R.W. Schanz, and Y.J. Meeks. "RUSTIC Code Documentation Vol. 1, Theory and Testing," Woodward-Clyde Consultants, Technical Report submitted to US EPA Environmental Research Lab, Georgia (1988).

21. Perlmutter, N.M. and M. Lieber. "Dispersion of Plating Wastes and Sewage Contaminants in Groundwater and Surface Water, South Farmingdale-Massapequa Area, Nassau County, New York," U.S. Geological Survey, Water Supply Paper 1879-G (1970).

22.   Wilson, J.L. and P.J. Miller. "Two-Dimensional Plume in Uniform Ground-Water Flow," *ASCE J. of the Hydraulics Div.,* 104 (HY4):503-514 (1978).

23.   Huyakorn, P.S. and S. Panday. "VAM3D-CG - Variably Saturated Analysis Model in Three-Dimensions with Preconditioned Conjugate Gradient Matrix Solvers," HydroGeoLogic, Inc., Technical Report submitted to Westinghouse Savannah River Lab., Aiken, South Carolina (1989).

24.   Kershaw, D.S. "The Incomplete Cholesky-Conjugate Gradient Method for the Iterative Solution of Systems of Linear Equations," *J. Comp. Physics,* 26:43-65 (1978).

25.   Anderson, D.V. "ICG3: Subprograms for the Solution of a Linear Symmetric Matrix Equation Arising from a 7, 15, 19 or 27 Point 3D Discretization," *Computer Physics Communications,* 38:51-57 (1983).

26.   Behie, A. and P.K.W. Vinsome. "Block Iterative Methods for Fully Implicit Reservoir Simulation," *Soc. Petrol. Engr. J.,* Oct 22:658-668 (1982).

6

COUPLING OF SPECIATION AND TRANSPORT MODELS

Charles J. Hostetler and Robert L. Erikson
Earth and Environmental Sciences Department
Pacific Northwest Laboratory
Richland, WA 99352

1.   INTRODUCTION

Understanding the temporal and spatial distribution of chemicals dissolved in groundwater is an important component of analysis in contaminant hydrogeology. Prediction of solute migration requires a quantitative understanding of solute mass transport and interphase mass transfer reactions. Developing such an understanding remains a challenging problem because of the complex nature of contaminant sources, sub-surface geologic materials, and waste-disposal practices.

The one-dimensional advection-dispersion-linear retardation (ADR) equation has been a common starting point for developing models to describe solute transport [1]. The importance of the classical description is partly because of the inclusion of three primary phenomena observed in a variety of field settings: the movement of solutes through groundwater systems, the dilution of solute concentrations, and the differential velocities of solutes. In the ADR equation, these processes are described using three adjustable parameters, the pore-water velocity (v), the coefficient of hydrodynamic dispersion (D), and the retardation factor (R). The importance of the

0-87371-277-3/93/$0.00 + $.50
©1993 by Lewis Publishers

ADR equation also derives from the availability of analytic solutions for a wide variety of initial and boundary conditions [2,3]

Although use of a constant retardation factor permits analytic solution of the transport equation in many cases, it has been recognized that retardation depends on a variety of solution and soil chemical properties, and that a constant retardation factor is inappropriate for describing solute transport in geologically complex systems (for example, [4,5]) In the area of reactive solute transport, improvements to the classical description have focussed on providing more mechanistic treatments of chemical reaction processes. Ion exchange, specific ion adsorption, and precipitation-dissolution reactions are the primary mechanisms that have been invoked to account for chemical retardation and attenuation.

In one-step chemical transport models, the chemical and transport equations are combined into a set of coupled, nonlinear partial differential equations which are solved using numerical techniques. Rubin and James [6] and Valocchi *et al.* [7] replaced the constant retardation factor with ion exchange equilibria to describe solute transport. In these investigations, the resulting set of coupled, nonlinear partial differential equations was solved numerically by the Galerkin finite element method. Miller and Benson [8] extended the treatment of ion exchange equilibria by including aqueous complexation reactions and individual ion activities in their model. Jennings *et al.* [9] used a similar treatment while replacing the constant retardation factor with solute complexation and surface complexation equilibria. A general framework for including solute complexation, adsorption, and solubility equilibria in the transport equation was presented by Kirkner and Reeves [10] and Reeves and Kirkner [11] examined numerical solution techniques for a variety of cases involving aqueous complexation and adsorption.

In two-step chemical transport models, the solute movement calculation is separated from the chemical reaction calculation. Grove and Wood [12] developed a model in which the transport equation (without retardation) was solved using a finite difference method for each solute. The solutes were allowed to react through solute complexation, solubility, and ion exchange equilibria calculated using a direct iteration method. Walsh *et al.* [13] employed a similar approach for solute movement, but relied on the Newton-Raphson method for calculation of solute complexation and solubility equilibria. Kirkner *et al.* [14] and Cederberg *et al.* [15] solved the transport equations using a finite element method, and solved solute complexation and surface complexation equilibria using the Newton-Raphson method. Narasimhan *et al.* [16] simulated transport using an integrated finite difference method, and used the geochemical code PHREEQE [17] as a module to solve solute complexation and solubility equilibria.

Kirkner and Reeves [10] point out that, although the methods currently employed to simulate solute transport are extendable to two- and three-dimensional domains, the nature and size of the governing equations require extensive computational resources for a multicomponent/multidimensional solute transport simulation. Simulation of an ensemble of one-dimensional streamtubes (for example [18]) also requires extensive computational resources and neglects transverse dispersive phenomena. The simulation of solute transport under transient flow conditions is not adequately addressed in current models. From a review of chemical transport models, Engesgaard and Christensen [19] concluded that many models were created for a specific purpose and that it is difficult to identify models with a wide range of applicability.

We have attempted to address these limitations by creating a framework for a generalized chemical transport model (CTM). In this paper, CTM is applied to one-dimensional transport of equilibrium governed solutes in a steady-state flow field. However, the implementation of the hydrologic and chemical process description is such that extension to multidimensional transport, transient flow fields, and rate limited reactions is straightforward. The purpose of this paper is to provide a description of the generalized framework of CTM and use results from CTM calculations to classify solute distribution profiles. Because these calculations employ a standard Fickian mass transport model and separate the effects of solubility from surface complexation, the results can serve as a benchmark in evaluation of the effects of alternate process descriptions. An important result of this work is that dominant attenuation mechanisms can, in some cases, be inferred from observations of solute concentration profiles.

2.   MODEL DESCRIPTION

The design criteria for CTM were based on the goal of providing a framework for comprehensive simulation of solute transport. The processes currently included in CTM are advection, diffusion, hydrodynamic dispersion, solute complexation reactions, precipitation/dissolution reactions, and surface complexation reactions. We sought to provide a computational capability for simulating transport under spatially varying initial solute concentrations and temporally varying solute boundary concentrations. Finally, we required that our numerical method be compact, computationally efficient, stable, both locally and globally mass conserving, and directly extendable to multidimensional problems.

The direct simulation method was chosen for CTM. Prior to the simulation, the simulation domain is divided into a number of bins of equal length. A linear operator representing the solute movement processes is constructed, and the initial conditions in each bin (i.e., the aqueous, solid, and adsorbed masses of each chemical component) are specified. Finally, the chemical reactions allowed to occur during the simulation are specified.

The two-step coupling algorithm for combining solute movement and chemical reaction was chosen for CTM. During each time step in the simulation, the distribution of mobile constituents (i.e., solutes) is governed by advection, diffusion, and hydrodynamic dispersion. This part of the calculation is the solute movement step. The chemical reaction step involves the reaction of the solution in each bin with the immobile reactive soil constituents (i.e., solid and adsorbed phases) in that bin. Thus, in each time step during the calculation, the solute distribution is modified by both transport and chemical processes.

2.1   SOLUTE MASS TRANSPORT

The solute mass transport processes represented in CTM are advection, diffusion, and hydrodynamic dispersion. Advection, the displacement of solutes because of bulk fluid displacement, and diffusion are described by the fluid flow field and molecular diffusion coefficients, respectively. In CTM, hydrodynamic dispersion is modeled as a result of variations from the mean pore water velocity caused by small scale heterogeneities in the surrounding porous medium [20]. This dispersion model contains the Fickian description as a special subcase.

For the initial version of CTM, transport is simulated in a one-dimensional streamtube with a uniform, steady flow field. If a pulse of conservative tracer is injected into the flow field at time t, the solute will move downstream and spread out. The resulting concentration distribution at a given time is a probability density function (pdf) for distance traveled [21]. The mean of the pdf (mean distance traveled by the tracer) is given by:

$$E\ [c(x)]\ =\ v\ t \tag{1}$$

where v is the pore water velocity and t is the time since injection. In the Fickian model of dispersion, probability density is normally distributed, with the variance of the concentration distribution given by:

$$Var\ [c(x)]\ =\ 2\ D\ t \tag{2}$$

where D is the coefficient of hydrodynamic dispersion (including, for convenience, the molecular diffusion coefficient):

$$D\ =\ \alpha_L\ v\ +\ D^* \tag{3}$$

α_L is the dispersivity, and D^* is the effective molecular diffusion coefficient [22]. Non-Fickian pdfs lead to non-normal solute concentration profiles (e.g. [20,23,24]) The pdf concept is readily extended to two- and three-dimensions.

2.1.1   Implementation

The Markov method, a direct simulation method, is used in CTM to implement the pdf-based description of solute movement. The simulation domain is discretized into a number of bins. In general, the bins contain equal water volume. For the special case of homogeneous, fully saturated media, the bins are of equal length. The number of bins is chosen according to the resolution required for the simulation. The total mass of each solute is expressed as a state vector, with each entry in the vector representing the solute mass in a particular bin. At each time step, the set of state vectors contains all the information available about the solute mass distribution in the system.

A linear operator (the Markov transition matrix) is used to represent solute mass transport processes in calculating the evolution of the state vectors as a function of time. The Markov transition matrix is used to project the state vectors forward in time via:

$$S^{(t+1)} \; = \; [T] \; S^{(t)} \tag{4}$$

where $S^{(t)}$ is a state vector at time t, and [T] is the Markov transition matrix. The elements of the transition matrix [T] describe solute mass transport in the system. In particular, T_{ij} is the fraction of solute mass in bin i at time t that is transported to bin j at time t+1.

The Markov transition matrix is constructed from the pdf defined for the system. The pdf is a spatially varying function depending on starting location (i.e., injection point) and time step length (Δt). The Markov transition matrix elements are obtained by integration of the pdf over bin volume:

$$T_{ij} \; = \; \int_{bin} pdf\,(x,i,\Delta t) \; dx \tag{5}$$

For the uniform, steady flow field simulated by CTM, the pdf is stationary (i.e., independent of spatial location). In particular, use of the Fickian description of dispersion leads to a pdf that is independent of starting position:

$$pdf\ (x,\Delta t)\ =\ (4\pi D\ \Delta t)^{-1/2}\ \exp\ [\frac{-(x\ -\ v\ \Delta t)^2}{4\ D\ \Delta t}] \qquad (6)$$

The mean and variance of the pdf are given by:

$$E\ [pdf(x)]\ =\ v\ \Delta t \qquad (7)$$

and

$$Var\ [pdf\ (x)]\ =\ 2\ D\ \Delta t \qquad (8)$$

respectively. The matrix elements are simply:

$$T_{ij}\ =\ \{\ erf\ [\frac{(x_d\ -\ v\ \Delta t)}{(4\ D\ \Delta t)^{1/2}}]\ -\ erf\ [\frac{x_u\ -\ v\ \Delta t)}{(4\ D\ \Delta t)^{1/2}}]\ \}\ /\ 2 \qquad (9)$$

where erf(z) is the error function evaluated at z [25]. x_u is the x-coordinate of the upstream bin boundary, and x_d is the x-coordinate of the downstream bin boundary.

The log normal pdf is similar to the normal pdf. However, the logarithm of the distance traveled is normally distributed:

$$pdf\ (x,\ \Delta t)\ =\ (4\pi\ x\ D\ \Delta t)^{-1/2}\ \exp\ [\frac{-(\ln(x)\ -\ v\ \Delta t)^2}{4\ D\ \Delta t}] \qquad (10)$$

The mean and variance of the log normal distribution are also related to the time step length, mean groundwater velocity, and dispersion coefficient by Equations (7) and (8). The matrix elements for the log normal pdf must be evaluated by numerical integration.

Using the Markov method, formulations of the pdf are limited only by the ability to integrate (numerically, if necessary) the pdf over each bin interval. As the dispersive properties of the sub-surface media are described by the pdf, a number of forms for the dispersive flux can be investigated using CTM. For example, Tompson [26] has developed a macroscopic mass balance description of the dispersive flux. The one-dimensional solutions of Tompson's Equation 18 [26], with Dirac delta boundary conditions, can be used directly as a pdf with the Markov method. One strength of the Markov method is the flexiblity of the description of mass transport processes.

The Markov method as implemented in CTM makes no distinction between one-dimensional and multi-dimensional systems. Implementation for multi-dimensional systems follows directly from the definition of the state vector and transition matrix. For the three-dimensional case, the bins become volume regions of space (either equal soil volume or equal water volume), and the state vectors hold the component masses in each bin. The bin numbering scheme is completely arbitrary. The pdf is integrated over each bin volume before the start of the simulation. This preprocessing step contains the majority of the extra computational effort involved in multi-dimensional calculations. Thus, the implementation during the simulation for the multi-dimensional case is equivalent to the implementation for the one-dimensional case. The state vector evolution is predicted using Equation (4). A second strength of the Markov method is the computational efficiency of the vector-matrix multiplication for calculating mass transport.

The Markov method is currently implemented in CTM for systems with a steady-state flow field. In this case, the transition matrix elements are independent of time and are evaluated prior to the transport simulation. For transport in systems with transient flow, if the flow equations are solved independently of the transport equations, the matrix elements can be calculated for each time step prior to the start of the transport calculation using Equation (5) with the pdf parameters varying with time. The appropriate matrix elements are then used with Equation (4) for each time step in the transport calculation.

2.1.2   Parameterization

A conceptual model for solute mass transport contains a physical parameter, two transport parameters, and two discretization parameters.

The physical parameter is the length of the streamtube, Equation (1). The transport parameters are the mean pore water velocity (v) and the dispersion coefficient (D). The discretization parameters are the length of each bin (Δl) and the length of each time step (Δt). These parameters are input to a preprocessor (PRE) that prepares the input files to CTM. One of the functions of PRE is to calculate the Markov transition matrix used by CTM.

To obtain the transport parameters for a particular laboratory experiment or field site, observed concentration distributions of a conservative tracer are required. The most direct method uses measurements of concentration as a function of distance at a fixed time. If several injection points are available, the variation in pdf with location can be investigated directly, otherwise the hypothesis of stationarity can be invoked. The resulting matrix elements can be obtained directly from spatial integration of the concentration distributions. Alternatively, the form of the pdf can be chosen and parameters obtained through regression. A third method of parameterization uses measurements of concentration as a function of time at a fixed location. In this method, the form of the pdf is chosen and the controlling parameters are obtained using regression analysis (e.g., [27]).

The discretization parameters are chosen to provide the resolution required for the calculation. The first step in choosing the discretization parameters is choosing the time step length. Two dimensionless numbers describe the spatial relationship of the pdf to the streamtube length. The displacement fraction is given by:

$$F_d = v \, \Delta t \, / \, l \qquad (11)$$

and represents the fraction of the streamtube length traveled by the mean pore water velocity front during one time step. The spreading fraction, given by:

$$F_s \;=\; (2\ D\ \Delta t)^{1/2} \;/\; l \qquad\qquad (12)$$

is the ratio of the standard deviation of the pdf to the streamtube length. CTM works in dimensionless variables (time steps and bins), so the derived parameters (F_d and F_s) represent the dimensionless velocity and dispersion coefficient, respectively. The time step length should be chosen to position the pdf such that no significant amount of probability mass lies outside the simulation domain. The ratio of the displacement fraction to the spreading fraction (F_d / F_s) provides a useful constraint on the time step length. This ratio should be greater than three for the normal distribution to ensure mass conservation to approximately three parts per thousand.

Using the two-step approach, discretization leads to errors in calculated solute distributions because solute mass may be transported over several bins before chemical reaction is calculated. The magnitude of the temporal discretization error is described by the local Courant number [28]:

$$C_r \;=\; v\ \frac{\Delta t}{\Delta l} \qquad\qquad (13)$$

which is the number of bin lengths traversed by a parcel of water moving at the mean pore-water velocity during one time step. Minimizing the local Courant number (within the constraints of computer time) is the consideration in choosing the bin length.

3.   INTERPHASE MASS TRANSFER

A focal point of chemical transport models is the use of specific chemical mechanisms to describe attenuation. Chemical attenuation is a result of mass transfer from the aqueous phase to an immobile phase. The interphase mass transfer reactions included in CTM are solid phase precipitation/dissolution and surface phase complexation reactions. In CTM, these reactions are calculated using a chemical equilibrium formulation. Mass transfer from the aqueous phase leads to the apparent differential velocities of solutes. As will be discussed in more detail below, solid phase attenuation reactions and surface phase attenuation reactions can lead to distinct types of solute transport behavior. The behavior, as recorded in solute profiles, can be used to help identify the important attenuation mechanisms operating in the field.

3.1 Implementation

The thermodynamic activities of the reactive solutes are required to accurately calculate the effects of attenuation reactions. Standard electrolyte theory and aqueous complexation reactions are used in CTM to calculate individual ion activities.

Molality was used for the concentration unit in CTM. The molality of the ith solute (m_i) is:

$$m_i = \omega \, n_i \, / \, n_w \qquad (14)$$

where n_i is the number of moles of the ith solute, ω is equal to 1000 divided by the molecular weight of water, and the subscript w indicates the component water.

The ionic strength (I) of the solution is given by a sum over all solutes:

$$I = 0.5 \sum_i m_i \, z_i^2 \qquad (15)$$

where z_i is the charge of solute i. The Davies equation [29] was used to calculate the individual ion-activities (γ_i) of charged solutes:

$$\log \gamma_i \; = \; -A \, z_i^2 \left(\frac{\sqrt{I}}{1 + \sqrt{I}} \; - \; 0.3 \, I \right) \tag{16}$$

where A is a constant (equal to 0.5092 at 25°C) related to the physical properties of water. For charge neutral solutes, activity coefficients are approximated using:

$$\log \gamma_i \; = \; 0.1 \, I \tag{17}$$

[30] and the activity of water by:

$$a_w \; = \; 1 \, - \, 0.017 \, \Sigma \, m_i \tag{18}$$

[31]. The activity of surface complexes is given by the Vanselow convention [32]:

$$a_m \; = \; \frac{n_m}{n_{js} + \Sigma n_m} \tag{19}$$

Each chemical component (basis species, indicated by the subscript, j) is associated with a mass balance equation:

$$T_j \; = \; n_j + \Sigma v_{ij} \, n_i + \Sigma v_{lj} \, n_l + \Sigma v_{mj} \, n_m \tag{20}$$

where T_j is the total number of moles of component j, the subscript i indicates aqueous complexes, l indicates solids, and m indicates adsorbates. Each species is represented by a mass action equation:

$$K_i \; = \; \frac{m_i \, \gamma_i}{\Pi \, (\gamma_j \, m_j)^{v_{ij}}} \tag{21}$$

$$K_1 \; = \; \frac{1}{\Pi \, (\gamma_j \, m_j)^{v_{ij}}} \tag{22}$$

$$K_m = \frac{a_m}{\Pi \, (\gamma_j \, m_j)^{\nu_{mj}}} \qquad (23)$$

When the mass action expressions are inserted into the mass balance equations, the result is a set of coupled, nonlinear, algebraic equations. The nonlinear equations are solved using Newton-Raphson iteration in a manner similar to Westall *et al.* [33] and Felmy *et al.* [30]. Mass balance on all components (including water) is kept within a tolerance specified by the user.

Alternate formulations for the adsorption of solutes on soils have been proposed: semi-empirical isotherms, ion exchange reactions, and surface complexation models with electrostatic terms. The surface complexation model, as implemented in CTM, can be used to describe competitive Langmuir adsorption, ion exchange reactions, and specific ion adsorption. The surface complexation model used in CTM does not include electrostatic terms. As pointed out by Krupka *et al.* [34], these terms are not necessary for reasonably accurate descriptions of adsorption of common ions on the surface of amorphous ferric hydroxide. Uncertainties in the thermochemical data and limitations of the chemical conceptual model introduce errors in solute concentrations far greater than those introduced by neglecting the electrostatic nature of oxide surfaces.

The mass balance equations are normally solved for each bin at each time step in the two-step algorithm. However, in CTM, advantage is taken of the fact that in some cases, the chemical composition of a number of bins may be sufficiently similar to be grouped together. An adaptive geochemical grid algorithm is used to group together bins having similar compositions (within user specified tolerance) so that only one chemical calculation is performed for the group per time step. Significant time savings are obtained, particularly at time steps when the solute fronts have either not penetrated far into the domain, or have penetrated far enough that homogeneous chemical conditions are behind the fronts.

As currently implemented in CTM, the geochemical model is restricted to equilibrium controlled reactions. This model is readily extended to kinetically controlled reactions using either of two methods. The first method relies on the introduction of explicit functions for the rates of various reactions. Implementation of this method was described in Furrer *et al.* [35] using an algorithm similar to the mass balance formulation used in CTM. This method is limited by the availability of appropriate rate expressions and the necessity to encode the numerous forms of the rate expressions that have been developed for individual reactions (e.g., [34]). In the second method, the half times of kinetically controlled reactions are estimated [34,36]. The half times are used in conjunction with the time step length to calculate the reaction progress during each time step.

3.2   Parameterization

For each of the chemical species in the code, the charge and gram-formula weight must be specified. For the derived species (i.e., aqueous complexes, solids, and adsorbates) the reaction stoichiometry and equilibrium constant must be specified. These data have been calculated for a number of components and derived species and are available on database files that can be read directly by the preprocessor code.

For additional components and species, the data can be added to the database files using a text editor. The most difficult step is to obtain an equilibrium constant for a new derived species. These data must be obtained using one of several techniques. The best technique is to calculate the equilibrium constant directly from the difference in free energies of formation of the derived species and the components. A number of tabulations of free energies of formation are available for this purpose. We relied primarily on Wagman *et al.* [37] to prepare the data in the database used with CTM. A second technique is to fit the equilibrium constants from experimental data. An example of this technique is the derivation of mass action expressions for the specific adsorption of solutes on the surface of hydrous ferric oxide [34]. A third technique is to estimate the equilibrium constants from empirical relationships. This technique was used to estimate equilibrium constants for ion exchange reactions on smectite surfaces [34].

In addition to the thermochemical data, CTM requires specification of the chemical initial and boundary conditions. The initial conditions are the concentrations of aqueous components, solids, and adsorbates in the streamtube. The initial conditions in CTM can be specified for each individual bin if desired. The boundary conditions are the concentrations of each component entering the streamtube. The boundary conditions can be specified for each time step if desired. The initial and boundary conditions are entered using the preprocessing code.

3. CHARACTERIZATION OF TRANSPORT BEHAVIOR

As discussed in the introduction, an important feature of solute transport is the apparent differential velocity of solutes. In the classical ADR equation, the retardation parameter is used to describe the ratio of the velocity of a conservative tracer relative to the velocity of a solute. In chemical transport models, and in CTM in particular, chemical attenuation reactions are used to describe interphase mass transfer, leading to different solute velocities.

The solute velocity is usually defined as the rate of change of location of the breakthrough front of the solute [1]. The breakthrough front is the location at which the concentration of a solute is one half of the injection concentration. It is important to realize that other velocities can be useful in analyzing solute transport. We refer to the rate of change of location of the position at which an aqueous concentration standard is exceeded as a contamination velocity. An analogous velocity exists for the rate of change of location of the position at which a soil concentration standard is exceeded. A fourth type of velocity is the rate of change of a precipitation or dissolution front.

For a homogeneous streamtube with constant boundary conditions and non-interacting solutes these velocities are constant. However, changes in the initial conditions, boundary conditions, and interactions between the aqueous and immobile phases can lead to changes in the apparent velocities of solutes as they migrate. In the constant retardation model, the solute velocity is a constant equal to the pore-water velocity divided by the retardation factor.

3.1   CALCULATED SOLUTE PROFILES

To illustrate the importance of chemical mass transfer reactions in determining apparent solute velocities, we have prepared an example calculation describing the transport of cadmium through a homogeneous soil column. The results of the CTM calculation are compared to results predicted from the classical ADR equation. To further illustrate the importance of the type of attentuation reaction (solubility versus surface complexation), we have performed calculations for the transport of cadmium using aqueous speciation and adsorption reactions only, aqueous speciation and solubility reactions only, and combined aqueous speciation, solubility, and adsorption reactions.

The general framework of the conceptual model is an acidic influent reacting with a neutralizing soil. This general scenario is often found at disposal sites in the western United States. Acidic leachates are commonly derived from leaching of coal combustion residues, uranium mill tailings, and mine tailings. Carbonate materials (that act to neutralize acidic solutions) are often found in arid or semiarid environments. As the leachate contacts the soil, the pH is buffered and trace metal-carbonate minerals can precipitate. In addition, the adsorption behavior of trace metals is a strong function of pH. Eventually, the buffering capacity of the soil will become exhausted when the leachate has dissolved all of the carbonate materials. The pH will drop to that of the influent, and the trace metals will become remobilized.

The conceptual model for mass transport was the same for all of the CTM calculations. The streamtube length was 50 meters, the mean pore water velocity was 5 m/yr, and the dispersivity was 0.1 m. A normal pdf was used. The streamtube was divided into fifty bins, each 1 m in length, and the length of each time step was one year. Thus, the displacement ratio was equal to 0.1 and the spreading ratio was equal to 0.02. Mass conservation from the pdf was three parts in

10^5. The local Courant number was five. Figure 1 shows the discretized pdf used in the transport simulations.

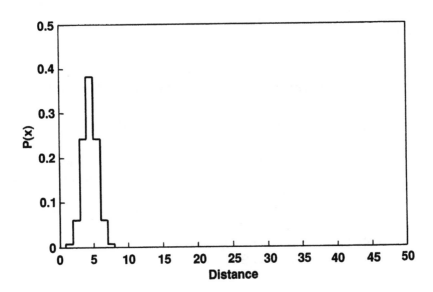

FIGURE 1. *Discretized Probability Density Function Used in All Example Transport Calculations Showing the Transport Probabilities (P(x)) Versus Distance in Meters.* The transport model conditions include a time step of 1 yr, mean pore water velocity of 5 m/yr, and dispersivity of 0.1 m for a normal pdf.

The components, adsorbates, and solids in the complete geochemical conceptual model are listed in Table 1. The equilibrium constants and reaction stoichiometries used in this calculation were taken from Krupka *et al.* [34]. A total of 47 aqueous complexes was selected from the database by the preprocessor for these calculations.

The discussion of the results will focus on the transport of cadmium. The attenuation reactions for cadmium included in the following example calculations are the adsorption of aqueous cadmium on amorphous ferric hydroxide (HFO) and the precipitation of the cadmium carbonate solid (otavite). Cadmium competes for the available adsorption sites with hydrogen, calcium, chromium, and sulfate. For cadmium, the midpoint of the adsorption edge is located at approximately pH 7.0, with almost no adsorption below pH 6.5 [34]. The other solutes competing for the available adsorption sites are also influenced by the solution pH.

TABLE 1.　　*Conceptual Model for Transport Calculations.*

Basis Species	Adsorbates	Solids
H_2O	$HFO\text{-}H^\circ$	$CaCO_3$ (calcite)
H^+	$HFO\text{-}H_2^+$	$CdCO_3$ (otavite)
Ca^{2+}	$HFO\text{-}Ca^+$	$Cr(OH)_3$
Na^+	$HFO\text{-}Cd^+$	$CaSO_4{\cdot}2H_2O$ (gypsum)
Cd^{2+}	$HFO\text{-}CrOH^+$	
$CrOH^{2+}$	$HFO\text{-}H_2CrO_4^-$	
CrO_4^{2-}	$HFO\text{-}H_2SO_4$	
CO_3^{2-}		
SO_4^{2-}		
Cl^-		
HFO^-		

The initial conditions for the geochemical conceptual model are listed in Table 2. The initial conditions represent a uniform, carbonate containing soil column with a moderate adsorptive capacity. The initial soil pH was 7.3. The initial aqueous solution in each bin was in equilibrium with both calcite and the amorphous ferric hydroxide surface.

The boundary conditions for the geochemical conceptual model are listed in Table 3. The leachate entering the soil column is acidic (pH = 3.3). The total cadmium in the leachate was constant at 1 mg/L. A twenty year time period was simulated (20 time steps). The adaptive geochemical grid was used in this calculation, which took approximately 12 cpu minutes on a VAX 11/780.

TABLE 2.   *Initial Conditions for Transport Calculations.*

Soil Properties
      bulk density = 1.6 g/cm^3
      saturated volumetric moisture content = 0.40
      buffering capacity = 0.006 wt % as calcite
      extractable Fe = 0.002 wt %
      saturated paste pH = 7.3
Solution Properties
      total Ca = 78.6 mg/L
      total Na = 0.2 mg/L
      total Cd = 0.01 µg/L
      total Cr = 0.01 µg/L
      total carbonate = 250 mg/L
      total sulfate = 4.8 mg/L
      total chloride = 0.03 mg/L

TABLE 3.   *Boundary Conditions for Transport Calculations.*

      pH = 3.3
      total Ca = 140 mg/L
      total Na = 122 mg/L
      total Cd = 1.0 mg/L
      total Cr = 1.7 mg/L
      total carbonate = 1.25 mg/L
      total sulfate = 584 mg/L
      total chloride = 3.5 mg/L

Figure 2 shows the arrival of chloride at the end of the streamtube over the simulated twenty year period. The solid line represents the values predicted by CTM and the dots are the values predicted by the ADR equation. Chloride is not associated with any attentuation reactions in our geochemical conceptual model. The CTM results are in good agreement with the classical transport equation for a conservative tracer. Note that the breakthrough time for a conservative tracer is ten years (breakthrough velocity of five m/yr) at the end of the 50 meter streamtube.

Figure 3 shows the arrival of carbonate, calcium, and pH at the end of the streamtube. The product of the calcium and carbonate activities is fixed by the presence of calcite in the streamtube. The total masses of calcium and carbonate are fixed by the speciation scheme. Because the leachate has a larger calcium concentration than the initial condition, the effluent chemistry changes after 9 years. This is the first indication of influent affecting the chemistry at the end of the streamtube. The initial soil pH was 7.3, and the pH of the soil in contact with the leachate was 8.0. The pH breakthrough occurs after ten years.

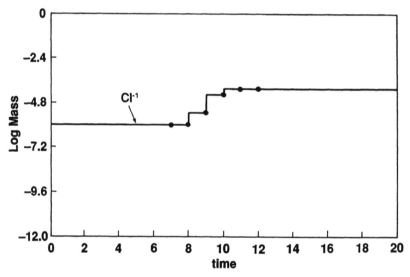

FIGURE 2.   ***The Arrival Concentration (log Mass, Molal) of Chloride in the Effluent as a Function of Time in Years.*** The solid line is the CTM result and the dots are the result obtained from an analytic solution of the advection-dispersion equation.

In Figure 4, the arrival of cadmium at the end of the streamtube is shown for five different calculations. The line labeled S shows the arrival of cadmium attenuated by solubility only, A by adsorption only, and SA by both adsorption and solubility. The lines labeled KD0 and KD2 were calculated using the advection-dispersion equation. The former shows the arrival of a conservative tracer and the latter shows the arrival of a solute retarded by a factor of three. Solubility reactions do not retard the first arrival of cadmium from the influent in this calculation. The effect of the solubility reaction is to lower the maximum total cadmium concentration observed. This solubility limit is a function of the major element chemistry (e.g., pH, ligand concentrations, stable solid phases). The agreement between the A and KD2 lines is not accidental; a retardation factor of three was estimated from a calculation of the equilibration of the influent solution with the initial soil materials. Because the major element geochemistry is fairly uniform ahead of the calcite dissolution front, the retardation factor for cadmium in this case is approximately constant. If the simulation covered a longer time period, both the A and KD2 lines would approach the influent value (shown by a dot on the right hand side of the figure), and the SA line would approach the S line. Note that the rise from background is much steeper for the solubility only run than for the either of the runs with adsorption. One effect of adsorption reactions is to lower the slope of the arrival curve.

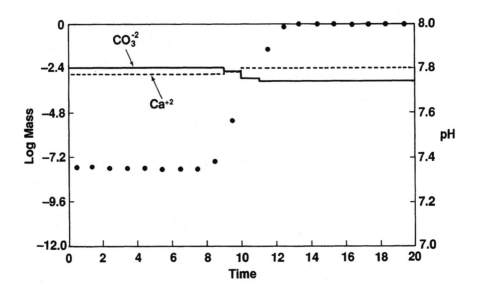

FIGURE 3. *The Arrival of Calcium and Carbonate Concentrations (log Mass, Molal) in the Effluent as a Function of Time in Years.* The dots represent the effluent pH (read from scale on right hand side of figure).

In Figure 5, some of the major element chemistry in the streamtube is shown at ten years. The chloride distribution shows breakthrough after 10 years, as discussed above. After ten years, calcite has been dissolved from the first 16 meters of the streamtube (dissolution frontal velocity of 1.6 m/yr). As a consequence, the pH in this region has fallen from 8.0 (the pH of the soil in contact with the leachate) to 3.3 (the pH of the leachate). At the 40 to 50 meter distance, the early pH breakthrough is shown, as discussed above. Eventually, the calcite dissolution front will migrate to the end of the streamtube and the pH in the effluent will fall from 8.0 to 3.3.

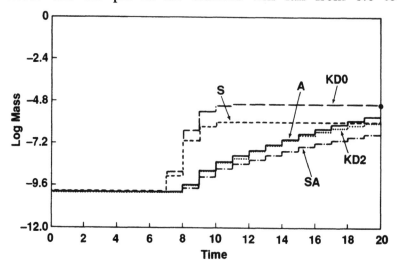

FIGURE 4.   *The Arrival of Cadmium Concentrations (log Mass, Molal) in the Effluent as a Function of Time in Years.*
The curves labeled KD0 and KD2 are from an analytic solution of the advection-dispersion equation using a retardation factor of 1 and 3, respectively. The curves labeled S, A, and SA are CTM results for the solubility only, adsorption only, and combined solubility and adsorption simulations described in the text. The dot on the right hand side of the figure represents the injection concentration of cadmium.

Figures 6a, b, and c show the spatial distribution of cadmium after ten years for the three runs. The influent composition is shown as a dot on the left hand side of each figure. In addition, the apparent local Kd is shown by the square symbols. The local Kd was calculated by dividing the mass of cadmium in the immobile phase by the mass of cadmium in the aqueous phase, and is expressed here in dimensionless units.

For the solubility only run (Figure 6a), cadmium has broken through at the end of the streamtube (also shown in Figure 1). The cadmium distribution is nearly flat in front of the calcite dissolution front. In this region, the cadmium concentration is solubility limited, as otavite is present between 16 and 21 m (interval over which the local Kd is nonzero). From 0 to 14 m, the buffering capacity of the soil has been exhausted and the cadmium is remobilized. The remobilized cadmium travels in a front directly behind the calcite dissolution front with a velocity of approximately 1.6 m/yr.

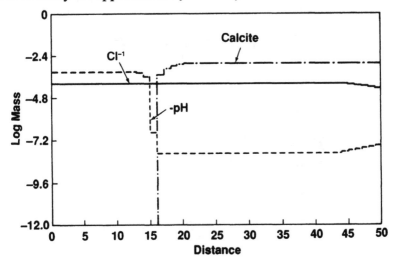

FIGURE 5. *The Distribution of the Concentrations (log Mass, Molal) of Chloride and Calcite, and the pH in the Streamtube Versus Distance (Meters) After Ten Years.* The pH scale is the negative of the scale on the left hand side of the figure.

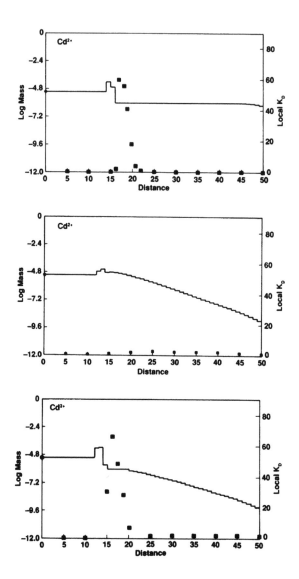

FIGURE 6. *Cadmium Concentrations (log Mass, Molal) in the StreamTube Versus Distance (Meters) After Ten Years for the (a) Solubility Only Calculation, (b) Adsorption Only Calculation, and (c) Combined Adsorption and Solubility Calculations.* The square boxes in each figure represent the local distribution coefficient (immobile Cd/mobile Cd) and are read from the right hand scale.

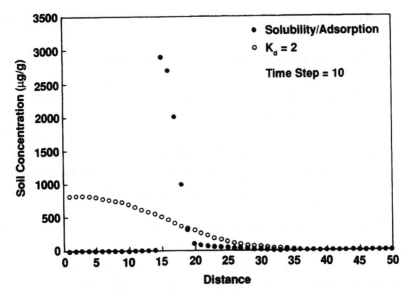

FIGURE 7.   *Cadmium Concentrations in the Soil Profile Versus*
Distance (Meters) After Ten Years for the Combined
Solubility and Adsorption Calculation (Filled Symbols).
The open symbols were calculated from the advection-dispersion
equation and the partition coefficient.

In Figures 6b (adsorption only) and 6c (both adsorption and
solubility) the local Kd values are near two for most of the streamtube
length.   Also note that adsorption reactions can attenuate the
concentration of solutes continuously along the streamtube, while
solubility reactions only attenuate concentrations to the solubility limit.
Figure 7 shows the soil concentration of cadmium after ten years
for the combined solubility and adsorption run. The filled symbols are
the results of the CTM simulation, and the open symbols are the results
calculated from the ADR equation, using a Kd of two. The presence
of cadmium in the soil as a discrete solid phase (in this case otavite)
and the sharpness of the calcite dissolution front leads to a large
increase in the predicted cadmium concentration at the pH front. This
is not predicted by the ADR equation. As cadmium concentrations of
3 µg/g in soil are detectable by standard analytical means, predictions
of this sort could be validated in the field.

3.2 SOLUTE PROFILES OBSERVED AT TWO FIELD SITES

To further illustrate the utility of the mechanistic description of solute attenuation, we now consider the observed distribution of cadmium at two field sites. Our goal is not to accurately predict solute concentration profiles using CTM; this is precluded by gaps in the site characterization and monitoring data. Instead, our approach is to use the qualitative discussion of the previous section to infer dominant attenuation mechanisms from the observed solute profiles at the sites. Both of these sites fit the general conceptual model of an acidic influent reacting with a neutralizing soil.

The first site is a fly ash disposal site located at the Michigan City Generating Station of the Northern Indiana Public Service Company. Site characterization data and monitoring data for this site were presented by Theis *et al.* [38]. The fly ash was disposed in two sets of settling ponds that were alternately filled and allowed to dry (in six to nine month cycles). During the monitoring period the first pond was being loaded with ash and the second pond was allowed to dry.

The leachate from the settling pond was weakly acidic. The pH in the settling pond was between 5.5 and 6.0. Cadmium concentrations in the pond water were not presented by Theis *et al.* [38]. However, from a study of the chemical characteristics of fossil-fuel combustion residues, it is reasonable to assume that cadmium concentrations at pH 5.5 are on the order of 5 μg/L [39]. The predominant anion in the pond water was sulfate.

The ponds were constructed over a clean, uniform sand of medium texture. The buffering capacity of this soil was not provided by Theis *et al.* [38], but it is reasonable to assume that the buffering capacity of the clean sand is very low. The porosity was assumed to be 0.5. Using a grain density of 2.65 g/cm^3 (that of quartz), we calculated a bulk density of 1.33 g/cm^3.

The hydrology of the site was complicated by the superposition of the pond on the regional groundwater system. Theis *et al.* [38] found a saddle point in the water table surface. Two monitoring wells (Numbers 13 and 18) were emplaced near this location.

Cadmium was measured in the monitoring network and found to be generally lower than 5 µg/L. This maximum value is close to the value we have assumed for the leachate. The pH in the monitoring network varied over a relatively narrow range, from 7.5 near the pond to 6.5 at a distance of 450 meters from the pond.

Cadmium concentrations were measured in several borings near the pond. The concentration of cadmium in the soil (at the elevation of the groundwater table) is relatively constant with distance. The maximum soil concentration was 0.66 µg/g (background was assumed to be 0.3 µg/g). Using an aqueous cadmium concentration of 5 µg/L, we calculated a dimensionless Kd of approximately 190 for the region from 250 to 450 meters from the pond.

The cadmium soil concentration profile observed at the Michigan City field site is qualitatively similar to the cadmium soil profile inferred from our Figure 6b. Our conclusion is that adsorption is the dominant attenuation mechanism for cadmium at this site. This conclusion gains additional support from a speciation analysis of the groundwater from Well 13, showing that otavite is undersaturated by over two orders of magnitude. In addition, we have calculated a Kd for cadmium using our adsorption model and the measured iron concentration in the soil [38]. The calculated value of 40 is within a factor of four of the measured value.

The second site is a uranium mill tailings disposal site located in the Gas Hills region of Central Wyoming. Site characterization data and monitoring data for this site were presented by Dames and Moore [40], Peterson et al. [41] and Erikson et al. [42]. Mill tailings were disposed in an evaporation pond over a twenty year period.

The leachate from the settling pond was highly acidic. The pH in the evaporation pond was 2.2. The source of the acidity was the sulfuric acid used to extract uranium from the crushed ore. The predominant anion in the pond water was sulfate. Cadmium concentrations in the pond water were approximately 300 µg/L.

The evaporation pond was constructed over an alluvial soil. The buffering capacity of this soil was approximately 0.1 weight percent as calcite. The porosity of the alluvium was 0.21 and the bulk density was

2.08 g/cm^3. The amount of amorphous ferric hydroxide in the soil is unknown.

Cadmium was measured in the monitoring network and found to vary between 100 to 300 µg/L at distances up to 250 m from the pond. At distances from 500 to 1000 m, cadmium concentrations dropped to 10 to 30 µg/L. The pH in the monitoring network varied over a large range, from 2.0 to 4.0 within 250 m of the pond to about 8.0 at distances greater than 500 m from the pond. Soil cadmium concentrations were not measured.

The cadmium aqueous concentration profile observed at the second field site is qualitatively similar to the calculated profile shown in our Figure 6a. Our conclusion is that the aqueous concentrations of cadmium are governed predominantly by the solubility of otavite at this site, although we cannot rule out the possiblity of adsorption reactions. Speciation analyses of the groundwaters in the monitoring network show that otavite is undersaturated near the pond where the pH is low. At greater distances from the pond, the saturation index for otavite ranges from 0.7 to 1.5.

4.   SUMMARY AND CONCLUSIONS

The chemical transport model (CTM) uses a two-step algorithm to model mass transport of aqueous solutes in the presence of chemically reactive sub-surface media. The CTM model offers an alternative to the classical ADR equation: advection and dispersion are modeled using a probability density function approach and retardation is modeled using chemical attenuation reactions. The description of retardation in terms of interphase mass transfer reactions leads to variable solute velocities. For situations involving complex trace metal/sediment interactions, the mechanistic treatment used in CTM (and other reactive solute transport models) represents an improvement over the lumped parameter approach used in the classical transport equation.

The CTM code is part of a solute transport simulation package that includes a preprocessor (PRE), a thermodynamic database, and a postprocessor (POST). The preprocessor accepts input from the user and creates the input files needed for CTM. The major functions performed by PRE include calculation of the Markov transport matrix, definition of the chemical reactions allowed to occur in the simulation, and specification of the initial and boundary conditions. The postprocessor uses output from CTM to prepare a graphical representation of solute profiles as a function of space or time. A tabular output file is also created by CTM.

The CTM model is numerically stable and efficient. For simulations of solute transport under steady flow, the Markov transition matrix is constant and is calculated prior to the start of the simulation. The mass transport calculation consists of a matrix/vector multiplication. The method works equally well for advection or dispersion dominated problems. The description of dispersion is limited by the user's ability to define an appropriate pdf. Extension of the model to calculate multidimensional mass transport is straightforward and efficient; the bins become multidimensional areas or volumes over which the pdf is integrated.

As demonstrated in the example calculations, solubility dominated attenuation does not lead to a delay in the first arrival of a solute at the end of the streamtube. The relationships among the solubility controlled concentration, the injection concentration, the background concentration, and the reference concentration determine the apparent solute velocity of a solubility controlled solute. For the example calculation, the injection concentration was higher than the solubility controlled concentration at the end of the streamtube during the early part of the simulation. The effect of the solubility reaction was to lower the concentration of cadmium to the solubility limit. As the soil buffering capacity was exhausted, the solution pH decreased, and the solubility limit increased. This led to a pulse of dissolved cadmium traveling behind the pH front. The local partition coefficient (cadmium mass in immobile phase/cadmium mass in aqueous phase) ranged from 0 where cadmium was not present in the solid phase assemblage to about 60 in the zone of otavite precipitation.

Adsorption dominated attentuation can lead to a delay in the first arrival of a solute above background levels. The delay is a function of the adsorption site density and the ability of the solute to compete for adsorption sites. Adsorption alone cannot lead to a permanent lowering of the solute concentration because the adsorption sites will eventually become saturated and the solute concentration will reach the injection level. In the example calculation, the local partition coefficient was almost constant for the adsorption dominated case.

The mechanistic treatment used in CTM relies on experimental and site characterization data to capture the nature of the solute soil interactions. The initial spatial distributions of the aqueous, solid, and adsorbed masses of each component must be specified. The possible chemical reactions, including potential solubility controls must be identified. The methods required to obtain this information are more complex than the batch or column tests traditionally used to estimate retardation factors. Application of chemical transport modeling methodology to field sites remains a challenging problem.

REFERENCES

1.   Freeze, R.A., and J.A. Cherry. *Groundwater* (Englewood Cliffs, NJ: Prentice-Hall, Inc., 1979), 604 pp.

2.   van Genuchten, M. Th., and W.J. Alves. "Analytical Solutions of the One-Dimensional Convective Dispersive Solute Transport Equation," *U.S. Dept. of Agric., Tech. Bull. No. 1661* (1982).

3.   Valocchi, A.J., and P.V. Roberts. "Attenuation of Groundwater Contaminant Pulses," *J. Hydr. Engineer.* 109:1665-1682 (1983).

4.   Reardon, E.J. "Kd's - Can They Be Used to Describe Reversible Ion Sorption Reaction in Contaminant Migration?," *Ground Water* 19:279-286 (1981).

5.   Serne, R.J., and A.B. Muller. "A Perspective on Adsorption of Radionuclides Onto Geologic Media," in *The Geological Disposal of High Level Radioactive Wastes* (Athens, Gr.: Theophrastus Publ., 1987).

6.   Rubin, J., and R.V. James. "Dispersion-Affected Transport of Reacting Solutes in Saturated Porous Media: Galerkin Method Applied to Equilibrium-Controlled Exchanges in Unidirectional Water Flow," *Water Resour. Res.* 9:1332-1356 (1973).

7.   Valocchi, A.J., R.L. Street, and P.V. Roberts. "Transport of Ion-Exchanging Solutes in Groundwater: Chromatographic Theory and Field Simulation," *Water Resour. Res.* 17:1517-1527 (1981).

8.   Miller, C.W., and L.V. Benson. "Simulation of Solute Transport in a Chemically Reactive Heterogenous System: Model Development and Application," *Water Resour. Res.* 19:381-391 (1983).

9.   Jennings, A.A., D.J. Kirkner, and T.L. Theis. "Multicomponent Equilibrium Chemistry in Groundwater Quality Models," *Water Resour. Res.* 18:1089-1096 (1982).

10.   Kirkner, D.J. and H. Reeves. "Multicomponent Mass Transport with Homogeneous and Heterogeneous Chemical Reactions: Effect of the Chemistry on the Choice of Numerical Algorithm 1. Theory," *Water Resour. Res.* 24:1719-1729 (1988).

11.   Reeves, H., and D.J. Kirkner. "Multicomponent Mass Transport with Homogeneous and Heterogeneous Chemical Reactions: Effect of the Chemistry on the Choice of Numerical Algorithm 2. Numerical Results," *Water Resour. Res.* 24:1730-1739 (1988).

12.   Grove, D.B., and W.W. Wood.   "Prediction and Field
       Verification of Subsurface-Water Quality Changes During
       Artificial Recharge, Lubbock, Texas," *Ground Water* 17:250-257
       (1979).

13.   Walsh, M.P., L.W. Lake, and R.S. Schecter.   "A Description of
       Chemical   Precipitation   Mechanisms   and   Their   Role   in
       Formation Damage During Stimulation by Hydrofluoric Acid,"
       in *Symposium on Oilfield and Geothermal Chemistry* (Dallas,
       TX: Society of Petroleum Engineers, 1982).

14.   Kirkner, D.J., T.L. Theis, and A.A. Jennings. "Multicomponent
       Solute Transport with Sorption and Soluble Complexation," *Adv.
       Water Resour.* 7:120-125 (1984).

15.   Cederberg, G.A., R.L. Street, and J.O. Leckie. "A Groundwater
       Mass   Transport   and   Equilibrium   Chemistry   Model   for
       Multicomponent Systems," *Water Resour. Res.* 21:1095-1104
       (1985).

16.   Narasimhan, T.N., A.F. White, and T. Tokunaga. "Groundwater
       Contamination From An Inactive Uranium Mill Tailings Pile, 2.
       Application of a Dynamic Mixing Model," *Water Resour. Res.*
       22:1820-1834 (1986).

17.   Parkhurst, D.L., D.C. Thorstenson, and L.N. Plummer.
       "PHREEQE - A Computer Program for Geochemical
       Calculations,"   *U.S. Geological Survey, Water Resources
       Investigations 80-96* (1980).

18.   Hostetler, C.J., R.L. Erikson, J.S. Fruchter, and C.T. Kincaid.
       Overview of the FASTCHEM™ Package: Application to
       Chemical Transport Problems," *EPRI Report EA-5870-CCM*
       Volume 1 (1989).

19. Engesgaard, P., and Th. H Christensen. "A Review of Chemical Solute Transport Models," *Nordic Hydrol.* 19:183-216 (1988).

20. Simmons, C.S. "A Stochastic-Convective Transport Representation of Dispersion in One-Dimensional Porous Media Systems," *Water Resour. Res.* 18:1194-1214 (1982).

21. Campbell, J.E., D.E. Longsine, and M. Reeves. "Distributed Velocity Method of Solving the Convective-Dispersion Equation: 1. Introduction, Mathematical Theory, and Numerical Implementation," *Adv. Water Resour.* 4:102-108 (1981).

22. Bear, J. *Dynamics of Fluids in Porous Media* (New York: Elsevier Science, 1972), 763 pp.

23. Gillham, R.W., E.A. Sudicky, J.A. Cherry, and E.O. Frind. "An Advection-Diffusion Concept for Solute Transport in Heterogeneous Unconsolidated Geological Deposits," *Water Resour. Res.* 20:369-378 (1984).

24. Devary, J.L., C.S. Simmons, and R.W. Nelson. "Groundwater Model Parameter Estimation Using a Stochastic-Convective Approach," *EPRI Report CS-3629* (1984).

25. Abramowitz, M., and I.A. Stegun. *Handbook of Mathematical Functions with Formulas, Graphs, and Mathematical Tables.* Applied Mathematics Series No. 55 (Washington, D.C.: U.S. Department of Commerce, National Bureau of Standards, 1966), p. 299.

26. Tompson, A.F.B. "On a New Functional Form for the Dispersive Flux in Porous Media," *Water Resour. Res.* 24:1939-1947 (1988).

27.   Parker, J.C., and M.Th. van Genuchten. "Determining Transport
       Parameters From Laboratory and Field Tracer Experiments,"
       Virginia Agricultural Experimental Station Bulletin 84-3 (1984).

28.   Huyakorn, P.S., and G.F. Pinder. *Computational Methods in
       Groundwater Flow* (Orlando, FL: Academic Press, Inc., 1983),
       473 pp.

29.   Davies, C.W. *Ion Association* (Washington, D.C.: Butterworths
       Publications, 1962), 190 pp.

30.   Felmy, A.R., D.C. Girvin, and E.A. Jenne. "MINTEQ - A
       Computer Program for Calculating Aqueous Geochemical
       Equilibria," *EPA Report EPA-600/3-84-032* (1984).

31.   Garrels, R.M., and C.L. Christ. *Minerals, Solutions, and
       Equilibria* (New York: Harper and Row, 1965), 450 pp.

32.   Vanselow, A.P. "Equilibria of the Base-Exchange Reactions of
       Bentonites, Permutites, Soil Colloids and Zeolites". *Soil Sci.*
       33:95-113 (1932).

33.   Westall, J.C., J.L. Zachary, and F.M.M. Morel. "MINEQL, a
       Computer Program for the Calculation of Chemical Equilibrium
       Composition of Aqueous Systems," *Tech. Note 18*, Dept. Civil
       Eng., Massachusetts Institute of Technology, Cambridge, MA
       (1976).

34.   Krupka, K.M., R.L. Erikson, S.V. Mattigod, J.A. Schramke, C.E.
       Cowan, L.E. Eary, J.R. Morrey, R.L. Schmidt, and J.M.
       Zachara. "Thermochemical Data Used by the FASTCHEM
       Package," EPRI Report EA-5872-CCM (1988).

35. Furrer, G., J. Westall, and P. Sollins. "The Study of Soil Chemistry Through Quasi-Steady-State Models: I. Mathematical Definition of Model," *Geochim. Cosmochim. Acta* 53:595-601 (1989).

36. Eary, L.E., and J.A. Schramke. "Rates of Inorganic Oxidation Reactions Involving Dissolved Oxygen and Applications to Geochemical Modeling," in *Chemical Modeling in Aqueous Systems II* (Washington, DC: American Chemical Society, 1990), pp. 379-396.

37. Wagman, D.D., W.H. Evans, V.B. Parker, R.H. Shumm, I. Halow, S.M. Bailey, K.L. Churney, and R.L. Nuttall. "The NBS Tables of Chemical Thermodynamic Properties. Selected Values for Inorganic and C1 and C2 Organic Substances in SI Units," *J. Phys. Chem. Ref. Data* 11(2) (1982).

38. Theis, T.L., J.D. Westrick, C.L. Hsu, and J.J. Marley. "Field Investigation of Trace Metals in Groundwater From Fly Ash Disposal," *J. Water Pollut. Control Fed.* 50:2457-2469 (1978).

39. Ainsworth, C.C., and D. Rai. "Chemical Characterization of Fossil Fuel Combustion Wastes". *EPRI Report EA-5321*, (1987).

40. Dames and Moore. "Detailed Seepage Investigation of Mill Waste Disposal Alternatives West Gas Hills, Wyoming for Federal American Partners," *Job 10500-004-06*, Salt Lake City, UT (1981).

41. Peterson, S.R., W.J. Martin, and R.J. Serne. "Predictive Geochemical Modeling of Contaminant Concentrations in Laboratory Columns and in Plumes Migrating from Uranium Mill Tailings Waste Impoundments," *USNRC Report NUREG/CR-4520* (1986).

42.   Erikson, R.L., C.J. Hostetler, and M.L. Kemner. "Mobilization and Transport of Uranium at Uranium Mill Tailings Disposal Sites: Application of a Chemical Transport Model," *USNRC Report NUREG/CR-5169* (1989).

7

RISK ASSESSMENT
OF METALS IN GROUNDWATER

Reva Rubenstein
U.S. Environmental Protection Agency
Washington, D.C. 20460

Sharon A. Segal
Clement International, Inc.
Fairfax, VA 22031

1.   RISK ASSESSMENT OF METALS

The health levels used for the risk assessment of metals in groundwater are either Maximum Contaminant Limits (MCLs), which are standards set for drinking water, or EPA verified oral Reference Doses (RfDs) for non-carcinogens and Risk-Specific Doses (RSDs) for carcinogens.

The Reference Dose (RfD) for non-carcinogens is the means by which EPA quantifies the potential health effects associated with chronic chemical exposure. The RfD addresses only systemic health effects (i.e. health effects other than cancer and gene mutations). These systemic health effects are considered to occur only after a threshold exposure has been exceeded. The RfD serves as a benchmark for deriving regulatory threshold levels to protect exposed populations from adverse health effects (other than cancer or gene mutations) due to the ingestion of toxicants. The RfD is defined as an estimate (with uncertainty spanning one or more orders of magnitude) of a daily

0-87371-277-3/93/$0.00 + $.50
©1993 by Lewis Publishers

exposure to the human population (including sensitive subpopulations) that is likely to be without an appreciable risk of deleterious effects over a lifetime of exposure. The RfD is ideally derived from the highest experimentally determined dose in humans or animals that is without adverse biological effect (No Observed Adverse Effect Levels - NOAEL) by consistent application of uncertainty factors that correct for the various types of data used, the extrapolation from animals to humans, and any other additional modifying factor based on professional judgment of the quality and adequacy of the entire database.

Although the RfD is useful as a reference point for estimating the potential risk associated with exposures to various chemicals, it should not be considered to represent a sharp demarcation between "safe" and "unsafe" exposures because of the uncertainties inherent in the RfD derivation. There is no simple way to estimate the magnitude of the uncertainty in any given RfD. Uncertainties arise from many sources and are related to the quality and completeness of the toxicity data, the uncertainty in the dose response information and the NOAEL, and the usual lack of chemical-specific comparison data on intra- and inter-species variability in response.

Exposures somewhat higher than the RfD are not likely to result in an increased risk of adverse effects, but there is no way of determining this likelihood quantitatively. The RfD methodology tends to be protective of the public health and is thus conservative in the face of uncertainty, or incomplete data. For example, the RfD assumes lifetime exposure, even though many real exposures occur less frequently and intermittently. Adverse effects are considered unlikely in most of the population if brief, small excursions above the RfD occur. (Unfortunately, there is no precise definition of "brief").

Particularly sensitive members of the population may suffer adverse responses at exposures below the RfD. Likewise, excursions above the RfD cannot always be expected to be without effect in sensitive subgroups and each case should be treated individually by examining factors such as toxicokinetics, metabolism, nature and severity of effect, shape of the dose response-curve, and specific characteristics of the sub-population exposed. Professional judgment is required.

1.1   How are RfDs Determined?

The first step in developing an RfD (or an MCL) is to critically evaluate the available database. The determination of the potential of a chemical to cause harm to humans following exposure is based ideally on human data. Often that is impossible and thus, experimental specially bred animals are used instead. A critical assessment of the available data must consider many factors including the quality of the data, the sensitivity of the studies, the number and appropriateness of the toxicological endpoints examined, the relevance of the animal testing species to humans, the completeness of reporting and appropriateness of the exposure regimen, the uptake and distribution of the chemical, the appropriateness of the dose level selection, the existence of corroborative data, and the availability and quality of human data.

Each study is then evaluated by considering whether the study design and execution follows acceptable standards for the given type of endpoint being measured. The experimental design determines the nature and extent of the results and sets permanent limits on the interpretation and analysis of the data.

Whether animal data can be used as the basis for a prediction of harm to humans is dependent on factors such as similar metabolism and susceptibility. If an animal species is known to metabolize a chemical differently from humans, how significant are the results of tests in that species to humans? Often such information is lacking, however, and the species which is most sensitive to the chemical is assumed to be appropriate for extrapolation of human risk.

The linchpin of toxicity testing is the subchronic study. The exposure period normally does not exceed 10% of the animal's lifespan -- three months for rodents, one year for dogs. Subchronic studies identify the type of toxicity induced by the chemical, the major histopathology and, if enough dose levels were chosen, the general shape of the dose-response curve.

As part of the evaluation of a dose-response relationship, a NOAEL (No Observed Adverse Effect Level), if possible, or a LOAEL (Lowest Observed Adverse Effect Level) may be determined. The NOAEL is defined as the highest experimental dose of a chemical at which there is no statistically or biologically significant increase in frequency or severity of a toxicological effect between an exposed group and its appropriate control. Adverse effects are defined as any effects that result in functional impairment and/or pathological lesions that may affect the performance of the whole organism or that reduce an organism's ability to respond to an additional challenge. In general, NOAELs for several different toxicological endpoints will differ. Everything else being equal, the critical end point is the one with the lowest NOAEL. In some instances, the NOAEL for the critical toxic effect is simply referred to as the NOEL (No Observed Effect Level). This latter term, however, is ambiguous because there may be observable effects that are statistically significant but are not considered to be of biological or toxicological significance and thus are not "adverse". This is often a matter of professional judgment.

The magnitude of the NOAEL is dependent on the population size under study and dose selection. Studies using a small sample size are generally less sensitive to low-dose effects than studies with larger number of subjects. In addition, if the interval between doses is large, then the NOAEL determined from that particular study may be lower than what would be seen in a study that used intermediate doses. In instances when a NOAEL cannot be demonstrated, a LOAEL may be used to evaluate critical toxic end points.

The major end points in subchronic (and chronic) studies are not mortality but nonlethal adverse effects, which can be defined by biochemical, hematological, or clinical measurements, including changes in body and organ weight, food, or water consumption, and histopathological examinations. The dose levels are selected to result in no toxicity at the low dose, none or slight toxicity at the intermediate-dose levels, and frank toxicity in the high-dose group, but without excessive mortality that prevents meaningful interpretation of the data. For both subchronic and chronic studies, the species selected should maximize the biological comparability between the experimental animal and humans with regard to metabolism, toxico-kinetics, and target organ toxicity (although this information is not usually available).

In general, the dose levels determined to be without adverse effect in a subchronic study, when corrected for length of study (see below), are protective for all other effects (except cancer) for approximately 90% of chemicals tested. More detailed information can be gleaned from a subchronic bioassay than any other testing protocol, and its cost is within reason (i.e., $100 thousand to $200 thousand).

Once the available database has been evaluated, the RfD is developed. Following an assessment of the data quality, a data array is constructed. This is most often easily accomplished by plotting the known effects vs doses for each duration of exposure. The critical study is then selected from this data array. The critical study from which the RfD is developed must be identified from among the studies that satisfy both toxicological and statistical criteria. In the absence of an appropriate critical study in humans, the first choice is to select the most relevant animal model with regard to comparability to humans, either because of similar toxicological responses or similar pharmaco-kinetic profiles. If no clear choice can be made based on these criteria, then data from the most sensitive species are chosen, based on the species that exhibits a toxic response at the lowest exposure level. From the critical study, the critical data are extracted. This is generally the highest NOAEL observed (i.e., the most sensitive end point) associated with an effect level. "Free-standing" NOAELs, i.e., NOAELs seen at the highest dose tested, are generally not used because it is not known where the threshold for toxicity lies. However, if several studies are available that collectively define NOAELs and LOAELs for a particular critical effect, then a "free-standing" NOAEL from a particular study can be selected as the critical data point. If no clear NOAEL has been established, a LOAEL can be used.

The next level of critical judgment in the development of RfDs is an assessment of the uncertainty inherent in the critical data and proper application of uncertainty and modifying factors (UFs and MFs). The UFs and MFs take the place of traditional "safety factors". These terms are more descriptive because they represent scientific uncertainties, and they avoid the risk management connotation of safety.

In the assessment of risk, the ideal situation is one in which the chemical, subject, and exposure conditions are identical to those for which the risk is estimated. In practice, departures from the ideal situation, such as the variability among subjects, the use of animal data, or differences in route, dose, duration, and frequency of exposure, require extrapolation from the testing situation to the "real world". UFs are applied to compensate for known deficiencies in the database, i.e., departures for what would constitute an ideal database. The UFs are applied to extrapolate from the type of study serving as the basis of the NOAEL and the human situation for which the risk is estimated.

The UFs are meant to account for processes that we postulate to be occurring, e.g. metabolic rate differences between species, genetic differences between human subgroups etc. One of the aims of the U.S. Environmental Protection Agency is to reduce the uncertainty in risk assessment by elucidating which of these types of processes can be quantified.

Much of the uncertainty inherent in the risk assessment process lies in two areas: (1) extrapolation from the experimental species from which the critical data were obtained to humans, and (2) the experimental design. A number of factors contribute to the differences between species. These include body size, lifespan, and metabolic rate; absorption, distribution, metabolism, and excretion; target organ susceptibility; rate of target organ maturity; age; sex; diet; anatomy and physiology; and route and timing of exposure. There is much debate over which method of expressing the dose gives the best measure of equivalence across species, and all methods try to account for differences in body size and surface area. Differences exist between species with regard to absorption from the same site, the distribution of a parent compound and its metabolites throughout the body, and the rate, routes, and sites of metabolism and excretion of the parent compounds and its metabolites. Differences in these processes can result in considerable differences in the effective dose level of active metabolites at a target organ. Furthermore, differences in susceptibility of target organs across species casts uncertainty on the appropriateness of other species as surrogates for humans.

Experimental design considerations also contribute to the uncertainty in risk assessment. For example, extrapolation from animal data at high doses to human exposure levels at low doses is problematic. These high doses may themselves produce altered physiological conditions which can qualitatively affect the expression of toxicity. Normal physiology, homeostasis, and detoxification or repair mechanisms, may be overwhelmed and toxicity, which otherwise might not have occurred, may be induced or promoted. If a qualitatively different toxico-kinetic profile is seen at such high doses, a toxic response at this dose may not be indicative of effects at low exposure levels. Other factors that could affect the outcome of the experiment, and thus contribute to the uncertainty associated with the data generated include distribution of animals among dose levels and total number of animals used; selection of dose levels; purity of drinking water and dietary considerations; animal husbandry; intercurrent disease; adherence to good laboratory procedures; proficiency of personnel conducting study; and statistical methods used.

Toxicologists have long utilized orders of magnitude estimations to quantitatively account for some of these factors. Standard UFs include a ten-fold factor to account for variation in sensitivity among human populations; an additional ten-fold factor to account for the uncertainty inherent in extrapolating animal data to humans; an additional ten-fold factor to account for the uncertainty when extrapolating from subchronic (less than lifetime) NOAELs to chronic (lifetime) NOAELs; and an additional ten-fold factor when extrapolating from a LOAEL to a NOAEL. Up to an additional ten-fold factor may also be applied when the animal data are incomplete, that is, data are not available for more than one species or a critical toxicity test (e.g., reproductive/developmental) is missing.

Within a given population, there is a statistical threshold level for a given effect. It is assumed that this threshold lies somewhere between the NOAEL and the LOAEL. The Uncertainty Factor (UF) for use of a LOAEL rather than a NOAEL accounts for not knowing where that threshold is. The concept of a threshold also assumes that everyone responds in the same manner to exposure to a given chemical.

However, this is not necessarily the case because of variability within any given population and the existence of sensitive sub-populations. The current approach for extrapolation from general to sensitive human populations includes a ten-fold uncertainty factor. Available data suggest that the ten-fold factor may, in some circumstances, account only for normal human variability and not sensitive individuals. Although examples of very broad variations in sensitivity have been summarized by Calabrese [1], there currently are not sufficient data for global refinement of this uncertainty factor. One approach is to consider all characteristics related to sensitivity that are common in the population, such as aging, pregnancy, or certain chronic diseases such as emphysema, and evaluate all available data to ascertain how much more sensitive these people may be than the average individual.

Professional judgment is required in determining the magnitude of each uncertainty factor since, in several instances, the guidance suggests use of "up to a ten-fold factor". Precedents exist in EPA risk assessments for the use of factors of less than ten, given availability of supporting evidence. For example, incomplete data is often assessed by applying a factor of three.

There are several factors that may introduce uncertainty but which are not specifically addressed by the standard uncertainty factors. For example, the sample size of the test population is not considered when applying standard uncertainty factors. This and other types of uncertainty could be accounted for by the application of an uncertainty factor greater than ten for interspecies variability. However, adjustment to the standard ten-fold uncertainty factor may be perceived as arbitrary and imply a mathematical precision (for example, an uncertainty factor of 125) that is not appropriate considering the underlying biology. The modifying factor (MF) then is applied to separate the "traditional" areas of uncertainty that have already been addressed by the standard uncertainty factors from other areas of scientific uncertainty that have not yet been addressed quantitatively or qualitatively.

In the ideal situation, an extensive database is available upon which to base a weight of evidence evaluation and from which a NOAEL for the critical endpoint from the critical study is derived. The uncertainty factors described above are applied to account for intraspecies and interspecies variation as well as account for inadequacies in the database. Therefore, theoretically, all of the uncertainty factors could be applied to relatively poor data. As an extreme example, chemical X has been tested in only one sex of one species, only in a subchronic study of questionable quality, and only a LOAEL was identified. Application of uncertainty factors as described above might then include factors of ten for intraspecies variations, ten for interspecies variation, ten for adjustment from a subchronic to a chronic exposure, ten for adjustment from a LOAEL to a NOAEL concept, and possibly an additional ten for an "incomplete" database. In such a case the total uncertainty factor applied to this data would be one hundred thousand. The Agency suggests that there are no cases which would justify so large an uncertainty factor. Instead, Agency guidance suggests that the minimum database for determination of an RfD might include only one subchronic bioassay of good quality that clearly defines the threshold range. If a composite uncertainty factor is greater than 3,000 to 5,000, it indicates that the underlying data are not sufficient to support an RfD calculation and may suggest directions for future research. The above discussion serves to emphasize the importance of a weight of evidence evaluation to include a qualitative characterization of the confidence level attached to the RfD and the rationale for that level of confidence.

Critical judgment is required to determine the level of confidence in the database and the RfD. Factors to be considered include whether similar results have been obtained by different investigators in replicated animal studies; if similar effects are observed across sex, strain, and species; if there is clear evidence of a dose-response relationship; if a plausible relationship exists between metabolic data, postulated mechanism of action, and the effect of concern; if similar toxic effects are exhibited by structurally related compounds; and if the critical effect is also observed in humans following exposure to the chemical. Thus, the more data that exists for a particular chemical, the more confidence one has in the database, and the more stable and, therefore, the health limit developed from that NOAEL will be less uncertain.

Once the database has been evaluated, the critical study and data selected, and the uncertainty factors identified, the RfD is calculated according to the following equation:

$$RfD = \frac{N(L)OAEL}{UF} MF \tag{1}$$

where

UF = an uncertainty factor to account for recognized deficiencies in the database, and

MF = a modifying factor based on professional judgment of the entire database.

The final number, or RfD, should be carried to only one significant figure. This reflects the imprecision inherent in the actual conduct of the experimental study and the use of UFs. For example, animals are generally fed *ad libitum*. Food consumption is determined by difference, i.e., the food container is weighed at specified intervals, and the amount of food consumed is assumed to be the difference between the latest value and the previous value. Whereas the scales used are generally accurate to one part in 10,000, the fact that food is often spilled and otherwise dispersed by the animal, renders this measurement imprecise. Furthermore, when the N(L)OAEL (which already contains some uncertainty, as discussed above) is divided by UFs that may range from 10-10,000, the final number can realistically only be accurate to one significant figure.

For toxic agents not considered to have carcinogenic potential and for which appropriate data are available, the setting of the Maximum Contaminant Limit Goal (MCLG) is based largely on the determination of the RfD. From the RfD, a Drinking Water Equivalent Level (DWEL) can be calculated. The DWEL represents a medium-specific (i.e., drinking water) lifetime exposure at which adverse, non-carcinogenic health effects are not expected to occur. The DWEL assumes 100% exposure from drinking water. The DWEL provides the non-carcinogenic health effects basis for establishing a drinking water standard. For ingestion data, the DWEL is derived as follows:

$$DWEL\ (ng/L)\ =\ \frac{RfD\ x\ (body\ weight\ -\ kg)}{drinking\ water\ volume\ (l/day)} \qquad (2)$$

where:

Body weight is assumed to be 70 kg for an adult, and drinking water volume is assumed to be 2 liters per day for an adult.

To calculate the MCLG, the DWEL must then be apportioned into amounts reflective of the relative source contribution. In determining the drinking water contribution to the total dose, an assessment of all source contributions must be considered with the relative percent from drinking water estimated. In general, EPA assumes that a 20% contribution from drinking water would be reasonably conservative and protective due to the wide range of environmental exposure distributions and because of age- and occupation-related differences. The MCLG is therefore generally calculated by multiplying the DWEL by 0.20, and thus, the MCLG is not equivalent to the RfD. For most metals, however, EPA does an exposure assessment to quantify contributions from other routes of exposure. Maximum Contaminant Limits (MCLs) currently exist for barium, cadmium, chromium, mercury, and selenium, and MCLGs have been proposed for lead and copper. The unusually small UFs reflect an attempt by the Agency to use human data where possible. These values are presented in Table 1.

Metal toxicology must be determined on a case-by-case basis. The pharmaco-kinetics of metals must be considered; metals in the ionic form often require active transport mechanisms to be absorbed from the gut, and are not absorbed by passive diffusion. Metal-induced toxicity is often a route-specific phenomenon. For example, chromium ingested in solution is not carcinogenic, whereas inhaled chromium(VI) dust is carcinogenic.

Metals are also a unique group because several of them (chromium, cobalt, copper, iron, manganese, molybdenum, selenium, and zinc) are essential elements either in trace or nontrace quantities. Therefore, each of these metals has three levels of biologic activity: Trace (or higher) levels required for optimum growth and development, homeostatic levels, and toxic levels. A careful evaluation of the available data must take into consideration those levels optimally required for normal function so that a health limit is not estimated that lies below the level considered essential.

TABLE 1.   *Current MCLs and MCLGs for Metals.*

Metal	*MCl/MCLG (mg/L)*	*UF*	*Critical Effect*
Barium	5[a]	2	Hypertension and cardiovascular effects (human)
Cadmium	0.005[a]	10	Renal dysfunction (human)
Chromium	0.1[a]	-	Cancer
Mercury	0.002[a]	1,000	Renal dysfunction (rats)
Selenium	0.05[a]	15	Loss of hair and fingernails, dermatitis, and muscular dysfunction (humans)
Copper	1.3[b]	2	Gastrointestinal effects (human)
Lead	0[b]	-	Cancer (rats)

[a]MCL (54 FR 22062-22160)
[b]MCLG (53 FR 31516-31578)

Finally, metals are a unique group because of their speciation. Different forms of a metal (different oxidation states, organic vs. inorganic salts, soluble forms vs. insoluble forms) will influence its behavior. For example, ten times more elemental mercury will cross the blood-brain barrier than mercuric or mercurous salts following inhalation exposure. Consequently, neurotoxic effects are more often observed following acute exposure to elemental mercury than other forms of mercury.

This paper examined major factors which contribute to uncertainty in the regulatory health limits for metals. Due to the nature of these contributing factors, it is unlikely that the uncertainty can be significantly reduced. In light of this, the health limits cannot be considered to encompass a precision greater than one significant figure. Uncertainty in health numbers used in groundwater modeling exercises may be similar in magnitude to uncertainty generated in other modeling data.

REFERENCES

1.   Calabrese, E.   "Pollutants and High Risk Groups." *The Biological Basis of Increased Human Susceptibility to Environmental and Occupational Pollutants.* (New York, NY: John Wiley and Sons, 1978).

2.   Barnes, D. and M. Dourson.   Reference Dose (RfD): "Description and Use in Health Risk Assessment." *Reg. Toxicol. Pharm.* 8:471-486 (1988).

3.   Dourson, M. and J. Stara.   "Regulatory History and Experimental Support of Uncertainty (Safety) Factors." *Reg. Toxicol. Pharm.* 3:224-238 (1983).

4.   Casarett and Doull's Toxicology. *The Basic Science of Poisons.* 3rd Ed: Klaassen, C.D., M.O. Amdur, and J. Doull, Eds. (New York, NY: Macmillan Publishing Company, 1986).

8

INFLUENCE OF REDOX ENVIRONMENT AND AQUEOUS SPECIATION ON METAL TRANSPORT IN GROUNDWATER:
Preliminary Results of Trace Injection Studies

James A. Davis, Douglas B. Kent, Brigid A. Rea
U. S. Geological Survey
Water Resources Division
Menlo Park, CA 94025

Ann S. Maest
RCG/Hagler, Bailly
PO Drawer O
Boulder, CO 80306

Stephen P. Garabedian
U.S. Geological Survey
Water Resources Division
Marlborough, MA 01752

1.   INTRODUCTION

Contamination of groundwater by metal ions may occur as a result of municipal or industrial landfills, disposal of liquid mining wastes or mining tailings, or ineffective containment of nuclear wastes. Reliable assessment of hazards or risks arising from groundwater contamination and the design of efficient and effective means of remedial action requires a capability to predict the behavior of dissolved

0-87371-277-3/93/$0.00 + $.50
©1993 by Lewis Publishers

solutes in groundwater. Quantitative prediction of metal movement requires an integration of the processes controlling transport, including chemical, physical, and biological processes.

Several processes interact to influence the transport of metal ions in groundwater. These include complexation reactions in water, redox-related processes, and reactions that result in the removal of metal ions from water, e.g. adsorption and precipitation. Metal ions generally are strongly adsorbed by the surfaces of minerals in a porous medium or in rock fractures, although their reactivity depends on the particular mineral assemblage and the composition of the groundwater [1]. The formation of an aqueous complex can have a significant effect on the tendency of a metal ion to adsorb [2-4]. When complexed with a strongly-binding ligand, a metal may be transported downgradient at an average velocity that is orders of magnitude faster than would be expected in the absence of ligands. For example, field studies have shown that metal radionuclides may be mobile in groundwater when complexed with organic ligands [5,6]. The ligand may be present naturally, e.g. fulvic acid or carbonate anions, or may be introduced as a contaminant. Changes in speciation, and therefore reactivity, can result from encountering different chemistries along a groundwater flow path. Many elements also may exist in more than one oxidation state, e.g. As, Cr, Cu, Fe, Hg, Mn, Mo, Se, U, V, and the actinide elements. The mobilities of the different oxidation states of these metals may differ by orders of magnitude.

We are using both field- and laboratory-scale experiments to study the transport of metals that undergo changes in speciation (and reactivity) involving complexation, adsorption, and electron-transfer reactions. Our objectives are to identify important processes affecting metal transport, to examine the relationships between chemical and physical factors affecting transport, and to stimulate the development of hydrogeochemical models that can predict the transport of metal ions in complex geochemical environments. In this paper, we present preliminary results of small-scale, natural-gradient tracer tests conducted in oxic and suboxic zones of a shallow, unconfined sand and gravel aquifer near Falmouth, Massachusetts (USA). The reactive tracers injected into the aquifer included zinc, as an EDTA complex, chromium(VI), and selenium(VI). An overview and some examples of significant processes likely to affect the transport of the injected tracers is given, along with a detailed description of the hydrologic and geochemical characteristics of the field site.

2.   BACKGROUND

2.1   Sorption of Metal Ions by Mineral Surfaces

The concentrations of many minor elements in natural waters often reflect undersaturation with respect to pure mineral phases of known composition and structure. Aqueous concentrations of these elements are usually thought to be controlled by adsorption-desorption (sorption) reactions [1,7,8]. The most common inorganic adsorbents in geological materials are the hydrous metal oxides (as discrete minerals or as coatings), carbonates, clays, and sands, while the most common organic adsorbents consist of detrital plant material and humic coatings on mineral surfaces. Many studies of metal sorption onto geological materials have concluded that iron and manganese hydrous oxides and organic materials are the predominant adsorbents [3,9-11], because of their charged, reactive hydroxyl groups, combined with their high specific surface area [1]. However, Fuller and Davis [8] recently showed that calcite dominated the sorption of cadmium by an aquifer sand.

Summaries of the adsorption behavior of zinc (Zn), chromium (Cr), and selenium (Se) are provided below because of their use as tracers in this study. In dilute aqueous solutions, Cr(VI) exists primarily as chromate (CrO_4^{2-}) and bichromate ($HCrO_4^{-}$) anions. These ions are adsorbed by the surfaces of many oxide minerals, especially those with high values of the zero point of charge, e.g. hydrous iron and aluminum oxides [12-14]. Cr(VI) is more weakly adsorbed by aluminosilicate minerals, such as montmorillonite [13,15] and kaolinite [16]. Less is known about its sorption behavior on soils and sediments, although existing studies confirm the importance of pH and the content of hydrous iron and aluminum oxides [11,17]. Cr^{3+} and its hydroxy complexes, like Fe^{3+}, are very strongly adsorbed by mineral surfaces, are only slightly soluble in water, and form solid solutions readily with ferrihydrite [15,18-20].

Oxide minerals are considered the most important adsorbents for selenium (Se) in soils and sediments [21], although the selective extraction techniques applied to make these findings may be subject to considerable errors [22]. The hydrous oxides of iron, aluminum, and manganese can adsorb both Se(IV) and Se(VI) under favorable conditions, especially in weakly acidic solutions at low ionic strength [14,23-25]. As will be discussed further below, Se adsorption is very dependent on its oxidation state.

Zinc (Zn), like many other transition metal ions, is strongly adsorbed by the surfaces of hydrous oxides, especially by iron and manganese oxides [26-28]. On iron oxides, adsorption increases markedly with increasing pH, from near zero sorption at weakly acidic pH values (e.g. pH 5.0) to essentially quantitative removal from water at pH values greater than 7.5 [27-29]. Adsorption on manganese oxides is less pH-dependent [30] but is stronger at low pH, with significant sorption occurring at pH 4.0 [26].

Complexation of metal ions by ligands can drastically alter adsorption by oxide surfaces [3,4,31]. Metal ion adsorption is often decreased by complex formation in solution, assuming that the complex formed has little reactivity with the mineral surface. For example, metal-EDTA complexes and metal-fulvate complexes are not adsorbed by the surfaces of silica and aluminosilicate minerals. In this case, from coordination chemistry considerations, the mineral surface and complexing ligand compete thermodynamically for metal ion coordination, and the resulting metal adsorption can be estimated based on straightforward calculations using the concentrations of the complexing ligand and surface sites. However, oxides with high points of zero charge, e.g. aluminum oxides, do adsorb metal-EDTA or metal-fulvate complexes in acidic solutions, resulting in a complicated pattern of increasing metal adsorption at low pH and decreasing metal adsorption at high pH [3,4]. The presence of Fe(III) in such systems leads to even more complex interactions, because of the strong binding of Fe(III) with these ligands and the pH dependence of Fe(III) solubility. For example, Bowers and Huang [31] have shown that adsorption of ZnEDTA complexes is strong on aluminum oxide in the pH range 4.0-6.0 in the absence of Fe(III). However, in the presence of Fe(III), Zn is only weakly adsorbed in this pH range because: 1) EDTA is preferentially complexed with Fe(III), causing the dissociation of $ZnEDTA^{2-}$ complexes and leaving Zn^{2+} as the principal species of dissolved Zn, and 2) adsorption of Zn^{2+} is highly pH-dependent, being relatively weak at pH 5.0 and becoming strong at pH values greater than 6.0. Clearly, metal adsorption in such complex systems can only be described by models that account for both aqueous and mineral surface speciation [1,2].

2.2 Redox Processes in Groundwater

Early diagenetic processes in marine sediments provide a conceptual framework for examining redox processes in groundwaters. The concept of redox reaction sequences, with distinct redox zonation, has proven useful in modeling the evolution of porewater chemistry and sediment composition in vertical profiles of marine sediments [32,33]. Each redox zone is dominated by a particular redox reaction. Berner [34] has classified the redox status of sedimentary environments as: 1) oxic (oxygen-rich), 2) post-oxic (oxygen-depleted, usually containing Fe(II)), 3) sulfidic (sulfide-rich), and 4) methanic (methane-rich). The observed sequence of vertical redox zones is explained in terms of a model in which organic matter is oxidized by the oxidant which will yield the greatest free energy change per mole of organic carbon oxidized [32]. When this oxidant is depleted, oxidation proceeds utilizing the next most efficient oxidant, until all oxidants are consumed or all oxidizable organic matter is depleted. Organic matter is usually assumed to be the only electron donor. Thus, oxygen is first consumed, then manganese oxides and nitrate are reduced, followed by iron oxide reduction, sulfate reduction, and finally methane is produced.

The analogy of such redox sequences to the evolution of groundwater quality along a sub-surface flow path has been recognized by many authors [35-37]. The analogy is based on the similarity of major chemical reactions and processes involving organic matter in an environment with limited exposure to oxygen [36]. In the case of groundwater systems, however, the redox zones are usually observed to occur in a lateral sequence along a groundwater flow path downgradient from the source of an organic-rich discharge. This may lead to the movement of iron and manganese in post-oxic reducing groundwaters, until these metals are deposited downgradient in narrow sub-surface bands, as more oxic waters mix with the metal-rich anoxic water [38].

Little is known about the manner in which sequential redox zones affect the transport of redox-sensitive metal contaminants, such as Cr and Se [17,39]. Cr(VI) is rapidly reduced to Cr(III) by Fe(II) in circumneutral solutions [40]. In addition to dissolved Fe(II), Cr(VI) may be reduced by the surfaces of mineral phases containing Fe(II), such as biotite or magnetite [15,41,42]. This reaction effectively limits the transport of chromium in iron-rich reducing groundwaters, since Cr(III), like Fe(III), is very strongly adsorbed by mineral surfaces and is of limited solubility at pH values greater than 6.0 [18-20]. Cr(VI) may also be reduced by reactions with dissolved or detrital organic material, e.g. oxalic acid, lactic acid, and various other carboxylic acids, alcohols, and aldehydes [43-45]. Complexation of Cr^{3+} by organic material, however, can increase its solubility and mobility [46]. The reduction of Cr(VI) by less reactive organic matter may be catalyzed by bacteria in either aerobic or anaerobic environments, in which Cr(VI) is used as a terminal electron acceptor [47]. Chromate-resistant strains of bacteria have been isolated from sediments contaminated with chromium [48].

The geochemical behavior of Se has been reviewed extensively in recent years [49,50]. In aqueous systems, Se may exist in one of four oxidation states: Se(VI) or selenate, Se(IV) or selenite, Se(O) or elemental selenium (solid), and Se(-II) or selenide. Thermodynamic analysis suggests that the latter two oxidation states should be found in reducing environments, selenite in mildly reducing environments, and selenate in oxidizing environments. Sulfur and Se share many chemical and biochemical properties [50], and selenate (like sulfate) is usually mobile in oxic environments and may be reduced by bacteria in anaerobic environments [51]. The chemical kinetics of conversion between selenate, selenite, and elemental selenium are slow and poorly understood; however, it has been recently shown that selenate can be reduced to elemental selenium by anaerobic bacteria independent of sulfate reduction [52]. Selenate transport behavior has been investigated by Benson and coworkers [53,54].

The physical size of redox zones and the ability to observe them in groundwater systems are highly dependent on a large number of complex processes, including the relative rates of redox processes and the rates of transport of reactants and products. Unlike marine sediments, which are commonly believed to be at redox equilibrium [34], an assumption of local chemical equilibrium in groundwater systems may not be justified. In addition, the rates of the chemical and physical transport processes may be affected by the heterogeneity of aquifers [55,56]. As noted by Barcelona *et al.* [57], many authors have emphasized the importance of horizontal (lateral) redox gradients along a groundwater flow path; however, vertical concentration and redox gradients in groundwater systems can be far more pronounced than horizontal gradients. Thus, the observation of oxidation-reduction boundaries or zones in groundwater systems will often require a spatially dense array of monitoring points [56,57].

2.3 Effects of Groundwater Composition on the Transport of Metals

Assessment of the long-term fate of toxic substances entering groundwater requires the identification and understanding of processes controlling the release, movement, and retention of individual solutes within the range of possible geochemical settings. A variety of empirical approaches have frequently been used to model sorption by mineral surfaces, including partition coefficients, isotherm equations (such as the Langmuir and Freundlich equations), and Kurbatov plots [1,58], but these empirical fitting parameters have little transfer value. Considerable progress has been made in the development of theoretical models for solute transport that include sorption processes [59-62]. However, there is a general lack of appreciation for the importance of water composition and metal speciation in determining the sorption behavior of metal ions and, hence, in affecting their mobilities in groundwater.

For example, consider the effect of sulfate concentration on the adsorption of Cr(VI) onto the surface of the mineral, ferrihydrite (Figure 1). Ferrihydrite is believed to be one of the more important adsorbents of metal and metalloid ions in low-temperature aquifer systems [1,7,9,12,28]. At low sulfate concentrations and in the pH range 4.0-7.0, Cr(VI) is strongly adsorbed by the ferrihydrite surface. However, as the sulfate concentration in water increases, Cr(VI) adsorption diminishes, due to competitive adsorption of the sulfate anion [14]. Thus, adsorption of Cr(VI) anions is highly dependent on both pH and sulfate concentration in water, as well as many other solutes, such as bicarbonate ion and dissolved silicate [12,63]. In this way, the mobility of Cr(VI) in aquifers may depend significantly on groundwater chemistry, being highly mobile under certain conditions and retarded under other conditions. Examples of this diversity in Cr(VI) transport behavior will be demonstrated in the field results presented later.

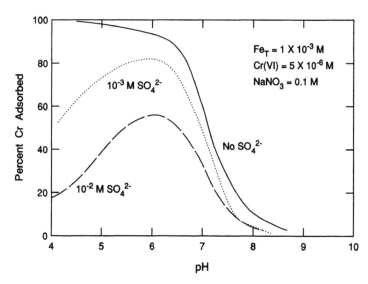

FIGURE 1.   *Adsorption of Cr(VI) by Ferrihydrite as a Function of pH and Sulfate Concentration. Data Described in Reference 101.*

Figure 2 illustrates the effect of pH, salt concentration, and oxidation state on the adsorption of Se anions by ferrihydrite (or goethite). Se(VI), present as the selenate (SeO_4^{2-}) anion, adsorbs extensively only at low pH and low ionic strength on the ferrihydrite surface [23]. Like Cr(VI), selenate mobility in aquifers can be dependent on groundwater chemistry, but conditions favoring its retardation are likely to occur less frequently, due to the low pH and low ionic strength conditions required for adsorption. Nonetheless, variability in Se(VI) mobility will be demonstrated in the field results presented in this study. In contrast, Figure 2 shows that Se(IV), present as the selenite (SeO_3^{2-}) anion, is strongly adsorbed by the ferrihydrite surface at circumneutral pH values. Se(IV) adsorption is independent of ionic strength [23]. Hence, changes in the oxidation state of Se may have significant effects on its solubility and reactivity with mineral surfaces, and may strongly influence its mobility. As mentioned earlier, a change in oxidation state for chromium, i.e. reduction to Cr(III), would greatly affect its mobility also, since Cr(III) is strongly adsorbed and weakly soluble at low pH. In this manner, changes in the redox environment of an aquifer, either temporal or spatial, can greatly change the mobility of some metal and metalloid ions.

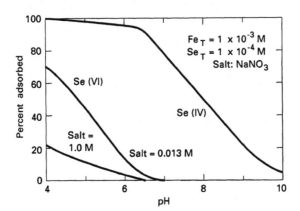

FIGURE 2. *Adsorption of Se(VI) and Se(IV) by Ferrihydrite as a Function of pH and Background Electrolyte Concentration.* Adsorption of Se(IV) is independent of background electrolyte concentration and only one curve is shown. Data described in reference 23.

The empirical sorption models that employ partition coefficients, or isotherms, are of limited utility in the cases of Cr, Se, or Zn, if the water chemistry of an aquifer is highly variable. This is because an empirical set of parameters must be derived for every set of aqueous conditions, e.g. various pH values, sulfate concentrations, and complexing ligand concentrations. A more fundamental, mechanistic approach that uses the surface ionization and coordination concept in conjunction with aqueous speciation models is now widely accepted for modeling the interactions of metal ions with mineral surfaces [1,2,5,64-67]. To date, only Cederberg *et al.* [68] have incorporated this type of approach within a hydrogeochemical transport model, although other models could be adapted to this approach for a limited number of reactions.

3.   FIELD SITE DESCRIPTION

Tracer-injection tests were carried out in a shallow, unconfined sand and gravel aquifer near Falmouth, Massachusetts (Figure 3). The aquifer is composed of about 100 meters of unconsolidated sediments that overlie a relatively impermeable bedrock [69]. The upper 30 meters of the aquifer are a permeable, stratified, glacial outwash. Beneath the sand and gravel outwash, the sediments are a fine-grained sand and silt [70]. The median grain size of the outwash is about 0.5 mm, and the outwash generally contains less than 1 percent silt and clay [71,72].

The hydrologic characteristics of the site are summarized in LeBlanc *et al.* [71]. The estimated average hydraulic conductivity of the sand and gravel is 110 m/day. Hydraulic tests made with a flowmeter in long-screened wells indicate that the hydraulic conductivity of the outwash varies one order of magnitude [73]. This variation results from the interbedded lenses and layers of sands and gravels that form the glaciofluvial deposit and that are evident in surface exposures at the site [71]. The effective porosity of the outwash is estimated to be about 0.39. The water table at the site is generally 3 to 7 m below land surface and slopes to the south at

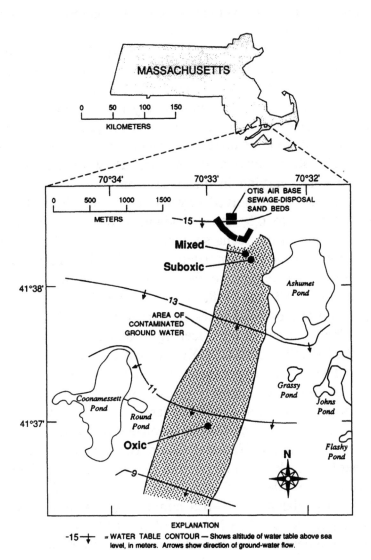

FIGURE 3. *Location of Field Sites for Natural-gradient Tracer Tests in a Shallow Sand and Gravel Aquifer near Falmouth, Massachusetts.* The oxic site (Site O) illustrates where the tracer test was conducted in the recharge zone. MIXED and SUBOXIC illustrate the locations of tests in the suboxic, sewage-contaminated zone. At the mixed site, oxic water with tracers was injected into the suboxic zone.

0.15m per 100 m. The average velocity of groundwater in the sand and gravel is 0.4 m/day.

Disposal of treated sewage to the aquifer at Otis Air Base has caused the formation of a contaminant plume [70] that underlies the test sites (Figure 3). The plume is about 23 m thick and more than 3 km long. Water in the plume contains very little or no dissolved oxygen and has a specific conductance as high as 350 µS/cm (microsiemens per centimeter). A zone of uncontaminated, aerobic groundwater, derived from precipitation, overlies the plume. Between the two zones is a transition zone, varying in thickness from 1 to 3 meters, containing sharp gradients in dissolved oxygen (DO) and nitrate [74,75], a pH gradient from 5 to 6.5, and a gradient in dissolved salts. Dissolved organic carbon in the recharge zone is 0.3 mg/l and between 2 and 4 mg/l in the sewage-contaminated zone [71,76]. The sewage effluent had a dissolved organic carbon concentration of 11 mg/l during sampling in 1983 and 1985 [76]. General aspects of the water composition of these two zones are summarized by LeBlanc et al. [71].

The locations of the tracer-injection tests are shown in Figure 3 as OXIC, MIXED, and SUBOXIC, hereafter referred to as sites O, M, and S. Two of the experiments (at sites M and S) were conducted within an array of densely-instrumented multilevel groundwater samplers (MLS) used in a previous large-scale tracer test [71,77]. This array of samplers (over 600 in number) is located about mid-way between the Otis Air National Guard Base sewage-disposal site and Ashumet Pond, a kettle-hole pond in the outwash plain (Figure 3). Site M was at MLS 10-15 and site S was at MLS 37-12 (see Garabedian [77] for an explanation of MLS numbering). A third experiment was conducted at site O (MLS M1 at well cluster F168), approximately 2.3 km downgradient of sites M and S, where an additional small array of multilevel samplers was installed.

The distribution of tracers in the aquifer was monitored by collection of water samples using the arrays of multilevel samplers (MLS) mentioned above and described in detail by LeBlanc et al. [71]. Each MLS consists of 15 color-coded polyethylene tubes (0.47 cm inside diameter). The cluster of tubes runs from land surface down the inside of a polyvinyl chloride (PVC) casing, and then out holes drilled at various depths. The open, downhole end of each tube is screened with a fine nylon fabric. The vertical spacing between sampling ports is generally constant for a given MLS, but varies from 25.4 cm to 50.8 cm for the MLS used in this study. Each tube of the MLS can be sampled by suction from land surface using a peristaltic pump.

Three natural-gradient tracer tests were initiated in July 1988, each characterized by a different set of conditions with respect to dissolved oxygen. At the oxic site (O), recharge water was withdrawn, mixed with the tracers, and pumped back into the recharge zone at the same depth interval (Figure 4). At the suboxic site (S), the same type of experiment was conducted with suboxic water. At the mixed site (M), oxic recharge water was mixed with the tracers and then pumped into the suboxic zone (Figure 4). Approximately 380 liters of groundwater was withdrawn for each tracer test from either two or three tubes of an MLS into an acid-washed polyethylene bag containing the tracers (as salts), and the "spiked" water was then pumped back into the same tubes from which groundwater was withdrawn (except for the mixed site, explained below). The tracers and their concentrations are shown in Table 1. Breakthrough of the tracers was monitored at several MLS downgradient from the injection MLS. Bromide ion (Br) was used as a nonreactive tracer of the injected water. EDTA was added to enhance the transport on Zn. The EDTA may be specifically involved in the transport of metals in some contaminated groundwaters [78]. However, its use here is solely to serve as a general model for complexing ligands and their role in enhancing metal transport.

4.   EXPERIMENTAL METHODS

Specific conductance was measured with a YSI Model 32 conductance meter and dip-type plastic cell (YSI #3417), which had a cell constant of 1.0 cm^{-1}. The temperature of the sample was measured simultaneously and used to correct the specific conductance to 25°C.

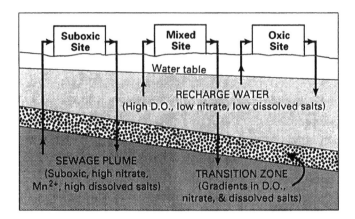

*FIGURE 4.   Schematic   Diagram   Showing   Recharge   and
Sewage-contaminated Zones of the Shallow Aquifer and the
Zones used for Withdrawing and Injecting Groundwater for
the Three Natural-gradient Tracer Tests.*

TABLE 1.   Concentration of Tracers Added to Groundwater.

TRACERS	CONCENTRATION (mM)
KBr	1.9
K_2H_2EDTA	0.5
$Zn(NO_3)_2 \cdot 6H_2O$	0.1
K_2CrO_4	0.1
Na_2SeO_4	0.06

Most pH measurements were obtained on 50 ml samples in open
bottles immediately after collection with Ross (Orion) combination
electrodes. Electrodes were pre-tested for their ability to measure pH
values in low ionic strength solutions using dilute H_2SO_4 solutions [79].
Each electrode was fitted with a silicone stopper that fits snugly into the
top of the sample bottle. An electrode was allowed to equilibrate for
3 to 5 minutes with the sample before the pH was recorded. The
electrodes were calibrated by equilibration in standard 0.05M potassium

hydrogen phthlate (ca. pH 4.0) and potassium hydrogen phosphate (ca. pH 7.0) buffers. Temperatures were measured as a matter of course, but the differences between samples and buffers never exceeded 5°C. Considering the variation in electrode slope with temperature, a 5°C difference between samples and standards results in a maximum error of 0.05 pH units. In some cases, pH values were obtained on samples collected in 400 cm^3 flow-through cells. The cells were located in-line between the well and peristaltic pump. The "open bottle" pH values averaged 0.05 pH units above the "flow-through cell" values; only 6 of 47 pH values differed by more than 0.1 pH units. The open bottle method was much faster than the flow-through cell method, thus, most of the pH values presented here were obtained with the open bottle method.

Dissolved oxygen (DO) concentrations were determined using Chemets (Chemetrix Inc, Calverton, VA). The application of Chemets to determining dissolved oxygen concentrations has been described by White *et al.* [80]. Briefly, a Chemet consists of a glass ampoule containing a small volume of reagent. The ampoule has been evacuated and partially back-filled with an inert gas and has a pre-scored capillary tube at one end. Various methods were used to collect water samples for DO measurement. The simplest was to use Norprene tubing (Cole-Parmer) to draw the water from an MLS tube through the peristaltic pump. Norprene tubing was found to be impermeable to atmospheric oxygen [81]. A cone-shaped flow-through sampler (included in the dissolved oxygen kit) was connected to the outflow end of the Norprene tubing, the Chemet was immersed in the sampler such that the capillary end was isolated from contamination by the atmosphere, and the Chemet was broken while water was flowing through the sampler at a rate of about 300 cm^3 per minute. Comparison of this method with other methods that were more rigorous in their exclusion of contamination of the sample with atmospheric oxygen proved that this method was effective [81].

Fe(II) was determined in the field using the ferrozine method [82]. Water samples were collected in 60 ml bottles captured in 400 ml flow-through cells placed in-line between the MLS and peristaltic pump. This was done to minimize contamination of the sample with atmospheric oxygen that might result in oxidation of Fe(II) before analysis. Nevertheless, many nearly-anoxic samples were contaminated with about 5 μM DO due to permeation of oxygen through the polyethylene tubing comprising the MLS during sampling (see Design of Field Experiments below). In fact, the only water samples yielding zero DO were those that had sufficient Fe(II) to give the sample an oxygen demand in excess of 5 μM DO.

The absence of detectable dissolved sulfide was determined by collecting water samples in flow-through cells [as described above for Fe(II)], withdrawing water from the cell using a syringe equipped with a stainless steel needle, and injecting the sample into a vacutainer containing 1 ml of 0.57M zinc acetate solution. The lack of ZnS precipitate indicated that dissolved sulfide was not present at detectable concentrations.

Water samples for metal analysis were collected in acid-washed, polyethylene bottles and acidified to pH 2.0 with HCl for preservation. The total concentrations of many elements (including Cr, Fe, Al, Ca, Mg, K) were determined by inductively-coupled plasma emission spectroscopy. Zinc was determined by flame atomic absorption spectroscopy. Total dissolved Se was determined by flameless atomic absorption spectroscopy with Zeeman background correction. Samples were diluted 1:1 for Se analysis with a matrix modifier solution containing 3 g/dm^3 palladium nitrate and 2 g/dm^3 magnesium nitrate [83]. Most water samples were collected during these experiments without filtration, and thus reported Fe, Mn, Al, and Zn background concentrations could reflect contamination by colloidal particles. A comparison of metal concentrations in selected filtered and unfiltered samples showed no significant difference, with occasional exceptions found for Fe, Al, and Mn.

Water samples for Br, Cl, and SO_4 analyses were collected in polyethylene bottles rinsed with MILLI-Q water and kept refrigerated until analysis. Samples were subsequently filtered in the laboratory, and bromide, chloride, and sulfate concentrations were determined by ion chromatography.

4.1 Laboratory Study of Zn Adsorption

Sub-surface material (also referred to as Otis sand) was obtained from a core taken adjacent to Site M. The core was obtained in a 5 cm diameter aluminum core liner, frozen immediately, and sent by express mail to the laboratory in Menlo Park, California. The core had thawed somewhat by the time of arrival, but was still quite cold when processed. Processing involved discarding the top two inches of core material, decanting excess groundwater, drying at room temperature in a laminar flow hood, and dry sieving through a 2 mm nylon screen to remove gravel and rock fragments. The sand was stored under ambient conditions in polyethylene bottles until used.

The physical characteristics and composition of the sub-surface material were not examined in detail, but appeared similar to previous descriptions of Otis sand [72,85]. Most of the sand falls in the 0.1 - 1.0 mm size range (fine to coarse sand). A few percent were greater than 1 mm (coarse sand and gravel), and a very small amount falls in the 0.002 - 0.1 mm size range (silt and very fine sand). Quartz, plagioclase, and orthoclase constituted about 95% of the minerals by weight; the remaining 5% was made up of accessory minerals, like biotite, hornblende, muscovite, and magnetite. The abundance of accessory minerals was greater in smaller size fractions. Organic material and clay minerals constituted less than 0.1% by weight. Hydroxylamine hydrochloride leaches of the sand used in the Zn adsorption experiments were performed using the method of Chao and Zhou [86]. Al, Fe, and Mn were extracted at concentrations of 2.5, 2.0, and 0.1 μmoles/g respectively.

Batch experiments were performed to study adsorption of Zn by sub-surface material collected at the field site. These experiments examined the amount of Zn adsorbed in the presence and absence of EDTA and the amount of Fe dissolved from the sub-surface material by EDTA. All experiments were performed in an electrolyte solution of 0.1M $NaNO_3$, at 22 ± 2°C, and 445 g of sand/dm^3 of solution. The extent of Zn adsorption was determined by measuring the removal of dissolved Zn from solution in the batch experiments. Ten grams of sand were placed in acid-washed polycarbonate centrifuge tubes. A solution containing 0.1M $NaNO_3$ and 5 µM Zn was prepared from reagent grade $NaNO_3$ and a Zn atomic absorption spectroscopy (AAS) standard. The solution was split into two stocks, one of which was spiked with a small quantity of carrier-free ^{65}Zn. For experiments with EDTA, an aliquot was added from a 0.1M EDTA volumetric standard (Baker) to yield a concentration of 6.4 µM EDTA in solution. A precise volume (22.5 cm^3) was pipetted into the centrifuge tube, the pH was adjusted with NaOH, and the suspension was rotated end over end for 18 hours. No attempt was made to prevent equilibration with the atmosphere. After the reaction period, the suspensions were centrifuged, solution was removed for analysis, and the pH was measured to an accuracy of ±0.03 pH units. The electrodes were calibrated with NBS buffers whose temperature never differed from that of the sample by more than 1°C. Activities of ^{65}Zn in water samples were determined by gamma spectrometry using an intrinsic germanium detector. For samples without ^{65}Zn, Zn and Fe concentrations were determined by flame AAS.

Some experiments were performed to determine the amount of Fe and Zn leached from the sand by EDTA. These experiments were performed using the same procedure described above except that no Zn was added to the stock solution. Fe and Zn concentrations in solution were determined by flame AAS.

5.   RESULTS AND DISCUSSION

5.1   Background Chemistry

As mentioned above, the upper aquifer at the site contains zones of varying chemical composition due to mixing of the sewage plume with recharge water. The recharge zone contains uncontaminated, oxic water with approximately 250 µM DO (8 ppm), low pH (4.0 to 5.0), and low concentrations of dissolved salts (specific conductivity of 30 µS/cm). The sewage-contaminated zone is nearly anoxic (referred to here as suboxic), containing DO concentrations less than 10 ppb, has a higher pH (6.5-7.0) and higher conductivity (350-400 µS/cm), and contains dissolved Mn(II) in concentrations as high as 10 μM. Dissolved sulfide was below detection in both zones. Dissolved Fe(II) was occasionally present above the detection limit, but was found only at a few points in the sewage-contaminated zone that is sampled by MLS in the large array at the field site. When found, the presence of Fe(II) was usually distinct but localized, with concentrations often exceeding 20 μM when Fe(II) was detected.   However, Fe(II) concentrations were usually very low in sampling ports adjacent to, above, and below any points at which it was found at elevated concentrations.   Gschwend and Reynolds [87] reported an Fe(II) concentration of 190 μM in a 5 cm diameter well (F343-57) located about 20 meters upgradient of Site M. That well is screened over a 60 cm interval that is several meters below the bottommost tube of the MLS at Site M.

Figures 5 and 6 show the gradients with depth of pH, specific conductivity, and dissolved Ca, Mn, nitrate, and DO at Sites M and S. The uppermost tubes of these MLS are located at depths within or near the recharge zone, but most of the tubes sample from the sewage-contaminated zone. The nitrate profiles at these sites (data shown for Site S only) show sharp increases within the transition zone and top of the sewage zone, but then decrease substantially to near zero at the center of the sewage-contaminated zone [75] (Figures 5 and 6). Very low concentrations of DO in the sewage zone (6 ppb, 0.2 μM) have been measured with a special MLS constructed with copper tubing located near Site S [81]. Measurements of DO in the sewage zone using MLS with polyethylene tubing consistently yielded values of 3 to 6 μM.   However, these values were in error, due to permeation of oxygen through the walls of the tubing in the vadose zone (approximately 6 meters in thickness).

The thickness of the recharge zone increases with distance downgradient from the Otis Air Base sewage disposal site as the sewage plume sinks [70].   The recharge zone within the densely-instrumented array of MLS at sites M and S is only about two meters thick or less (Figures 5 and 6). Since it was desired to conduct one of the tracer tests under oxic conditions, a small array of MLS was installed at Site O, where the recharge zone is much thicker (about 6

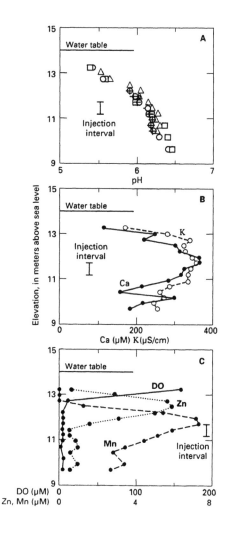

FIGURE 5. *Vertical Profile of pH, Specific Conductivity, and Dissolved Oxygen, Ca, and Mn at the Mixed Site.*

A: circles, pH at MLS 10-15 using flow-thru cells, April 1988; squares, pH at MLS 12-11 using flow-thru cells, April 1988; triangles, pH at MLS 10-16 using bottles, July 1988; + signs, pH at MLS 10-15 using bottles, July 1988.
B: Data from MLS 10-15 in April 1988.
C: Data from MLS 10-15 in July 1988.

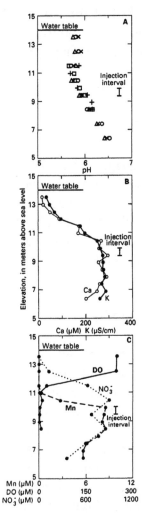

FIGURE 6. *Vertical Profile of pH, Specific Conductivity, and Dissolved Oxygen, Ca, Mn, and NO_3^- at the Suboxic Site.*

A: Triangles, X's, pH at MLS 38-12 using cells, April 1988; circles, pH at MLS 37-12 using cells, April 1988, + signs, pH at MLS 37-12 using bottles, July 1988.

B: Data from MLS 37-12, April 1988.

C: All data from MLS 37-12: Mn, April 1988; dissolved oxygen identical in April 1988 and September 1989; NO_3^-, September 1989.

m). Most of the tubes of the MLS at this site sample water from the recharge zone, although the bottom tubes showed increases in specific conductivity (Figure 7), perhaps indicating the top of the transition zone. The pH increases with depth from values of 4.6-4.7 at the top tube to 5.5 for the bottom tubes.

5.1.1   The Oxic Site

The MLS designated as F168-M1 (Site O) was used for the experiment conducted with recharge water, initiated on July 9, 1988 [88]. Three tubes (vertical separation of 0.38 meters) were used to obtain groundwater characterized by oxic conditions, low pH (5.1), and low concentrations of background salts (Table 2). The vertical interval used for sampling groundwater and injecting the tracers is indicated in Figure 7.

5.1.2   The Suboxic Site

MLS 37-12 (Site S) was used to conduct a tracer experiment within the sewage-contaminated zone, commencing on July 10, 1988 [88]. Two tubes (vertical separation of 0.51 m) were used to obtain groundwater characterized by very low concentrations of dissolved oxygen, higher pH (6.1), and greater concentrations of background salts than found in the recharge water (Table 2). The water contained 8 μM dissolved Mn but no detectable Fe(II), indicating mildly reducing conditions. The vertical interval for groundwater withdrawal and tracer injection is indicated in Figure 6. The groundwater and tracers were mixed in the polyethylene bag after it had been purged twice with nitrogen. The water contained about 25 μM DO when it was reinjected with the tracers.

5.1.3   The Mixed Site

MLS 10-15 (Site M) was used to conduct a tracer test on July 8, 1988, in which oxic recharge water with tracers was injected into the sewage-contaminated zone [88]. The same three tubes of MLS F168-M1

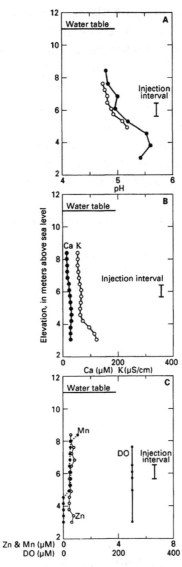

FIGURE 7. *Vertical Profile of pH, Specific Conductivity, and Dissolved Oxygen, Ca, Zn, and Mn at the Oxic Site (MLS F168-M1).*

A: Filled circles, pH using cells, April 1988; open circles, pH using bottles, July 1988.
B: Data from April, 1988.
C: Data from April 1988.

that were employed subsequently for the oxic injection were used to withdraw oxic water from the recharge zone (Table 2). The oxic water was mixed thoroughly with the tracers in the polyethylene bag and then was transported to MLS 10-15 (Site M), where it was pumped into three tubes (vertical separation of 0.25 m, interval shown in Figure 5) located within the suboxic sewage-contaminated zone. In this case, dissolved oxygen was also an injected tracer that was monitored downgradient. In addition to the difference in dissolved oxygen, the injected water contained lower concentrations of major salts, e.g. sulfate and calcium, than the ambient groundwater.

TABLE 2.   *Background Chemistry of Injected Groundwater.*

	CONCENTRATIONS IN RECHARGE WATER (μM)	CONCENTRATIONS IN SEWAGE ZONE (μM)
CATIONS	SITE O	SITE S
Na	406	1348
Ca	27	274
Mg	57	181
K	10	133
pH	5.1	6.1
ANIONS		
Cl	363	1380
SO$_4$	102	304
NO$_3$	trace	780
		highly variable
Alkalinity	--	467

5.2   Speciation of the Metal Tracers

There are several ways in which the speciation of the injected tracers could affect their transport (Figure 8). Changes in the speciation of these elements may cause significant differences in sorption behavior and mobility. Zn^{2+}, as a free cation, may adsorb extensively onto minerals present in the porous medium. The ZnEDTA^{2-} complex may also adsorb, but is expected to be considerably more mobile, based on a study of CdEDTA^{2-} mobility in a sandy aquifer [8,89]. Changes in the distribution of dissolved zinc between Zn^{2+} and the ZnEDTA^{2-} complex can result from dispersion-induced dilution of EDTA or competition for EDTA between Zn^{2+} and other cations (e.g. Fe, Mn, Ca) [31]. Such changes in aqueous speciation would be expected to influence the mobility of Zn. The reduction of Cr(VI), a weakly sorbing anion, to Cr^{3+}, a strongly sorbing cation, can cause immobilization of Cr. Similarly, reduction of Se(VI) to Se(IV), a strongly sorbing anion, or to elemental Se, can lead to immobilization of Se. Under certain limiting conditions, determined by aqueous speciation and kinetic considerations, the Cr(III)-EDTA complex may form (Figure 8), increasing the mobility of Cr(III).

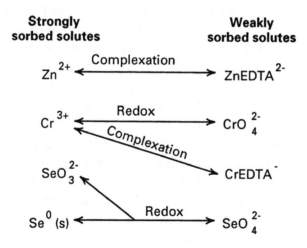

FIGURE 8. *Schematic Diagram Showing Important Reactions Affecting the Speciation of Zn, Cr, and Se in the Tracer Tests. For Each Element, the Reactivity with Mineral Surfaces May Vary from Strong Adsorption Tendencies to Weak, Given the Range of Possible Species.*

The redox status at Sites M and S within the sewage-contaminated zone are poised at mildly reducing, suboxic conditions. At the elevations sampled by the MLS, the conditions appear to fall between the oxic and iron-rich (post-oxic) classes of redox status given by Berner [34] in that the water is characterized by very low concentrations of dissolved oxygen yet contains very little Fe(II). Smith and Duff [74] have shown that denitrification activity in the sewage-contaminated zone is limited by the availability of organic carbon. Thus, the sequence of redox zonation processes appears to have been halted (or appreciably slowed) at a stage of partial denitrification and manganese reduction (Figures 5 and 6) due to depletion of readily-oxidized organic carbon. Almost all dissolved oxygen has been consumed in the sewage-contaminated plume before the groundwater reached sites M and S. Because the rate of vertical dispersion is very small and the groundwater flow rate is relatively high [90], dispersion (including molecular diffusion) of dissolved oxygen from the recharge water into the sewage-contaminated zone is negligible. This situation affects the reactants and products of all the redox processes, and produces the very large vertical gradients in solutes observed across the transition zone between the recharge water and the sewage plume boundary (Figures 5 and 6). The presence of high concentrations of Fe(II) observed by Gschwend and Reynolds [87] suggest that the redox sequence may have proceeded to more reducing conditions at lower elevations.

It is instructive to compare the equilibrium constants of half redox reactions to see what might be expected for Cr and Se in such a mildly reducing environment. As discussed by Morel [91], such a comparison can be made more easily by fixing the concentrations of certain reactants, e.g. H^+, at typical values, which allows the calculation of effective (or apparent) stability constants for specific half reactions. Table 3 shows the values of effective equilibrium constants for several important half redox reactions at pH 6.0, a typical value for the sewage-contaminated zone at Sites M and S. Each reaction is normalized for the exchange of one electron, and the equations are positioned in order of decreasing equilibrium pe_{eff} at pH 6.0. Thus, in theory, any reductant on the right-hand side should be capable of reducing any oxidant on the left-hand side that is higher up in the table [91]. Recall, however, that organic matter, shown as "CH_2O" in Table 3, was assumed to be the only reductant (or electron donor) in the model of Froelich *et al.* [32].

TABLE 3. *Effective Equilibrium Constants for Selected Redox Reactions at pH 6.0.*

EQUATION	$pe^{0\ c}$	pe_{eff} (pH=6)
$1/4\ O_2(g)+H^++e^-=1/2\ H_2O$	20.75	14.75
$1/5\ NO_3^-+6/5\ H^++e^-=1/10\ N_2(g)+3/5\ H_2O$	21.05	13.85
$1/2\ MnO_2(s)+2\ H^++e^-=1/2\ Mn^{2+}+H_2O$	20.8	8.8
$1/2\ NO_3^-+H^++e^-=1/2\ NO_2^-+1/2\ H_2O$	14.15	8.15
$1/6\ SeO_4^{2-}+4/3\ H^++e^-=1/6\ Se(s)+2/3\ H_2O$	14.8	6.8
$1/3\ HCrO_4^-+7/3\ H^++e^-=1/3\ Cr^{3+}+4/3\ H_2O$	20.2	6.2
$1/2\ O_2(g)+H^++e^-=1/2\ H_2O_2$	11.5	5.5
$Fe(OH)_3(s)+3\ H^++e^-=Fe^{2+}+3H_2O$	16.0	-2.0
$1/8\ SO_4^{2-}+5/4\ H^++e^-=1/8\ H_2S(g)+1/2\ H_2O$	5.25	-2.25
$1/4\ HCO_3^-+5/4\ H^++e^-=1/4"CH_2O"+1/2\ H_2O$	1.8[a]	-5.7[b]

[a] Based on "CH_2O" equal to one sixth of a glucose molecule
[b] Based on the activity of bicarbonate ion equal to 1 mM
[c] Constants from Morel [91].

Following the argument that organic matter is oxidized by the oxidant yielding the greatest energy, the order of the half redox reactions in Table 3 shows the order in which oxidants should be consumed for the conditions of the sewage plume. Dissolved oxygen should be utilized first, followed by denitrification, manganese oxide reduction, etc. Se(VI) reduction (to elemental Se) [52] and Cr(VI) reduction occur at nearly the same pe_{eff}, after the consumption of nitrate and manganese oxides, but before the reduction or iron hydroxide [88]. Note that this theoretical sequence of redox reactions is consistent with the results of Oremland *et al.* [52], who showed that dissolved oxygen, nitrate and manganese oxide inhibited the reduction of selenate to elemental selenium, but iron oxide and sulfate did not. White *et al.* [54] also reported that selenate was not reduced in the presence of nitrate in a field study of selenate mobility. Thus, it is not apparent that Cr and Se will be reduced under the mildly-reducing conditions within the sewage-contaminated zone. Moreover, the depletion of organic carbon within this zone could minimize the role of dissimilatory microbial processes in the reduction of these elements.

5.3   Zinc Adsorption Experiments with Otis Sand

Results of the Zn adsorption experiments were computed as the percent of Zn removed from solution versus pH. For experiments utilizing ^{65}Zn, this was determined by the equation: $100(A_o-A_s)/A_o$, where A_o and A_s are the activities of the original stock solution and the centrifugate solution, respectively. The experiments performed without ^{65}Zn showed that the total amount of Zn involved in adsorption-desorption reactions was greater than 5 μM, due to release of Zn from the sand. In order to compute the amount of Zn adsorbed, it was necessary to evaluate the total dissolved Zn concentration in the absence of adsorption. A total Zn concentration of approximately 8 μM in the experiments yielded optimal agreement between calculated adsorption as a function of pH in all experiments (with and without ^{65}Zn).

The sand used in Zn adsorption experiments was taken from an interval of the top of the core just below the water table, near the bottom of the recharge zone (cf. Figure 5); results are presented in Figure 9. In the absence of EDTA, Zn adsorption increased from about 15% adsorbed at pH 4.8 to 100% at pH 7.0 (Figure 9a). The observed strong dependence on pH is typical for cation adsorption onto oxide minerals and results from the release of protons that accompanies the chemical reaction between cations and surface sites [27,28,64,92].

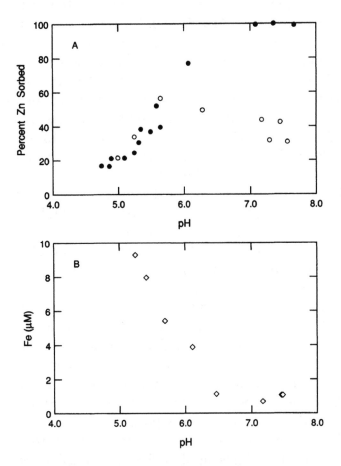

FIGURE 9. *A: Zinc Adsorption on Otis Sand as a Function of pH and EDTA Concentration. Filled Circles, no EDTA Added.* Open circles, 6.4 μM EDTA added. Zn added, 5 μM. Total surface-reactive Zn approximately equal to 8 μM. Sand:water ratio was 445 grams/dm³.

B: Dissolution of Fe from Otis Sand as a Function of pH after 18 Hours of Reaction in 0.1 M NaNO₃ Solution Containing 6.4 μM EDTA.

The effect of adding EDTA to the system results in complex adsorptive behavior for Zn. As can be seen in Figure 9a, Zn adsorption was not affected by 6.4 μM EDTA at pH <5.6. This result is surprising because the ZnEDTA complexes are stable at pH values less than 3 under these conditions. At pH >6.0, Zn adsorption was significantly decreased in the presence of the EDTA. The change in the response of Zn adsorption to EDTA addition with increasing pH is related to two factors: 1) variable dissolution of Fe from the sand by EDTA with increasing pH, and 2) competition between Fe and Zn for complexation with EDTA. Figure 9b shows the measured concentrations of dissolved Fe as a function of pH in a solution containing EDTA in separate experiments performed without added Zn. Dissolved Fe increased markedly as the pH decreased below 6.5.

These effects can be understood by examining the pH dependence of Zn speciation in this system. Equilibrium speciation calculations were performed with HYDRAQL [93]. The following assumptions were made for the calculations: 1) EDTA remained completely in the solution phase, i.e. no EDTA or metal-EDTA complexes were adsorbed, and 2) dissolved Fe was present completely as Fe(III), i.e. no Fe(II) was dissolved. The results are shown in Table 4. At pH values greater than 6.0, most of the dissolved Zn was calculated to be present as the ZnEDTA^{2-} complex. This species has little tendency to adsorb on the surfaces of the porous medium. Zn adsorption decreased in the pH range 6.0 to 8.0 (Figure 9a) as the ZnEDTA^{2-} complex became increasingly stronger. However, at pH < 5.5, most of the dissolved Zn was present as Zn^{2+}, because sufficient Fe(III) had dissolved from the sand to cause dissociation of the ZnEDTA^{2-} complex. Thus, at pH < 5.5, EDTA had a negligible effect on Zn adsorption, because the complexing ligand was bound to Fe(III), leaving Zn speciation unaffected.

While the results above are only preliminary, important points have been illustrated with these limited data. The experiments and theoretical calculations show the significance of speciation in determining the adsorption of Zn by the sub-surface material. Secondly, they show that it is important to consider the dissolution of trace minerals in the porous medium when interpreting the results of metal adsorption experiments in the presence of complexing ligands. Finally, the results illustrate the inadequacy of the distribution coefficient approach in modeling ligand-enhanced metal transport in groundwaters [68,94]. In this system, a large, multidimensional matrix of distribution coefficients would be needed as a function of pH, EDTA concentration, and Fe concentration. Conversely, the site-binding adsorption model coupled with an equilibrium speciation model [1,8,95] is better suited for this type of application to transport modeling.

TABLE 4.   *Calculated Major Species of Dissolved Zn(II) and Fe(III) in Adsorption Experiments with Otis Sand in 0.1M NaNO$_2$ with 6.4 µM EDTA.*

pH	Zn^{2+} µM	$ZnEDTA^{2-}$ µM	$FeEDTA^-$ $+FeOHEDTA^{2-}$ µM
4.0	8.0(100)[a]	<0.001 (0)[a]	6.4
5.0	6.5 (100)	0.002 (0)	6.4
5.5	4.7 (99)	0.06 (1)	6.3
6.0	1.4 (37)	2.4 (63)	4.0
6.5	<0.001 (0)	4.6 (100)	1.0
7.0	<0.001 (0)	4.8 (100)	0.7

[a] Percent of total dissolved Zn(II)

5.4   Results of Natural-Gradient Tracer Tests

The distribution of Br in time and space reveals the velocity, general shape, and mode of spreading the tracer cloud. Groundwater velocities determined from the peaks of Br breakthrough curves were in good agreement with the previous hydrologic studies [71]. Given the porosity (0.39) and injected volume of 0.38 m^3, the tracer clouds would occupy an aquifer volume of approximately 1 m^3 without mixing with ambient water. At each site, analysis of Br in samples taken within one hour of completing the injection showed that the tracer clouds spread vertically (both up and down) approximately 0.4 m. Br was undetectable at MLS adjacent to the injection sites, indicating that the clouds spread less than one meter laterally during injection; lateral spreading was minimal even after the cloud had travelled 22 meters downgradient. Most of the spreading occurred in the direction of groundwater flow, in agreement with the observations made during a large-scale tracer test at the site [90]. Br breakthrough curves at various depths at selected MLS illustrate vertical heterogeneity in hydraulic conductivity, as has been demonstrated more comprehensively at this site by Garabedian [77,90] and Hess *et al.* [73].

Cr transport showed a marked dependence on chemical characteristics of the transport zone. At the oxic site, Cr was retarded relative to Br at two meters downgradient from the injection MLS (Figure 10a). The transport of Cr was probably substantially influenced by adsorption of Cr(VI) onto the surfaces of minerals, especially hydrous oxides of iron and manganese. Adsorption of Cr(VI) is greatly enhanced by the low pH and low concentrations of other anions present in the recharge zone. Especially important are the low concentrations of sulfate and phosphate, anions that compete effectively for sorption sites on oxide mineral surfaces (Figure 1). At the mixed site, where groundwater with DO was injected into the mildly-reducing sewage-contaminated zone, Cr was attenuated but not retarded at four meters downgradient (Figure 10b). The attenuation was due to reduction of Cr(VI) to Cr(III), which was irreversibly sorbed by the surfaces of minerals in the porous medium [88,96]. Even though DO was transported conservatively in this zone, reduction of Cr(VI) to Cr(III) can occur if an appropriate reductant is present, such as detrital organic matter or minerals containing Fe(II). Essentially all Cr(VI) had been reduced after the tracer cloud had traveled 22 meters downgradient. Unlike the oxic site, Cr(VI) transport was only slightly retarded in the sewage-contaminated zone [88], and the peak in Cr concentration occurred at virtually the same time as the peak in Br concentration (Figure 10b). This occurred because adsorption of Cr(VI) was inhibited by the higher pH and greater concentrations of sulfate and phosphate ions within the sewage-contaminated zone (relative to the recharge zone). Thus, Cr(VI) as a species was transported conservatively within the sewage plume, but its reduction led to significant attenuation of

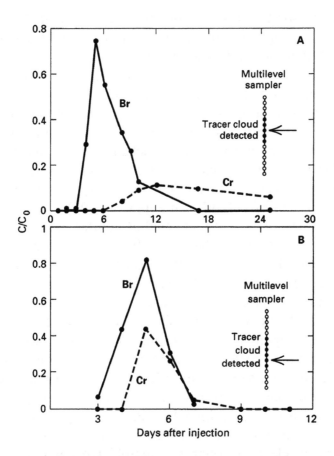

FIGURE 10. *A: Breakthrough Curves for Br and Cr in the Recharge Zone at MLS F168-M2 at an Elevation of 6.1 Meters, Near the Vertical Center of the Tracer Cloud.* The sampling point was approximately 2 meters downgradient from the injection well (the oxic site).

B: Breakthrough Curves for Br and Cr During the Tracer Test at the Mixed Site (Oxic Tracer Cloud Injected in the Suboxic Zone). Water samples taken from MLS 12-14 at an elevation of 11.0 meters, near the vertical center of the tracercloud, approximately 4 meters downgradient of the injection well (the mixed site).

total Cr transport. Results from the tracer test at the suboxic site exhibited Cr attenuation similar to the mixed site, with slight retardation of Cr(VI) transport (results not shown).

More detailed tracer tests conducted in 1989 have confirmed these hypotheses and extended the findings [88]. These experiments have shown that the reductant for Cr(VI) is not present as a dissolved species. Dissolved oxygen transport in this zone is essentially conservative, implying that the reductant does not react rapidly with DO. The mobile Cr species was only Cr(VI) that had not yet been reduced. The mechanism of Cr reduction appears to be by reaction with the surfaces of Fe(II)-bearing minerals [96].

Se transport was also affected by aquifer chemistry. At the oxic site, Se was retarded relative to Br at the sampler two meters downgradient (Figure 11a). We assume that all dissolved Se was present as Se(VI). As in the case of Cr(VI), the low pH and low concentrations of other sorbing anions favor the adsorption of Se(VI), cf. Figure 2. At the mixed and suboxic sites, however, Se transport was conservative (Figure 11b; data for mixed site not shown). Apparently, reduction did not occur under the mildly reducing conditions present in the sewage-contaminated zone. Abiotic reduction of Se is known to be slow, and microbial catalysis of the reaction apparently did not occur, perhaps because of inhibition by nitrate or manganese oxides [52]. However, another reason could have been the depletion of organic carbon as an electron donor for microbial respiration. The mobile species was probably Se(VI), which did not adsorb at the higher pH and greater concentrations of competing anions present in the sewage-contaminated zone, in comparison to conditions in the recharge zone. More discussion of these results is given in Kent *et al.* [88].

Transport of Zn was also affected by chemical characteristics of the injection sites [97]. At the oxic site, Zn was retarded relative to Br at MLS F168-M2, about two meters downgradient (Figure 12). The peak in Zn concentration occurred 3 to 4 days after the peak concentration of EDTA passed by this sampling point. Associated with the peak in EDTA were high concentrations of Fe and Al (the EDTA concentration was estimated from metal concentrations above

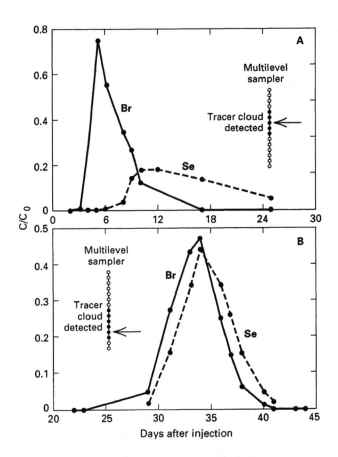

FIGURE 11. *A: Breakthrough Curves for Br and Se in the Recharge Zone at MLS F168-M2 at an Elevation of 6.1 Meters, Near the Vertical Center of the Tracer Cloud.* The sampling point was approximately 2 meters downgradient from the injection well (the oxic site).

B: Breakthrough Curves for Br and Se in the Suboxic Zone at MLS 39-11b at an Elevation of 9.2 Meters, near the Bottom Part of the Tracer Cloud. The sampling point was approximately 10 meters downgradient from the injection well (the suboxic site).

background levels, see Figure 12 caption). As was found in the laboratory study of Zn adsorption by Otis sand, dissolution of Fe(III) from the sand surface would be expected in this pH range (ambient pH 5.0) in the presence of EDTA [97]. This would cause dissociation of the ZnEDTA^{2-} complex, releasing Zn^{2+}, which is confirmed by the separation of the Zn and EDTA peaks (Figure 12). Zn^{2+} sorption on the sand at pH 5.0 was weak (Figure 9a) but sufficient to cause the retardation of Zn observed in the tracer test.

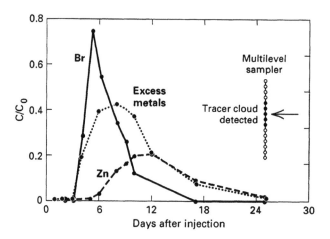

FIGURE 12. **Breakthrough Curves for Br, Zn, and "Excess Metals" in the Recharge Zone at MLS F168-M2 at an Elevation of 6.1 Meters, Near the Vertical Center of the Tracer Cloud.** The sampling point was approximately 2 meters downgradient from the injection well (the oxic site). "Excess metals" was defined as the sum of the concentrations of Zn, Fe, Al, Mn, and Ca above their background concentrations, as a semi-quantitative indicator of EDTA concentration.

The results also suggest that EDTA or its metal complexes may have been weakly adsorbed, since the estimated peak in EDTA concentration was retarded relative to Br. EDTA adsorption on iron and aluminum oxides has been observed in laboratory studies [4,98,99]. However, dissolution of Fe(III) can decrease the adsorption of EDTA [31].

Analysis of Zn transport at the suboxic and mixed sites is complex, due to significant contamination of the sand and gravel within the sewage-contaminated zone [97,100]. Significant concentrations of Zn were desorbed from the sand surfaces as the tracer clouds passed through the sewage-contaminated zones at both Sites M and S. Elevated concentrations of dissolved zinc (6 µM) were observed within this zone before the tracer tests, indicating that significant quantities of Zn may be adsorbed. Transport of Zn was slightly retarded relative to Br at these sites, illustrating the importance of complexation in metal transport.

6.   CONCLUDING REMARKS

The redox environment, adsorption-desorption processes, and aqueous speciation significantly influenced the transport of Zn, Cr, and Se in natural-gradient tracer tests conducted in a shallow sand and gravel aquifer contaminated by sewage effluent. In the uncontaminated recharge zone, retardation of Zn (injected as a metal complex) occurred when it was displaced from EDTA by Fe(III) dissolved from minerals in the porous medium. Similar processes could occur in pristine or contaminated systems containing fulvic acids or synthetic compounds that are known to complex metals. Cr reduction occurred in the mildly reducing, suboxic zone, even though injected dissolved oxygen appeared to be transported conservatively. This resulted in attenuation of the Cr breakthrough curve downgradient, but the mobile species, Cr(VI), was only slightly retarded. Significant retardation of Cr(VI) and Se(VI) was observed only within the recharge zone, where the aqueous composition (low pH and low concentrations of competing sorptive anions) provided the appropriate conditions that favor sorption on oxides.

Modeling the transport of many metals will require that aqueous speciation be considered, because speciation changes caused by complexation and oxidation-reduction reactions are of primary importance in determining the reactivity of these elements with the porous medium via sorption or precipitation processes. Important speciation changes may occur along a flow path, as solutes from a waste discharge mix with ambient water or encounter different geochemical regimes downgradient. The physical processes that lead to macrodispersion and fluid mixing may affect aqueous speciation of metals via dilution of complexing ligands or the addition of new solutes from dissolving mineral phases, each of which may have direct or indirect effects on the sorption of metal ions. Significant advances are being made in the development of theoretical models for coupling flow with chemical and microbial processes occurring in the groundwater environment. However, the application of these models to metal transport in groundwaters is still limited by inadequate consideration of adsorption-desorption reactions within the models and the availability of relevant kinetic and thermodynamic data to describe metal sorption processes in the groundwater environment. Further advances in this area will require detailed laboratory and field investigations combined with additional refinement of modeling approaches.

REFERENCES

1. Davis, J. A., and D. B. Kent "Surface Complexation Modeling in Aqueous Geochemistry," in *Mineral-Water Interface Geochemistry*, Reviews in Mineralogy, vol. 23, M. F. Hochella and A. F. White, Eds, (Washington, D.C.: American Mineralogical Society, 1990). p. 177-260.

2. Hsi, C-K. D., and D. Langmuir, "Adsorption of Uranyl onto Ferric Oxyhydroxides: Application of the Surface Complexation Site-binding Model," *Geochim. Cosmochim. Acta* 49:1931-1942 (1985).

3.   Davis, J. A. "Complexation of Trace Metals by Adsorbed Natural Organic Matter," *Geochim. Cosmochim. Acta* 48:679-691 (1984).

4.   Bowers, A. R., and C. P. Huang "Adsorption Characteristics of Metal-EDTA Complexes onto Hydrous Oxides," *J. Colloid Interface Sci.* 110:575-590 (1986).

5.   Killey, R. W. D., J. O. McHugh, D. R. Champ, E. L. Cooper, and J. L. Young "Subsurface Cobalt-60 Migration from a Low-level Waste Disposal Site," *Environ. Sci. Tech.* 18:148-157 (1984).

6.   Cleveland, J. M., and T. G. Rees "Characterization of Plutonium in Maxey Flats Radioactive Trench Leachates," *Science* 212:1506-1509 (1981).

7.   Fuller, C. C., and J. A. Davis "Influence of Coupling of Sorption and Photosynthetic Processes on Trace Element Cycles in Natural Waters," *Nature* 340:52-54 (1989).

8.   Fuller, C. C., and J. A. Davis "Processes and Kinetics of Cd^{2+} Sorption by a Calcareous Aquifer Sand," *Geochim. Cosmochim. Acta* 51:1491-1502 (1987).

9.   Tessier, A., F. Rapin, and R. Carignan "Trace Metals in Oxic Lake Sediments: Possible Adsorption onto Iron Oxyhydroxides," *Geochim. Cosmochim. Acta* 49:183-194 (1985).

10.   Lion, L. W., R. S. Altmann, and J. O. Leckie "Trace Metal Adsorption Characteristics of Estuarine Particulate Matter: Evaluation and Contribution of Fe/Mn Oxide and Organic Surface Coatings," *Environ. Sci. Tech.* 15:660-666 (1982).

11.   Zachara, J. M., C. C. Ainsworth, C. E. Cowan, and C. T. Resch
       " Adsorption of Chromate by Subsurface Soil Horizons," *Soil Sci. Soc. Am. J.* 53:418-428 (1989).

12.   Zachara, J. M., D. C. Girvin, R. L. Schmidt, and C. T. Resch. "Chromate Adsorption on Amorphous Iron Oxyhydroxide in the Presence of Major Groundwater Ions," *Environ. Sci. Tech.* 21:589-594 (1987).

13.   Honeyman, B. D. "Cation and Anion Adsorption at the Oxide/Solution Interface in Systems Containing Binary Mixtures of Adsorbents: An Investigation of the Concept of Adsorptive Additivity," PhD Thesis, Stanford University, Stanford, CA (1984).

14.   Davis, J. A., and J. O. Leckie "Surface Ionization and Complexation at the Oxide/Water Interface. 3. Adsorption of Anions," *J. Colloid Interface Sci.* 74:32-43 (1980).

15.   Rai, D., J. M. Zachara, L. E. Eary, C. C. Ainsworth, J. E. Amonette, C. E. Cowan, R. W. Szelmeczka, C. T. Resch, R. L. Schmidt, D. C. Girvin, and S. C. Smith "Chromium Reactions in Geologic Materials," Electric Power Research Institute Report EA-5741 (1988).

16.   Zachara, J. M., C. E. Cowan, R. L. Schmidt, and C. C. Ainsworth "Chromate Adsorption on Kaolinite," *Clays Clay Min.* 36:317-326 (1988).

17.   Stollenwerk, K. G., and D. B. Grove "Adsorption and Desorption of Hexavalent Chromium in an Alluvial Aquifer near Telluride, Colorado," *J. Environ. Qual.* 14:150-155 (1984).

18.   Rai, D., B. M. Sass, and D. A. Moore "Chromium(III) Hydrolysis Constants and Solubility of Chromium(III) Hydroxide," *Inorg. Chem.* 26:345-349 (1987).

19. Schultz, M. F., M. M. Benjamin, and J. F. Ferguson "Adsorption and Desorption of Metals on Ferrihydrite: Reversibility of the Reaction and Sorption Properties of the Regenerated Solid," *Environ. Sci. Tech.* 21:863-869(1987).

20. Sass, B. M., and D. Rai "Solubility of Amorphous Chromium(III)-Iron(III) Hydroxide Solid Solutions," *Inorg. Chem.* 26:2228-2232 (1987).

21. Chao, T. T., and R. F. Sanzolone "Fractionation of Soil Selenium by Sequential Partial Dissolution," *Soil Sci. Soc. Am. J.* 53:385-392 (1989).

22. Gruebel, K. A., J. A. Davis, and J. O. Leckie "The Feasibility of Using Sequential Extraction Techniques for Arsenic and Selenium in Soils and Sediments," *Soil Sci. Soc. Am. J.* 52:390-397 (1988).

23. Hayes, K. F., C. Papelis, and J. O. Leckie "Modelling Ionic Strength Effects on Anion Adsorption at Hydrous Oxide/Solution Interfaces," *J. Colloid Interface Sci.* 125:717-726 (1988).

24. Balistrieri, L. S., and T. T. Chao "Selenium Adsorption by Goethite," *Soil Sci. Soc. Am. J.* 51:1145-1151 (1987).

25. Ryden, J. C., J. K. Syers, and R. W. Tillman "Inorganic Anion Sorption and Interactions with Phosphate Sorption by Hydrous Ferric Oxide Gel," *J. Soil Sci.* 38:211-217 (1987).

26. Zasoski, R. J., and R. G. Burau "Sorption and Sorptive Interaction of Cadmium and Zinc on Hydrous Manganese Oxides," *Soil Sci. Soc. Am. J.* 52:81-87 (1988).

27. Benjamin, M. M., and J. O. Leckie "Multiple-Site Adsorption of Cd, Cu, Zn, and Pb on Amorphous Iron Oxide," *J. Colloid Interface Sci.* 79:209-221 (1981).

28. Kinniburgh, D. G. "The H^+/M^{2+} Exchange Stoichiometry of Calcium and Zinc Adsorption by Ferrihydrite," *J. Soil Sci.* 34:759-768 (1983).

29. Balistrieri, L. S., and J. W. Murray "The Adsorption of Cu, Pb, Zn, and Cd on Goethite from Major Ion Seawater," *Geochim. Cosmochim. Acta* 46:1253-1267 (1982).

30. Loganathan, P., R. G. Burau, and D. W. Fuerstenau "Influence of pH on the Sorption of Co^{2+}, Zn^{2+}, and Ca^{2+} by a Hydrous Manganese Oxide," *Soil Sci. Soc. Am. J.* 41:57-62 (1977).

31. Bowers, A. R., and C. P. Huang "Role of Fe(III) in Metal Complex Adsorption by Hydrous Solids," *Water Res.* 21:757-764 (1987).

32. Froelich, P. N., G. P. Klinkhammer, M. L. Bender, N. A. Luedtke, G. R. Heath, D. Cullen, P. Dauphin, D. Hammond, B. Hartman, and V. Maynard. "Early Oxidation of Organic Matter in Pelagic Sediments of the Eastern Equatorial Atlantic: Suboxic Diagenesis," *Geochim. Cosmochim. Acta* 43:1075-1090 (1979).

33. Wilson, T. R. S., J. Thomson, S. Colley, D. J. Hydes, N. C. Higgs, and J. Sorenson. "Early Organic Diagenesis: The Significance of Progressive Subsurface Oxidation Fronts in Pelagic Sediments," *Geochim. Cosmochim. Acta* 49:811-822 (1985).

34. Berner, R. A. "A New Geochemical Classification of Sedimentary Environments," *J. Sedimentary Petrology* 51:359-365 (1981).

35. Champ, D. R., J. Gulens, and R. E. Jackson "Oxidation-Reduction Sequences in Ground Water Flow Systems," *Canad. J. Earth Sci.* 16:12-23 (1979).

36. Baedecker, M. J. and Back, W. "Modern Marine Sediments as a Natural Analog to the Chemically Stressed Environment of a Landfill," *J. Hydrol.* 43:393-414 (1979).

37. Edmunds, W. M., A. H. Bath, and D. L. Miles. "Hydrogeochemical Evolution of the East Midlands Triassic Sandstone Aquifer," *Geochim. Cosmochim. Acta* 46:2069-2081 (1982).

38. Liu, C. W., and T. N. Narasimhan "Redox-Controlled Multiple-Species Reactive Chemical Transport 2. Verification and Application," *Water Resour. Res.* 25:883-910 (1989).

39. Griffin, R. A., A. K. Au, and R. R. Frost "Effect of pH on Adsorption of Chromium from Landfill Leachate by Clay Minerals," *J. Environ. Sci. Health* Part A, A12:431-449 (1977).

40. Eary, L. E., and D. Rai "Kinetics of Chromate Removal from Aqueous Wastes by Reduction with Ferrous Iron," *Environ. Sci. Tech.* 22:972-977 (1988).

41. Eary, L. E., and D. Rai, "Kinetics of Chromate Reduction by Ferrous Ions derived from Hematite and Biotite at 25°C," *Amer. J. Sci.* 289:180-213 (1989).

42. White, A. F., "Heterogeneous Electrochemical Reactions Associated with Oxidation of Ferrous Oxide and Silicate Surfaces," in *Mineral-Water Interface Geochemistry*, Reviews in Mineralogy, vol. 23, M. F. Hochella and A. F. White, Eds. (Washington, DC: American Mineralogical Society, 1990), p. 467-509.

43. Haight, G. P., G. M. Jursich, M. T. Kelso, and P. J. Merrill "Kinetics and Mechanisms of Oxidation of Lactic Acid by Chromium(VI) and Chromium(V)," *Inorg. Chem.* 24:2740-2746 (1985).

44. Benson, D. *Mechanisms of Oxidation by Metal Ions* (New York: Elsevier Scientific Publishing Co., 1976), p. 149-214.

45. Stollenwerk, K. G., and D. B. Grove "Reduction of Hexavalent Chromium in Water Samples Acidified for Preservation," *J. Environ. Qual.* 14:396-399 (1985).

46. Douglas, G. S., and J. G. Quinn "Geochemistry of Dissolved Chromium-Organic Matter Complexes in Narragansett Bay Interstitial Waters," in *Aquatic Humic Substances*, I. H. Suffet and P. MacCarthy, Eds., Adv. Chem. Series 219, (Washington, D.C.: American Chemical Society, 1989), p. 297-322.

47. Bopp, L. H., and H. L. Ehrlich "Chromate Resistance and Reduction in Pseudomonas fluorescens strain LB300," *Arch. Microbiol.* 150:426-431 (1988).

48. Bopp, L. H., A. M. Chakrabarty, and H. L. Ehrlich "Chromate Resistance Plasmid in Pseudomonas fluorescens," *J. Bacteriol.* 155:1105-1109 (1983).

49. Jacobs, L. W. "Selenium in the Agricultural Environment," *Soil Sci. Soc. Am. Spec. Pub.* #23, 233p. (1989).

50. Doran, J. W. "Microorganisms and the Biological Cycling of Selenium," *Adv. Microbial Ecol.* 6:17-32 (1982).

51. Zehr, J., and R. S. Oremland "Reduction of Selenate to Selenide by Sulfate-Respiring Bacteria: Experiments with Cell Suspensions and Estuarine Sediments," *Appl. Environ. Microbiol.* 53:1365-1369 (1987).

52. Oremland, R. S., J. T. Hollibaugh, A. S. Maest, T. S. Presser, L. G. Miller, and Charles W. Culbertson, "Selenate Reduction to Elemental Selenium by Anaerobic Bacteria in Sediments and Culture: Biogeochemical Significance of a Novel, Sulfate-Independent Respiration," *Appl. Environ. Microbiol.* 55:2333-2343 (1989).

53. Benson, S. M. "Characterization of the Hydrologic and Transport Properties of the Shallow Aquifer under Kesterson Reservoir, Merced County, California," PhD Thesis, University of California, Berkeley, CA, 1988.

54. White, A. F., S. M. Benson, A. W. Yee, H. A. Wollenberg, and S. Flexer "Groundwater Contamination at Kesterson National Wildlife Refuge, CA. Part II. Geochemical Parameters Influencing Selenium Mobility," *Water Resour. Res.* 27:1085-1098 (1991).

55. Sudicky, E. A. "A Natural Gradient Experiment on Solute Transport in a Sand Aquifer: Spatial Variability of Hydraulic Conductivity and its Role in the Dispersion Process," *Water Resour. Res.* 22:2069-2082 (1986).

56. Back, W. and M. J. Baedecker. "Chemical Hydrogeology in Natural and Contaminated Environments," *J. Hydrol.* 106:1-28 (1989).

57. Barcelona, M. J., T. R. Holm, M. R. Schock, and G. K. George. "Spatial and Temporal Gradients in Aquifer Oxidation-Reduction Conditions," *Water Resour. Res.* 25:991-1003 (1989).

58.   Honeyman, B. D. and J. O. Leckie, J. O. "Macroscopic
       Partitioning Coefficients for Metal Ion Adsorption: Proton
       Stoichiometry at Variable pH and Adsorption Density," in
       Geochemical Processes at Mineral Surfaces, J. A. Davis and K.
       F. Hayes, Eds. (Washington, D.C.: American Chemical Society,
       ACS Symposium Series 323, 1986), p. 163-190.

59.   Grove, D. B. and K. G. Stollenwerk. "Modeling the
       Rate-Controlled Sorption of Hexavalent Chromium," *Water
       Resour. Res.* 21:1703-1709 (1985).

60.   Valocchi, A. J. "Spatial Moment Analysis of the Transport of
       Kinetically Adsorbing Solutes Through Stratified Aquifers,"
       Water Resour. Res. 25:273-279 (1989).

61.   Brusseau, M. L., R. E. Jessup, and P. S. C. Rao. "Modeling the
       Transport of Solutes Influenced by Multiprocess
       Nonequilibrium," *Water Resour. Res.* 25:1971-1988 (1989).

62.   Lewis, F. M., C. I. Voss, and J. Rubin. "Solute Transport with
       Equilibrium Aqueous Complexation and Either Sorption or Ion
       Exchange: Simulation Methodology and Applications," *J.
       Hydrol.* 90:81-115 (1987).

63.   Anderson, P. R., and M. M. Benjamin "Effects of Silicon on the
       Crystallization and Adsorption Properties of Ferric Oxides,"
       Environ. Sci. Tech. 19:1048-1053 (1985).

64.   Davis, J. A., and J. O. Leckie "Surface Ionization and
       Complexation at the Oxide/Water Interface. 2. Surface
       Properties of Amorphous Iron Oxyhydroxide and Adsorption of
       Metal Ions," *J. Colloid Interface Sci.* 67:90-107 (1978).

65.   Hayes, K. F., and J. O. Leckie "Modelling Ionic Strength Effects
       on Cation Adsorption at Hydrous Oxide/Solution Interfaces," *J.
       Colloid Interface Sci.* 115:564-572 (1987).

66. Catts, J. G., and D. Langmuir "Adsorption of Cu, Pb, and Zn onto Birnessite (δ-MnO_2)," *J. Applied Geochem.* 1:255-264 (1986).

67. LaFlamme, B. D., and J. W. Murray "Solid/Solution Interaction: The Effect of Carbonate Alkalinity on Adsorbed Thorium," *Geochim. Cosmochim. Acta* 51:243-250 (1987).

68. Cederberg, G. A., R. L. Street, and J. O. Leckie. "A Groundwater Mass Transport and Equilibrium Chemistry Model for Multicomponent Systems," *Water Resour. Res.* 21:1096-1104 (1985).

69. Oldale, R. N. "Seismic Investigations on Cape Cod, Martha's Vineyard, and Nantucket, Massachusetts, and a Topographic Map of the Basement Surface from Cape Cod Bay to the Islands," *U. S. Geological Survey Prof. Paper 650-B* p. B122-B127 (1969).

70. LeBlanc, D. R. "Sewage Plume in a Sand and Gravel Aquifer, Cape Cod, Massachusetts," *U. S. Geological Survey Water-Supply Paper 2218* (1984).

71. LeBlanc, D. R., S. P. Garabedian, K. M. Hess, L. W. Gelhar, R. D. Quadri, K. G. Stollenwerk, and W. W. Wood "Large-Scale Natural-Gradient Tracer Test in Sand and Gravel, Cape Cod, Massachusetts: 1. Experimental Design and Observed Tracer Movement," *Water Resour. Res.* 27:895-910 (1991).

72. Barber, L. B.II, "Geochemical Heterogeneity in a Glacial Outwash Aquifer: Effect of Particle Size and Mineralogy on Sorption of Nonionic Organic Solutes," PhD dissertation, Univ. of Colorado, Boulder, CO, 1990.

73.   Hess, K. M., S. H. Wolf, M. A. Celia, "Large-scale Natural Gradient Tracer Test in Sand and Gravel, Cape Cod, Massachusetts. 3. Hydraulic Conductivity Variability and Calculated Macrodispersivities," *Water Resour. Res.* 28:2011-2027 (1992).

74.   Smith, R. L., and J. H. Duff "Denitrification in a Sand and Gravel Aquifer," *Appl. Environ. Microbiol.* 54:1071-1078 (1988).

75.   Smith, R. L., R. W. Harvey, J. H. Duff, and D. R. LeBlanc "Importance of Closely Spaced Vertical Sampling in Delineating Chemical and Microbiological Gradients in Groundwater Studies," *J. Contamin. Hydrology* 7:285-300 (1991).

76.   Barber, L. B., E. M. Thurman, M. P. Schroeder, and D. R. LeBlanc "Long-Term Fate of Organic Micropollutants in Sewage-Contaminated Groundwater," *Environ. Sci. Tech.* 22:205-211 (1988).

77.   Garabedian, S. P. "Large-Scale Dispersive Transport in Aquifers: Field Experiments and Reactive Transport Theory," PhD Thesis, Massachusetts Institute of Technology, Cambridge, MA (1987).

78.   Jacobs, L. A., H. R. von Gunten, R. Keil, and M. Kuslys "Geochemical Changes Along a River-Groundwater Infiltration Flow Path: Glattfelden, Switzerland," *Geochim. Cosmochim. Acta* 52:2693-2706 (1988).

79.   Busenberg, E., and L. N. Plummer "pH Measurement of Low-Conductivity Waters," *U. S. Geological Survey Water-Resources Investigations Report 87-4060* (Reston, VA: U.S. Geological Survey (1987).

80. White, A. F., M. L. Peterson, and R. D. Solbau "Measurement and Interpretation of Low Levels of Dissolved Oxygen in Ground Water," *Groundwater* 28:584-590 (1990).

81. Smith, R. L., U. S. Geological Survey, Denver, CO, personal communication, 1989.

82. Gibbs, M. M. "A Simple Method for the Rapid Determination of Iron in Natural Waters," *Water Res.* 13:295-297 (1978).

83. Schlemmer, G., and B. Welz, "Palladium and magnesium nitrates, a more universal modifier for graphite furnace atomic absorption spectrometry" *Spectrochim. Acta* 41B:1157-1165 (1986).

84. Zapico, M. M., S. Vales, and J. A. Cherry "A Wireline Piston Core Barrel for Sampling Cohesionless Sand and Gravel Below the Water Table," *Ground Water Monitoring Review* 7:74-82 (1987).

85. Coston, J. A., C. C. Fuller, and J. A. Davis, "The Search for a Geochemical Indicator of Lead and Zinc Sorption in a Sand and Gravel Aquifer, Falmouth, Massachusetts, USA, *Water/rock Interaction* WRI-7 1, p. 41-44. Edited by Y. Kharaka and A. Maest, Balkema, Rotterdam, The Netherlands (1992).

86. Chao, T. T. and L. Zhou "Extraction Techniques for Selective Dissolution of Amorphous Iron Oxides from Soils and Sediments," *Soil Sci. Soc. Am. J.* 47:225-232 (1983).

87. Gschwend, P. M., and M. D. Reynolds "Monodisperse Ferrous Phosphate Colloids in an Anoxic Groundwater Plume," *J. Contam. Hydrol.* 1:309-327 (1987).

88.   Kent, D. B., J. A. Davis, J. A., L. D. Anderson, B. A. Rea, and T. D. Waite,"Transport of Chromium and Selenium in the Suboxic Zone of a Shallow Aquifer: Influence of redox and Adsorption Reactions," *Water Resour. Res.*, submitted.

89.   Davis, J. A., C. C. Fuller, J. O. Leckie, and D. Blowes "Surface Reactions and Mobility of Cd(II) After Injection in a Shallow Aquifer," *EOS* 66:273 (1985).

90.   Garabedian, S. P., D. R. LeBlanc, L. W. Gelhar, and M. A. Celia "Large-Scale Natural-Gradient Tracer Test in Sand and Gravel, Cape Cod, Massachusetts: 2. Analysis of Spatial Moments for a Nonreactive Tracer," *Water Resour. Res.*, 27:911-924 (1991).

91.   Morel, F. M. M. *Principles of Aquatic Chemistry* (New York: John Wiley & Sons, 1983), p. 311-376.

92.   Davis, J. A., and K. F. Hayes, "Geochemical Processes at Mineral Surfaces: An Overview," in *Geochemical Processes at Mineral Surfaces*, J. A. Davis and K. F. Hayes, Eds., ACS Symposium Series 323, (Washington, D.C.: American Chemical Society, 1986), p. 2-18.

93.   Papelis, C., K. F. Hayes, and J. O. Leckie "HYDRAQL: A Program for the Computation of Chemical Equilibrium Composition of Aqueous Batch Systems including Surface-Complexation Modeling of Ion Adsorption at the Oxide/Solution Interface," Technical Report No. 306, Dept. of Civil Engineering, Stanford University, Stanford, CA (1988).

94.   Reardon, E. J. "K_d's - Can They Be Used to Describe Reversible Ion Sorption Reactions in Contaminant Migration?" *Ground Water* 19:279-286 (1981).

95. Waite, T. D. "Mathematical Modeling of Trace Element Speciation," in *Trace Element Speciation: Analytical Methods and Problems*, G. Batley, Ed. (Boca Raton, FL: CRC Press, 1989), p. 117-184.

96. Anderson, L. C. D., J. A. Davis, and D. B. Kent, "Batch Experiments Characterizing the Reduction of Cr(VI) using Suboxic Material from a Mildly Reducing Sand and Gravel Aquifer," *Environ. Sci. Technol.*, submitted.

97. Kent, D. B., J. A. Davis, B. A. Rea, and L. D. Anderson, "Ligand-enhanced Transport of Strongly Adsorbing Metal Ions in the Ground-water Environment," *Water-rock Interaction WRI-7 1*, pp. 805-808, Edited by Y. Kharaka and A. Maest, Balkema, Rotterdam, The Netherlands (1992).

98. Bowers, A. R. and C. P. Huang, "Adsorption Characteristics of Polyacetic Amino Acids onto Hydrous γ-Al$_2$O$_3$," *J. Colloid Interface Sci.* 105:197-215 (1985).

99. Rubio, J. and E. Matijevic, "Interactions of Metal Oxides with Chelating Agents, 1. β-FeOOH-EDTA," *J. Colloid Interface Sci.* 68: 408-421 (1979).

100. Rea, B. A., D. B. Kent, D. R. LeBlanc, and J. A. Davis, "Mobility of zinc in a sewage-contaminated aquifer, Cape Cod, Massachusetts" in US Geological Survey Toxic Substances Hydrology Program-Proceedings of the technical meeting, edited by G. E. Mallard, and D. A. Aronson, US Geological Survey Water Resourc. Invest. Rept. 91-4034, 88-95 (1991).

101. Leckie, J. O., M. M. Benjamin, K. Hayes, G. Kaufmann, and S. Altmann "Adsorption/Coprecipitation of Trace Elements from Water with Iron Oxyhydroxide," Electric Power Research Institute Report EPRI-RPP-910 (1980).

9

METAL-ORGANIC INTERACTIONS IN SUBTITLE D LANDFILL LEACHATES AND ASSOCIATED GROUNDWATERS

Peter A. Gintautas, Kristina A. Huyck,
Stephen R. Daniel, and Donald L. Macalady
Department of Chemistry and Geochemistry
Colorado School of Mines
Golden, CO 80401

1.   INTRODUCTION

Sanitary landfills provide an amazing variety of feedstocks and processes for the production, microbial and abiotic, of an extensive suite of chemical compounds. Those of appropriate solubility characteristics will comprise the solute of leachates produced as water percolates through the landfills. Among expected constituents of leachates are organic compounds of environmental concern (e.g., benzene, tetrachloroethylene), as well as toxic metal ions (e.g., Pb^{2+}, Hg^{2+}). In addition, however, and the particular focus of this study, a variety of organic compounds can be expected in the leachates which afford the potential for interaction with toxic metal ions and, hence, influence on their mobilities.

The objective of this research is a test of the hypothesis that metal-organic interactions are important in the mobilization of toxic metals from subtitle D landfills. The tool chosen for evaluation of this

0-87371-277-3/93/$0.00 + $.50

hypothesis was the MINTEQ computer program, which is available from the U.S. Environmental Protection Agency, Environmental Research Laboratory, Athens, Georgia 30613.

Use of MINTEQ to test the hypothesis required availability of two bodies of information: 1) identity and concentrations of potential ligands in landfill leachates, and 2) thermodynamic (both formation constants and redox potentials) data for calculation of speciation of the metals of interest (Sb, As, Ba, Cd, Cr, Cu, Hg, Ni, Pb, Se, Ag, Tl, and Zn).

The approach taken consisted of analyses of landfill leachates for organic constituents with potential for interaction with metal ions, and inclusion in the MINTEQ database of all pertinent data available in the literature, as well as additional required thermodynamic data determined, or estimated, in this study.

The oxidation-reduction (redox) conditions in landfill leachates and receiving groundwaters affect metal-organic interactions by at least two mechanisms. First, the oxidation state of several metals on the OSW list, for example, arsenic, selenium, antimony, chromium and several other important competing metals (metals not on OSW list, but which compete with the OSW metals for ligands), e.g., iron and manganese, are dependent upon the redox conditions of the aqueous matrix. Each of the oxidation states has, of course, different metal-complexation constants, and, therefore, can be expected to be mobilized by anthropogenic organic compounds to an extent critically dependent upon the redox conditions. Perhaps the metal most dramatically affected by redox conditions is iron. Ferrous iron is quite soluble at most groundwater pH's so the complexation equilibria for iron(II) are quite important under reducing conditions. The general insolubility of iron(III) makes iron a much less important factor in oxidized waters.

Redox conditions also influence metal-organic interactions through the organic ligands, although such effects are generally less dramatic than effects on the metal. Organic groups sensitive to redox conditions in natural waters include the important quinone/hydroquinone functionality, which is present both in natural organic matter and in anthropogenic compounds [1]. Another functional group of interest is nitro which, when reduced to the amine form, can produce dramatically altered complexation equilibria for some metals, as illustrated by the following formation constants, measured under the same conditions.

(4-aminophenylthio)acetate - Ag^+ $\log K_1 = 3.22$ [2]

(4-nitrophenylthio)acetate - Ag^+ $\log K_1 = 1.98$ [2]

Thus, specification of redox conditions in leachates, and, perhaps, more importantly, in receiving groundwaters, is critical for estimates of metal transport from landfills into aquifers. Such specification, if it is to be useful, must clearly go beyond provision of the traditional platinum redox electrode potential. The inadequacy of this parameter is well-documented [3], but the alternative steps necessary for useful redox information are less well-established.

Two conceptually different considerations apply. First, the redox state of the system, as defined by the (equilibrium) distribution of oxidized and reduced forms of each of the redox-active species present, determines the redox potential, Eh, of the solution [4]. Eh thus represents the redox parameter (for electrons) analogous to the acid-base parameter (for protons), pH. Eh, in principle, defines the oxidation state distribution for each metal and organic compound in the leachate/groundwater systems. Methods for determination of accurate and/or kinetically relevant distributions of redox-active species in aqueous systems require detailed analytical information and have only recently been described [5-7].

The resistance of a solution to changes in redox potential, called the poising capacity, is another important consideration. It provides a conceptual parameter analogous to alkalinity (buffer capacity) in acid-base chemistry and, in principle, provides the means by which one can determine whether the receiving groundwater has sufficient poising capacity (or, alternatively, concentration of redox-active species) to substantially alter the redox conditions of the influent leachates. Very dilute near-surface groundwater systems, for example, are typically oxidized [6] but often have very low concentrations of oxidized species. Methods for specification of poising capacities are non-trivial, and detailed proposals for quantification of this parameter have only recently been outlined [8]. Thus the Eh of the leachate/groundwater mixture is assumed to be dominated by the Eh of the receiving groundwater, at least at some large distance downgradient from the infiltration zone. In addition, redox potentials for organic compounds are either unknown or not present in the database, so redox transformations of natural and anthropogenic organic components of the leachates (or groundwaters) are also ignored.

The above considerations have not been previously applied specifically to considerations of metal-organic interactions. For the work described in this paper, knowledge of the concentrations and redox state of oxygen, iron, sulfur and organic carbon (TOC) probably represents necessary and sufficient information for the estimates of metal-organic complexation which are the basis for the predictive models generated. Detailed discussion of the rationale for selection of these parameters is not critical to the considerations of this paper and will be presented elsewhere. Because the models are inherently based on the assumption of equilibrium, kinetic considerations are also beyond the scope of these investigations.

Compositional data for landfill leachates was obtained via analysis in our laboratory of leachate samples collected for us by the Athens Environmental Research Laboratory (USEPA) at six subtitle D landfills across the United States. The sites varied widely in climate, geology, age, and perhaps in the materials which were placed in the landfills. The sites are identified here only in terms of the abbreviations for the states in which they are located.

Characterization of the samples in terms of system parameters such as pH, Eh, conductivity, alkalinity, and TOC, provided useful definition of conditions in the various leachates and convenient indicators of sample integrity during shipment and storage. Samples were handled in an inert atmosphere box and, where possible, measurements were made in the box. The resulting data are presented in Table 1.

TABLE 1. *As-received Leachate Characterization Data.*

	FL	NJ	OR	TX	UT	WI
pH	8.05	7.14	8.24	6.94	7.38	7.29
Eh (mV vs SCE)	-5	49	+133	+126	-152	-114
D.O. (mg/L)	<0.5	<0.5	<0.5	<0.5	<0.5	<0.5
TOC (mgC/L)	675	416	262	188	105	1156
TIC (mgC/L)	814	747	383	309	809	523
COND. (μS)	17000	5300	6200	2900	13500	7300

Specific organic compound analysis was achieved by extraction, concentration, and GC/MS analysis. Because only compounds with potential for interaction with metal ions were of interest, analysis was not exhaustive but focused on Lewis bases. Slightly modified portions of the EPA Master Analytical Scheme for Organic Compounds in Water [9] were used for these analyses.

1.1 Modeling Approaches

The MINTEQ database was compiled by various people, at various points in the history of the model. Because of this, the first step in compiling an accurate database was to review the equilibrium constants already present in the database. These constants were checked, and duplicates were removed. In addition, values for complexes of organic ligands were compared with those values found in Smith and Martell, *Critical Stability Constants* [10]. The values for complexes of inorganic ligands were not reviewed. The constants compiled by Smith and Martell have been thoroughly screened; their criteria used for including a constant are described in detail elsewhere, and so will not be discussed here [10]. Those values in MINTEQ that differed from the values found in Smith and Martell were replaced with the latter. In addition, the literature was searched using both electronic and hard copy resources. Additional equilibrium constants found were reviewed and either included in the database, or discarded as of no interest at the current time. All of the constants for organic ligands included in the most recent version of MINTEQ were measured at

298K. In addition, the values recorded are for zero ionic strength. Those values reported in the literature at non-zero ionic strength were corrected using the Davies [11] equation before inclusion in the database.

It is important to note that a certain amount of subjectivity arises when evaluating the accuracy of one reported stability constant over another. An effort was made to follow the criteria for evaluation as established by Smith and Martell [10] in order to maintain consistency in the values contained in the MINTEQ database.

The literature search and database modification demonstrated major holes in the required data. Thermodynamic data for many potentially important metal-organic interactions have not been reported. For example, aliphatic amines are among the compounds expected in a typical landfill leachate [12]. Many stability constants for complexes of aliphatic amines with Ag^+ have been reported in the literature, (for example, [10]). Interactions of amines with other metal ions such as Cu(II), Ni(II), and Zn(II) may be much weaker; few of these constants are found in the literature. Another example of a group of compounds that would typically be found in a landfill leachate are normal aliphatic carboxylic acids [12]. More stability constants have been measured for this class of compounds than for the aforementioned amines. However, the absence of constants for some metal ions, such as silver(I), is troublesome. That a particular constant has not been reported does not ensure its insignificance.

Procedures for expanding these types of databases were outlined by Martell *et al.* in 1988 [13]. Two approaches to filling the gaps that occur in the stability constant data were considered. The first approach is measurement of all the missing constants. Several researchers have acknowledged that this is no small task, and may even be beyond current analytical capabilities [14-17]. The second approach is to estimate the needed constants using readily available data. Several attempts at predicting stability constants have been reported.

Harris [15] used a structure-activity relationship based on the structure of the organic ligand to predict formation constants for nickel, cadmium, zinc, and iron complexes. A separate relationship was developed for each metal ion. The model was developed by assigning each functional group in the ligand a number, or "adjustable parameter", that reflected its effect on the metal-ligand interaction. A separate model was developed for those ligands expected to display chelating tendencies, and ligands containing strongly electron-donating or electron-withdrawing functional groups were omitted. In addition, ligands where steric effects were expected to come into play were excluded. Harris also notes "... in a few cases stability constants were excluded because they were clearly inconsistent with the bulk of the data." [15]. Rather good correlations for nickel, cadmium and zinc, and a rather poor correlation for iron were found. The advantage of this type of model is obvious. In order to predict a stability constant, only the structure of the organic ligand must be specified. Unfortunately, only models for nickel, cadmium and zinc have so far been developed with reasonable accuracy. Disadvantages include the previously mentioned exclusion of some functional groups and that "... the parameters in this correlation are presented as empirical parameters with no direct theoretical definition." [15].

Niebor and McBryde [18] reported a linear free energy relationship (LFER) for prediction of formation constants from constants for metal ions with similar ligands, or from constants for ligands with similar metal ions. They noted that, in many cases, there is a linear correlation "... between proton dissociation constants of a series of related ligands and the stability constants of the corresponding metal ion complexes." When this linear relationship exists, it is found between ligands that are very similar [19]. Then, using Hammett [20] type relationships, one can use a "parent molecule" to predict the stability of a complex of a related ligand.

Another structure-activity relationship is exemplified by a study by Smith, Martell, and Motekaitis [16]. For carboxylate ligands, they found in most cases the ratio of log constants for Mg^{2+}/Be^{2+} is about 0.38 [16]. Using this relationship, they were able to predict the stability constants for Be^{2+} from data for Mg^{2+} complexes. A similar methodology was used to predict the constants for Sc^{3+}, Y^{3+}, Am^{3+} and Th^{4+} from corresponding La^{3+} stability constants. The limitation lies in the need for substantial data to generate an estimate. "No estimates [can be] made for systems for which baseline data on related complexes are not available." [21,16,17].

The fourth type of model is based on "hard-soft" interactions. The concept of hard-soft interactions has been outlined by several authors, [22-24], and thus will not be reviewed here. Brown, Sylva and Ellis [25] developed a hard-soft model for predicting hydrolysis constants of metal ions using formal charge and ionic radius of the metal ion to describe its hardness. The model developed works well for most of the complexes tested, but a problem does arise when the ionic radius, or the coordination number on which the radius depends [26], of the metal ion has not been determined. In later papers [14,27], Brown and others tried to extend this type of model to inorganic ligands other than hydroxide but it is not clear whether it can be extended successfully to organic ligands. Again, substantial data are needed to establish the "electroncity" of the interaction on which this model is based. In some cases, there is a "large difference between numbers obtained using this model, and corresponding stability constants that were determined experimentally" [14].

The hard-soft approach is most appealing. It provides for systematic variation in the complex forming tendencies of both ligands and metal ions. An application of this concept using a minimum of readily available or easily measured data would be most desirable. Stability constants for hydroxy complexes, which reflect the acidity of the metal ion toward the hard base OH^-, are available for all metal ions of interest here. Acid ionization constants, which reflect the basicity of a ligand toward the hard acid H^+, are available for acids corresponding to most ligands of interest. Softness of the ligand may be described in terms of the extent of polarization of the ligand when bonded to H^+, which is evident in the residual polarity of the H^+ to ligand bond. This polarity is reflected in several physical properties of the resulting acids including dielectric constant and index of refraction. The latter parameter is particularly attractive due to its ready availability. In addition, molar refractivity (and hence index of refraction) can be readily estimated from structure [28] when an experimental value is unavailable. Predictive models using these data were investigated and are discussed below.

Several metal-organic interactions are being measured in our lab in order to verify the accuracy of algorithms we have developed to predict metal organic interactions, as well as to provide additional data for further refining of the algorithms. Many methods are available to determine formation constants for complexes of metal cations with organic ligands. A summary of these methods has been previously published, and so will not be discussed here, [13,29]. Interactions of metal anions with organic molecules have been less extensively studied but merit consideration.

Three of the metals of interest in this study - arsenic, selenium, and antimony - exist as anions in solution under normal environmental conditions. Arsenate (As(V)) is the primary form of arsenic in aerobic soils within the pH range of 4 to 7 [30] and in aquatic systems [31,32]. In alkaline soils, selenium exists primarily as selenate (Se(VI)) [31], in acidic soils [31] and aqueous systems [32], primarily as selenite (Se(IV)). Selenium exists in aqueous solutions as Se(VI) only under conditions of vigorous oxidation [32].

It has been shown that arsenite, borate and tellurate form esters with polyol complexes in aqueous solutions [33], (see Figure 1 for an example of such an interaction). In addition, arsenic acid has been reported to form complexes with catechol [34], and ester type linkages between phosphate and phenols are also known [35]. By analogy, one might expect similar interactions between arsenate, selenite, selenate, and antimonate and phenols, although no definite evidence of this is found in the literature.

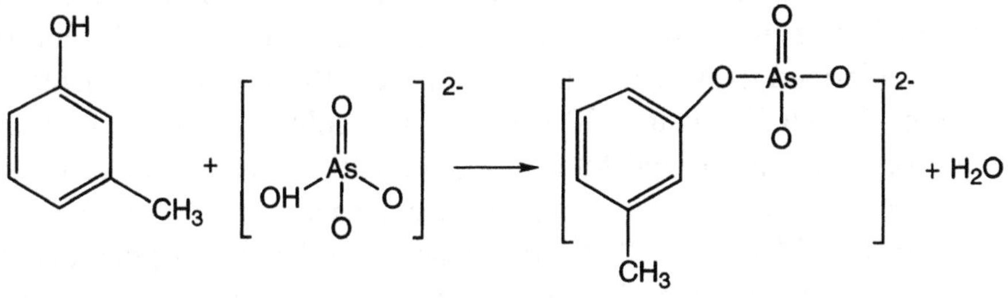

FIGURE 1. *Ester Formation Between m-Cresol and Arsenate.*

2.   EXPERIMENTAL METHODOLOGY

2.1   Chemicals and Instrumentation

All chemicals used for preparation of analytical solutions, or as standards, were reagent grade chemicals used without further purification. Deionized water was used to prepare aqueous solutions. Extraction solvents for organic analysis included dichloromethane (Baker, pesticide residue grade), methanol (Baker, pesticide residue grade), and methyl t-butyl ether (MTBE) (B&J, pesticide residue grade). Methanolic BF_3 for derivatizations was purchased from Supelco and stored at -10°C. Anhydrous salts for drying extracts were stored in a 110°C oven and cooled immediately before use.

Measurement of pH was performed using an Orion SA250 pH meter equipped with Corning Model 476540 pH electrode, the electrode was calibrated with standard pH 4.01 and 7.00 buffers. Solution potential (Eh) was measured using an Orion SA520 meter equipped with Ross double junction reference electrode and either a platinum or wax impregnated graphite probe. Electrode preparation and further method description has been reported elsewhere [36]. Conductivity was measured using a Cole Palmer conductivity meter. Dissolved oxygen was determined using an Orion Model 97-08 O_2 electrode. Dissolved oxygen, pH, Eh, and conductivity of leachate and groundwater samples were measured in a Forma inert atmosphere box containing an atmosphere of 5% H_2 in N_2. Metal determinations on leachates were performed using an Instrumentation Laboratories 951 atomic absorption spectrophotometer with graphite furnace accessory. GC/MS analyses were performed using either a Spectrel GC/MS or ELQ 400 triplequad system (EXTREL) using 30-meter DB-5 column (J&W Scientific). Total organic carbon and total inorganic carbon were measured using instruments manufactured by Coulometrics, Inc. (Model 5011 Coulometer, Model 5120 Total Carbon Combustion Apparatus, and Model 5030 Carbonate Carbon Apparatus). The copper ion selective electrode used was a h·nu Model 30-29-00. The measurements on the cation interactions were carried out in a thermostatted, water-jacketed beaker. The spectrophotometric measurements were performed using a Cary 219 visible-UV spectrophotometer. Equipment and chemicals used in the determination of oxygen, sulfate, sulfide, nitrate, and iron are described elsewhere [37-40].

2.2 Organic Ligand Analysis

Analysis of organic ligands in leachate samples (and in analogous surrogate standards prepared from reagent grade compounds) were performed via modified versions of portions of the EPA Master Analytical Scheme for Organic Compounds in Water [9]. An aliquot (ca. 750 mL) of sample was adjusted to pH 10 with NaOH, saturated with NaCl, and spiked with 10 µL of Supelco 6-component deuterated internal standard (Supelco 4-8902). A single batch extraction with 75 mL dichloromethane was used to obtain base/neutrals. The deuterated standards were found to be completely removed from the aqueous phase in this single extraction. The dichloromethane extract was dried over anhydrous Na_2SO_4 and then evaporated to ca. 2 mL using a Kuderna-Danish apparatus with 3-ball Snyder condenser followed by N_2 blowdown to 0.3 mL. The underivatized concentrate was analyzed via GC/MS.

The aqueous phase remaining after dichloromethane extraction was slowly acidified to pH 1 with 1 M HCl; substantial CO_2 gas production occurred during acidification of leachates. The acidified sample was extracted twice with 250 mL portions of methyl t-butyl ether (MTBE). The two extracts were combined, dried via passage through a column containing approximately 10 grams each of anhydrous Na_2SO_4 and anhydrous $MgSO_4$. The dried extract was evaporated to ca. 2 mL using a Kuderna-Danish apparatus with 3-ball Snyder condenser and then taken to dryness via N_2 blowdown. Two 1-mL aliquots of 14% BF_3-methanol were then sequentially added to the evaporated sample and transferred to a 15-mL vial. The resulting solution was heated at 100°C for 3-5 minutes (capped vial), cooled for 10 minutes, and treated with 10 mL saturated aqueous NaCl. Ten µL each of 20,000 ppm solutions of 2-fluoro-1,1'-biphenyl and 4-fluoro-2-iodotoluene in dichloromethane were added as standards. The solution was extracted twice (5 minutes shaking, 5 minutes settling) with 1-mL aliquots of dichloromethane; each aliquot was analyzed separately via GC/MS. When ligands were found in the second aliquot, concentration in the sample was calculated from the sum of material found in the two aliquots. Analysis of surrogate standards provided the basis for calculation of ligand concentrations from normalized (on the basis of the internal standards) areas of total ion chromatogram (TIC) peaks. Blanks were analyzed by the identical procedure. Where possible, compound identities were established via both library search matching and retention time comparisons for standards.

A separate extraction/ analysis procedure was used for very hydrophilic compounds. Separate 50 mL aliquots of leachate were loaded onto columns of AG-1 anion exchange resin (bed volume ca. 10 mL) in the OH$^-$ form. The resin, originally in the chloride form, had been exhaustively extracted with methanol followed by acetonitrile in a Sohxlet apparatus and then converted to the OH$^-$ form after slurrying into columns. Unretained components were rinsed through the column with approximately three bed volumes of deionized water and the anions were then displaced and collected via elution with 1:1 acetone-water solution of $NaHSO_4$ (1 M).

The resulting eluate was adjusted to pH 11 with 1 M K_2CO_3 and rotoevaporated to near dryness at 50°C. Approximately twice the remaining liquid volume of 2,2-dimethoxypropane was added. The resulting mixture was refluxed for ca. 20 minutes and then evaporated to dryness at ca. 80°C. The resulting solid was derivatized with benzyl bromide (5% in acetone) by heating for 12 hours at 60°C. The resulting solution phase was analyzed by GC/MS. Blanks were analyzed by the identical procedure.

2.3   Determination of Redox Characteristics

Groundwater characteristics, which determine the redox transformations predicted for influent leachates, assuming sufficient poising capacity, include dissolved oxygen, ferrous iron, sulfide, sulfate, and dissolved organic carbon concentrations, along with pH, temperature, and chemical oxygen demand. Field methods used in this work for oxygen, iron, pH, and temperature have been previously described [7,41]. Laboratory methods for sulfide [37], sulfate, total organic carbon, and chemical oxygen demand [38-40] are well known and will not be further described here.

Determinations of electrode redox potentials are not recommended, though they have apparent utility in many situations [42,7]. Procedures for such determinations, along with descriptions of situations for which such determinations may be useful, are also described in these references.

2.4   Stability Constant Determination

A solution of copper (for example) was titrated with the ligand of interest. Solution pH was maintained constant via addition of NaOH and Cu^{2+} activity was measured according to the procedure outlined by Ramamoorthy et al. [43]. Additionally, we maintained an ionic strength of 0.01 M during the titration using KNO_3 solution.

Phenol-anion interactions were investigated by titrating aqueous solutions of anion (selenite, selenate, antimonate, or arsenate), of phenol (m-cresol, catechol, or pentachlorophenol), and of anion-phenol mixtures with standard NaOH. Resulting pH titration curves were recorded. Spectrophotometric analysis similar to that described by Johnson and Pilson [44] was also employed. In these analyses, the phenol was extracted into isooctane both in the presence and absence of anion. The concentration of the phenol was determined spectrophotometrically (UV). Direct synthesis of an arsenate-catechol complex was attempted [34]. The analytical techniques employed were described in detail by Huyck et al. [45].

3.   RESULTS AND DISCUSSION

3.1   Leachate Characterization

The as-received characterization data for the six leachates studied (Table 1) reveal similarities and marked differences. All six samples were nearly neutral and strongly buffered by carbonate species (see total inorganic carbon (TIC) data) and, to a lesser extent, by organic acids, ammonia and silica species. No attempt was made to accurately measure the very low dissolved oxygen levels present in these samples. Measurement was simply to ensure that gross oxidation or contamination of the samples had not occurred in shipping. Conductivity in each of the samples is quite high. Variations between samples are mostly a result of differing major ion inorganic constituents and, to a much lesser extent, differences in concentrations of ionic organic species.

The six leachates vary from moderately low Eh (OR, TX) to low (FL, NJ) to extremely low (UT, WI). The OR sample was collected from a surface leachate pond and thus was continuously exposed to an oxygen supply. It may be assumed that the Eh of the leachate reaching groundwater at this site could be much lower. All other samples were collected from sub-surface collection systems directly and the Eh of these samples is more likely representative of *in situ* conditions. An order of magnitude difference in TOC between the most organic rich (WI) and the least organic rich (UT) samples was observed. It is interesting to note that the leachates are two to three orders of magnitude higher in TOC than most groundwaters [46].

Eh, pH, conductivity and carbon content, both inorganic and organic, were measured on aliquots of leachate from all six sites during the period of sample storage at varying time intervals. Upon receipt, an aliquot of each leachate was transferred under inert atmosphere into a 125 mL septum bottle. A syringe was then used to remove samples for periodic analysis.

Total organic carbon (TOC) measured in the samples is the most direct and reliable parameter for determining microbial or abiotic sample alteration. Two contrasting sample preservation histories are seen in Figures 2 and 3. In Figure 2 (OR), TOC and total inorganic carbon (TIC) remain relatively constant over eight months, whereas Figure 3 (WI) shows a continuous decline in TOC and concomitant increase in TIC over the monitoring interval. This could be due to leakage of the septum seal, as the carbon content change is accompanied by decrease in conductivity and increase in Eh.

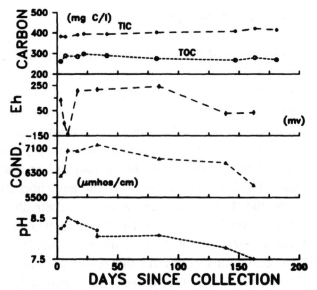

FIGURE 2. *Variation of Oregon Leachate Characteristics During Storage.*

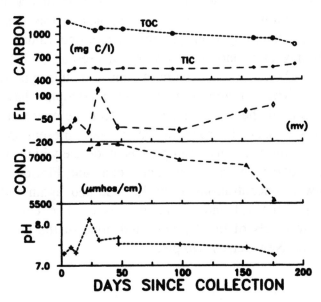

FIGURE 3. *Variation of Wisconsin Leachate Characteristics During Storage.*

3.2   Organic Ligand Analysis

Three groupings of the leachates on the basis of carboxylic acid content are evident in the data (Table 2). The first is those samples with predominantly short chain aliphatic (FL, NJ, TX) carboxylic acids. The second group of samples contain primarily aromatic acids and/or cyclohexane carboxylic acid (WI, OR). The third observed type, the Utah sample, has very low concentrations of organic, primarily aromatic, acids. This sample could be considered an end member of the second type (WI, OR). We have not found any short chain aliphatic acids in this sample, but are not convinced of their absence because Utah leachate has consistently posed extraction efficiency problems. The sample forms strong emulsions with both dichloromethane and MTBE. These emulsions could be the result of humic-like organic material in the sample or other, at present undetermined, matrix effects in this sulfide-rich sample.

The organic acids present in the leachates analyzed in this study are not dissimilar to acids identified in English landfill leachates [47], Canadian landfill leachates [48] or in U.S. landfill leachates [49,12]. The concentrations measured in the samples of this study, however, often vary considerably from previously reported values.

The presence and concentration, if present, of 1,2 or 1,3 or 1,4 benzene dicarboxylic acids is one example of this type of variability (Table 2). The Oregon and Wisconsin samples have very high concentrations of the 1,2 acid but no detectable concentrations of the 1,3 and 1,4 benzenedicarboxylic acids were present in the Oregon sample, while the Wisconsin sample has moderately high concentrations of the other two isomeric acids. The Texas and Florida samples have slightly lower concentrations of the 1,2 acid and, again, only one of the two (TX) has the 1,3 and 1,4 acids present. The Utah sample only contains low levels of the 1,2 acid and none of the these acids were present in the New Jersey sample.

TABLE 2. **Organic Ligands in Landfill Leachates.** P = Present, but concentration not quantified

	Concentration Found (mg/L)					
	UT	NJ	FL	WI	OR	TX
Carboxylic Acids:						
Acetic	P	P	P	P	P	P
Propanoic	23.80	4.09			1.13	22.69
Methylpropanoic	3.70	2.08			P	1.23
Butanoic	7.07	0.19			P	0.39
Dimethylpropanoic				0.02		
3-Methylbutanoic		3.06	1.35			4.65
Pentanoic		1.71	0.01	P		0.12
Methyl butanedioic						P
3,3-Dimethylbutanoic			0.01	P		
2,3-Dimethylbutanoic		P	0.01	P		
2-Methylpentanoic		0.17	0.04			0.11
3-Methylpentanoic		0.02	0.01	0.02		
4-Methylpentanoic		0.17	0.02			0.47
Hexanoic		0.06	0.01	P		0.01
2-Ethylbutanoic			0.04			0.07
4-Methylhexanoic		0.08				0.10
5-Methylhexanoic		0.03	P			
Heptanoic		0.70	1.16	0.49		0.55
2-Ethylpentanoic				0.03		0.01
Cyclohexane Carboxylic		0.11	0.03	5.85		0.76
3-Cyclohexene Carboxylic				0.03		
3-Cyclohexene-1,2-dicarboxylic				0.02		0.03
2-Ethylhexanoic		0.08	P	0.20		0.19
Palmitic	0.16			0.10		
Benzoic	P	0.05	0.09	0.02	0.49	1.37
Phenylacetic	P	0.03	0.41	10.54	1.41	1.38
Toluic	P		0.04	0.12		0.06
Phenylpropanoic				0.13		0.08
alpha-oxophenylpropanoic	0.13		0.22		0.22	0.08
1,2-Benzenedicarboxylic	0.06		0.95	4.40	5.37	0.90
1,3-Benzenedicarboxylic				2.04		0.43
1,4-Benzenedicarboxylic				1.22		0.25
Phenylbutanoic				0.06		
2-[4-Chloro-o-tolyloxy]propanoic				P		P
Phenols						
Phenol					P	
Cresol			P	P	P	P
3-Ethylphenol				P		
4-t-Butyl phenol	P		P	P		

The presence of only the 1,2 benzene dicarboxylic acid in the Utah sample is somewhat puzzling as it has the most diverse composition of aromatic compounds (non-acidic) of any of the samples, including many di- and tri- substituted benzene hydrocarbons. The conditions in the landfill may not be conducive to acid formation as it is, in terms of measured Eh, the most reduced of all the samples. Also, the near absence of any aromatic acids in the New Jersey site also is difficult to interpret at present.

Another interesting comparison between the samples is a lack of correlation of total carboxylic acid content with TOC. The acids reported in Table 2 account for less than 1% of the TOC in the UT leachate but 19% of TOC in the TX leachate. The unfortunate implication is that TOC analysis does not afford an indication of complexing capacity of leachates.

Although previous studies have reported the presence of amines [12] (aliphatic as well as heterocyclic) in landfill leachates, none were found in these samples. Analyses of surrogate standards clearly demonstrated the adequacy of our analytical methods for such compounds. Undefined matrix effects could perhaps impact extraction efficiency. Amine concentrations are particularly significant with respect to mobilization of "soft" metal ions, for example, Ag^+.

The analytical methods used for organic ligand determinations remain under development. We noted loss of long-chain (hydrophobic) fatty acids in the pH 10 extractions into dichloromethane. Similar losses from complex matrices were previously reported by Norwood *et al.* [50]. No concentration data are reported in Table 2 for acetic acid and values for propanoic acid are probably low. Our analyses of surrogate standards reveal low recoveries of these volatile compounds. The developers of the Master Analytical Scheme noted low recoveries of internal standards in both NOVA and VOSA procedures when applied to wastewaters as opposed to drinking water matrices used in the method development [9].

3.3 Redox Determinations

Detailed discussion of redox investigations conducted as part of this research is beyond the scope of this paper and will be presented elsewhere. Data which summarize findings relevant to considerations of metal/organic interactions are given in Tables 4 and 5. The data represent two distinct groundwater types. The first, shown in Table 3, are for a dilute near surface groundwater system, with low concentrations of redox-active species (i.e., low poising capacity). Note that oxygen is present in all well depths lacking measurable ferrous iron, and that nitrite (from a contaminant plume) is also anticorrelated with Fe(II) concentrations. These data, typical of those found for several other wells at the same site (Cape Cod, MA, USA), indicate that oxygen may dominate the abiotic redox chemistry of portions of the groundwater system. Also, presence of redox-active species, such as nitrite, may signify (biotic) redox processes which dominate the system. DOC in these wells is <2 mg C/L and measurable sulfide is absent. Where Fe(II) is present in this system (Table 4), redox potentials calculated from the Fe(III)/Fe(II) couple correspond (to within ± 15mV) to the potential recorded at a Pt electrode (corrected to the standard hydrogen electrode, SHE). Such correlation is not universal, as will be seen from the data discussed below.

The second data set (Table 5) was obtained from a system which contains relatively high concentrations of inorganic and organic species. Sulfate levels in these wells, for example, are typically 40-100mg sulfate/L, and DOC's as high as 80 mgC/L have been observed. Determinations of dissolved oxygen, ferrous iron, sulfide, sulfate, DOC and COD for these wells enabled calculation of redox potentials with respect to a variety of redox couples. From these calculations, no consistent correspondence between any of the calculated redox potentials and any of the observed electrode potentials emerges. Of the relevant redox couples, only the Fe(III)/Fe(II) couple has been shown to be electroactive [3,4,36,51] With the possible exception of the O_2/H_2O_2 and certain S^{2-}/S° couples, none but the iron has been indicated as a reversible, electroactive couple relevant to groundwater regimes. The data in Table 5 confirm that even the iron couple, however, cannot be represented by the Pt electrode potential.

TABLE 3.   *Estimations Derived From the Model.*

Compound	log K_1 MOH	dielectric constant	estimated log K_1 ML	actual log K_1 ML
Pb-Propanoate	7.78	3.35	2.35	2.64
Ni-Propanoate	4.70	3.35	0.96	0.78
Zn-Propanoate	4.38	3.35	0.91	1.01
Mg-Propanoate	4.38	3.35	0.59	0.54
Cu-Isobutyrate	6.36	2.64	2.10	2.17

TABLE 4.   *Groundwater Redox Parameters.* Wells near Otis Air Force Base, Cape Cod, MA, USA; Sampled July, 1988. Uncertainties represent the maximum range of (3 to 5) replicate analyses for Fe(II), O_2, and NO_2^-; estimated 90% confidence intervals from accumulated experience for pH and electrode potentials. Depth is from land surface to well screen. E_{Pt} is mV *vs.* SHE.
[1]. Estimated from data from nearby wells at similar depths. nd - not detectable.

Well #	Depth (m)	T,$^{\circ}C$ ±0.2	pH ±0.05	Fe(II) mg/L	O_2 µg/L	NO_2^- mg N/L	E_{Pt} - mV ± 5
343	11.0	12.1	6.23	<0.05	20±5	0.05	355
343	17.4	13.2	6.46	16±1	<0.50	<0.025	95
343	24.1	12.6	6.60	21±1	<0.50	<0.025	80
343	30.2	13.0	5.56	<0.05	9±3	<0.025	371
16-17	9.0	12.0	n.d.	<0.007	>100	<0.025	310
16-17	9.60	12.8	6.23	<0.007	>100	0.15	275
16-17	10.36	16.5	6.26	0.10	40±8	0.40	260
16-17	10.61	15.4	6.24	0.10	50±8	0.70	279
16-17	10.91	15[1]	6.25	<0.007	25±5	0.60	280
16-17	11.13	15[1]	6.25	0.12	25±5	0.06	240
16-17	11.40	15[1]	6.45	4.3±0.2	<0.5	<0.025	141
16-17	11.65	15[1]	6.43	4.8±0.2	<0.5	<0.025	140

TABLE 5.   ***Groundwater Redox Parameters.*** II. Wells in Tennessee
Park, near Leadville, CO, USA; Sampled July, 1989. Nitrite is
undetectable (< 0.025 mg N/L) in all wells. Uncertainties
represent the maximum range of (3 to 5) replicates for Fe, O_2,
SO_4, and TOC, duplicates for sulfide and COD. Others are
estimated 90% confidence intervals. Depth is below surface of
land. TOC - total organic carbon. COD - chemical oxygen
demand. E_{Pt} is mV *vs* SHE. [1] - no analysis, sulfide smell absent.

Well #	Depth (m)	T,°C ±0.2	pH ±0.05	Fe(II) mg/L ±10%	O_2 µg/L	S^{2-} mg S/L ±10%	SO_4^{2-} mg/L ± 5%	TOC mg C/L ±10%	COD mg O_2/L	E_{Pt} mV ±10
1	1.50	6.5	4.98	5.2	>1000	0.06	71	3.0	5±1	259
3	2.06	7.5	5.09	9.0	<0.5	0.02	94	3.7	2±2	233
4	2.06	6.0	5.50	<0.05	>1000	0[1]	84	3.4	12±4	412
5	2.08	7.0	5.52	18.2	<0.5	0.26	28	24.7	47±5	216
6	2.08	6.5	5.46	19.5	<0.5	1.26	30	44.0	104±2	137
7	2.08	6.5	5.63	10.0	13±3	0.06	68.5	13.3	18±2	300
8	2.08	6.5	5.37	10.6	1.0±0.5	0.30	119	13.7	18±2	73
9	3.56	6.5	4.54	0.15	1000±100	0[1]	1.5	8.5	4±1	427
10	2.97	6.0	6.10	17.5	<0.5	0.21	44	21.6	40±2	53
11	3.56	7.5	5.74	3.4	<0.5	0.35	55	14.6	20±2	153
12	3.56	7.5	5.29	<0.05	>1000	0[1]	82	11.6	4.5±2.5	313
13	2.97	7.5	5.68	15.1	0.5±0.5	0.20	67	4.0	26.6±1	51
14	2.97	8.0	5.51	5.5	<0.5	0[1]	81	3.1	9.6±2	367

Thus, it is less than obvious what experimental information can be
used to allow a computer model such as MINTEQ to accurately
represent the metal/organic interactions which pertain to a landfill
leachate as it infiltrates into a groundwater system. Our
recommendation, based upon limited data such as that illustrated above,
is that a calculated Eh which pertains to the ferrous iron and pH levels
in the receiving groundwaters be used wherever possible to determine
the redox conditions which govern the metal/organic interactions. This
procedure has been detailed elsewhere [42].

In the absence of measurable Fe(II), redox potentials
corresponding to the presence of oxygen or sulfide should be used. In

our opinion, exact potentials are, in such redox regimes, not important. For example, the presence of measurable oxygen, even at one to several micrograms per liter, corresponds to equilibrium conditions more oxidized than any of the metal oxidation states relevant to this study.

Alternatively, the presence of sulfide indicates redox conditions reducing enough to convert any of the metals considered here to as low an oxidation state as can be achieved in an aqueous environment. (The presence of sulfide as an inorganic precipitant and complexing agent is quantitatively important, however, and this makes accurate sulfide concentration data important beyond redox considerations). Thus, predictions of metal-organic interactions, within the framework of the present equilibrium computer codes, are susceptible to perturbations due to changes in redox potentials. However, limited knowledge of the redox conditions of the system, within the three broadly defined redox regimes defined above, should enable one to define and quantify such perturbations to an accuracy well within the limitations prescribed by the equilibrium assumptions, the lack of poising capacity information, and the incomplete nature of the database.

3.4   Stability Constant Determination

A value for pH and free Cu^{2+} was obtained for each titration point. Assuming a 1:1 complex for Cu-butyrate (as is established in the literature) [4], and using the titration curve illustrated in Figure 4, in addition to a log K_a value of 4.82 [10], a value of 2.26 ± 0.09 (corrected to zero ionic strength) was obtained for the log of the formation constant for Cu-butyrate at 25°C. This agrees well with the literature value of 2.14 at zero ionic strength [10]. A similar calculation was done for 3-methylvaleric acid with copper(II). Using the information from the titration curve in Figure 5, a value of 1.8 ± 0.2 (corrected to zero ionic strength) was obtained for the log of the formation constant for copper (II) with 3-methylvaleric acid.

A representative number of these types of determinations is necessary in order to determine the accuracy of models proposed. However, determination of all of the missing metal-organic interactions would be impractical and unnecessary.

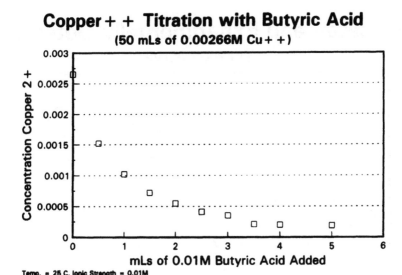

Temp. = 25 C, Ionic Strength = 0.01M
pH = 5.15 +/- 0.08

FIGURE 4. *Titration of a Copper Solution with Butyric Acid.* 50 mL of 0.00266 M Cu^{2+} at 25°C, μ = 0.01 M, pH = 5.15±0.08.

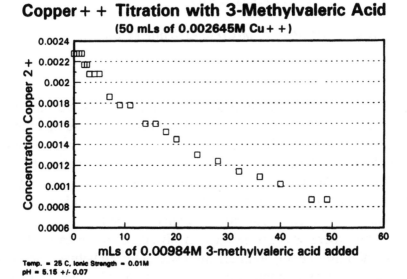

Temp. = 25 C, Ionic Strength = 0.01M
pH = 5.15 +/- 0.07

FIGURE 5. *Titration of a Copper Solution with 3-Methylvaleric Acid.* 50 mL of 0.02645 M Cu^{2+} at 25°C, μ = 0.01 M, pH = 5.15±0.07.

Simulated and Experimentally Determined
Titration curves

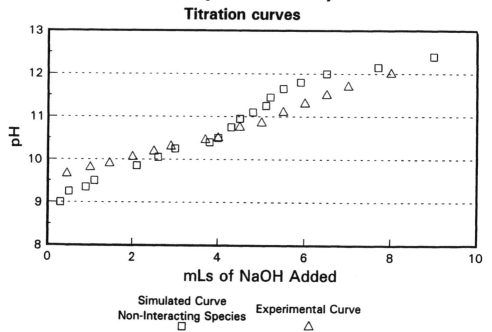

FIGURE 6.   Titration of m-Cresol-Selenite Mixtures.

The results of the studies of arsenate, antimonate, and selenate with m-cresol, catechol, and pentachlorophenol indicate that there is no significant interaction between these anionic metal species and these three phenols. Reactions were allowed three to four days to come to equilibrium. The generalization that arsenate, selenate, and antimonate react to no significant degree with phenolic compounds under normal aqueous environmental conditions seems justified.

While there was no interaction observed in the case of catechol and pentachlorophenol, an interaction was detected between selenite and m-cresol. The experimental curve for titration of a m-cresol-selenite mixture (Figure 6) differs significantly from that calculated for non-interacting m-cresol and selenite (using constants obtained from direct titration, or from the literature [52]). The differences cannot be removed with adjustment of the respective K_a values; nor can they be adequately modeled using an ester type complexation. These observations generally were confirmed via extraction/spectrophotometric measurements. It is possible that there are kinetic problems associated with the formation of a complex in this system or the components may interact in a fashion that is not well described by the esterification model. Further investigation of the phenomena continues in our laboratory.

3.5 Stability Constant Estimation

As discussed earlier, the need for methods of approximating unknown formation constants using readily available data is great. A recent paper by Brown *et al.* [25], used size and formal charge of the metal ion to predict formation constants for hydroxy-metal constants. In this paper the authors state that "...the problem of predicting the hydrolysis constants of metal cations is inseparable from that of determining the 'effective ionic radius'" [25]. This is indeed true and it points out an interesting concept. Upon examination of the literature [2,52], it is apparent that the first formation constant with hydroxide has been measured for a majority of the metals of interest to the environmental community. We decided to use the stability of the first hydroxy complex as relating both inherent acidity and hardness of the metal ion. Using pK_a data and stability constant data for metal carboxylate complexes from the literature, no useful correlation of stability constants with pK_a and hydroxy complex stability constants was found.

Two approaches to including a measure of ligand hardness were considered; use of dielectric constant of the carboxylic acid, or of its refractive index. A large number of dielectric constants have been measured [53] and others can be easily determined [54]. Figure 7 presents a schematic of the method used to predict metal ion-n-carboxylate formation constants from pK_1 for the metal hydroxy interaction and dielectric constant. Additional stability constants not used in developing the correlation were subsequently found to fit the

Step 1   Obtain log K1 value for metal of interest with OH⁻.

Step 2   Insert value found in Step 1 into the following equation:
$$Y = 0.216(\text{Step } 1)+0.511.$$

Step 3   Obtain dielectric constant for ligand of interest.

Step 4   Insert value found in Step 3 into one of the following three equations depending on the value found in Step 1:
$$Y = .09(\text{Step } 3)+0.8 \text{ if Step } 1<2.0$$
$$Y = -.01(\text{Step } 3)+2.0 \text{ if Step } 1>5.0$$
$$Y = .08(\text{Step } 3)+0.9 \text{ if } 2.0<\text{Step } 1<5.0.$$

Step 5   Add values obtained in Step 2 and Step 4.

Step 6   Insert value found in Step 5 into the following equation:
$$\text{Log K1} = 0.8246(\text{Step } 5) - 1.0795.$$

FIGURE 7.   Method of Predicting Metal-carboxylic Acid Interactions.

constants correlation reasonably well. In addition, using the dielectric constant for butyric acid [53] and K_1 for copper(II)-hydroxy complex [52], the value of pK_1 for copper(II)-butyrate complex was estimated as 2.10. Subsequent experimental measurement of the constant yielded 2.26 ± 0.09.

Few dielectric constants have been reported for amines; therefore, more readily available refractive indices are more useful here. These correlations are in a developmental, evolutionary stage. We are now considering three parameter correlations. As additional constants are measured, the expanding data will be used to further refine the correlations and improve estimates of constants for inclusion in MINTEQ. Irrespective of the specific correlation used to estimate stability constants for 1:1 complexes, the method of Brown [14] can be used to then estimate stepwise stability constants for higher order (i.e., 1:2, 1:3, etc.) complexes.

4.   CONCLUSIONS

The landfill leachates analyzed contained significantly varying concentration levels of organic ligands.   The higher concentrations found in some of the samples (e.g., NJ or TX) can significantly affect speciation.   For example, in the NJ leachate, approximately 10% of dissolved Pb(II) would be in the form of aliphatic carboxylate complexes.

Several additional considerations impact the significance of these results on metal ion mobilization and point to future directions of our research.   Only a single leachate sample collection was performed at each site.   The composition of leachate at a given site must be expected to vary [55], but the magnitude of temporal variation in concentration or distribution of organic ligands is not known.   The six sites studied also probably do not completely define variance arising from landfill location, age, climate and related conditions.

In addition, as leachates migrate, a variety of degradation reactions, both microbial and abiotic, will modify the organic components present.   If the system becomes more oxidizing, the introduction of additional Lewis base sites via oxidation may increase complexing capacity. This, in turn, would lead to metal solubilization from aquifer materials some distance downgradient.   Finally, toxic metal solubilization can occur both as a result of complexation (with organic ligands present) and as a result of release of toxic metals upon solubilization of sorbent phases (for example, ferric oxides) [55,56].

DISCLAIMER
Although the research described in this article has been supported by the United States Environmental Protection Agency through cooperative agreement CR-814290-01-0 to the Colorado School of Mines, it has not been subjected to Agency review and, therefore, does not necessarily reflect the views of the Agency and no official endorsement should be inferred.

REFERENCES

1. Macalady, D.L., P.G. Tratnyek and T.J. Grundl. "Abiotic Reduction Reactions of Anthropogenic Organic Chemicals in Anaerobic Systems: A Critical Review," *J. Cont. Hydrol.* 1:1-28 (1986).

2. Perrin, D.D., Ed., *Stability Constants of Metal-Ion Complexes Part B* (Pergamon Press, 1979), p.619, 623.

3. Lindberg, R.D. and D.D. Runnells. "Groundwater Redox Conditions: An Analysis of Equilibrium State Applied to Eh Measurements and Geochemical Modeling," *Science* 225:925-927 (1984).

4. Stumm, W. and J.J. Morgan. *Aquatic Chemistry 2nd Edition* (J. Wiley and Sons, New York, 1981) Chapter 7.

5. Barcelona, M.J., T.R. Holm, M.R. Shock and G.K. George. "Spatial and Temporal Gradients in Aquifer Oxidation-Reduction Conditions," *Water Resour. Res.* 25:991-1003 (1989).

6. Macalady, D.L., K. Walton-Day, V.T. Tate and M.W. Brooks. "Field Methods for Measurement of Ground Water Redox Chemical Parameters," *Ground Water Modeling Reviews,* Fall 1990, 81-89 (1990).

7. Macalady, D.L., D. Langmuir, T.J. Grundl and A.E. Elzerman. "Use of Model-Generated Fe^{3+} Ion Activities to Compute Eh and Ferric Oxyhydroxide Solubilities in Anaerobic Systems," in, *Chemical Modeling in Aqueous Systems III*, D. Melchoir and R. Bales, Eds. (American Chemical Society Symposium Series 416, Washington D.C., 1989), pp. 350-367.

8. Scott, M.J. and J.J. Morgan. "Energetica and Conservative Properties of Redox Systems," in, *Chemical Modeling in Aqueous Systems II,* D. Melchoir and R. Bales Eds., (American Chemical Society Symposium Series 416, Washington D.C., USA, 1990) pp. 368-378.

9. Pellizzari, E.D., L.S. Sheldon, J.T. Bursey, L.C. Michael, R.A. Zweidinger and A.W.Garrison. "Master Analytical Scheme for Organic Compounds in Water," EPA/600-S4-85/008 (July, 1985).

10. Smith, R.M. and A.E. Martell. *Critical Stability Constants* (Plenum Press, New York, Vol.1; 1974, Vol.2; 1975, Vol.3; 1977, Vol.4; 1976, Vol.5; 1982).

11. Brown, D.S. and J.D. Allison. *MINTEQA1, an Equilibrium Metal Speciation Model: User's Manual* (Environmental Research Laboratory Office of Research and Development U.S. Environmental Protection Agency, Athens, GA, 30613) p. 22.

12. Khare, M. and N.C. Dondero. "Fractionation and Concentration of Volatiles and Organics on High Vacuum Systems Examination of Sanitary Landfill Leachate," *Environ. Sci. Tech.* 11:814-819 (1977).

13. Martell, A.E., R.J. Motekaitis and R.M. Smith. "Structure-Stability Relationships of Metal Complexes and Metal Speciation in Environmental Aqueous Solutions," *Environ. Tox. Chem.* 7:417-434 (1988).

14. Brown, P.L. "Prediction of Formation Constants for Actinide Complexes in Solution," *Talanta* 36:351-355 (1989).

15. Harris, W.R. "Structure-Reactivity Relation for the Complexation of Ni, Cd, Zn, and Fe," *J. Coord. Chem.* 13:17-30 (1983).

16. Smith, R.M., A.E. Martell and R.J. Motekaitis. "Prediction of Stability Constants. I. Protonation Constants of Carboxylates and Formation Constants of their Complexes with Class A Metal Ions," *Inorg. Chim. Acta* 99:207-216 (1985).

17. Smith, R.M., R.J. Motekaitis and A.E. Martell. "Prediction of Stability Constants. II. Metal Chelates of Natural Alkyl Amino Acids and their Synthetic Analogs," *Inorg. Chim. Acta* 103:73-82 (1985).

18. Niebor, E. and W.A.E. McBryde. "Free-energy Relationships in Coordination Chemistry. I. Linear Relationships Among Equilibrium Constants," *Can. J. Chem.* 48:2248-2564 (1970).

19. Niebor, E. and W.A.E. McBryde. "Free-energy Relationships in Coordination Chemistry. II. Requirements for Linear Relationships," *Can. J. Chem.* 48:2565-2573 (1970).

20. Cary, F.A. and R.J. Sundberg. *Advanced Organic Chemistry* (Plenum Press, New York, 1984) pp. 179-189.

21. Smith, R.M. and A.E. Martell. "Critical Stability Constants, Enthalpies and Entropies for the Formation of Metal Complexes of Aminopolycarboxylic Acids and Carboxylic Acids," *Sci. Tot. Environ.* 64:125-147 (1987).

22. Huheey, J.E. *Inorganic Chemistry; Principles of Structures and Reactivity* (Harper and Row Publishers, Inc, 1983) pp. 312-324.

23. Pearson, R.G. "Hard and Soft Acids and Bases," *J. Am. Chem. Soc.* 85:3533-3539 (1963).

24. Parr, R.G. and R.G. Pearson. "Absolute Hardness: Companion Parameter to Absolute Electronegativity," *J. Am. Chem. Soc.* 105:7512-7516 (1983).

25. Brown, P.L., R.N. Sylva, and J. Ellis. "An Equation for Prediction the Formation constants of Hydroxy-metal Complexes," *J. Chem. Soc. Dalton Trans.* 723-730 (1985).

26. Shannon, R.D. and C.T. Prewitt. "Effective Ionic Radii in Oxides and Fluorides," *Acta Cryst.* B 25:925-946 (1969).

27. Brown, P.L. and R.N. Sylva. "Unified Theory of Metal Ion Complex Formation Constants," *J. Chem Res., Synop.* 4-5 (1987).

28. Shriner, R.L., R.C. Fuson, D.Y. Curtin. *The Systematic Identification of Organic Compounds,* Fourth Edition (John Wiley and Sons, New York, 1956) pp.48-51.

29. Martell, A.E. and R.J. Motekaitis. *Determination and Use of Stability Constants* (VCH Publishers, Inc, New York, 1988).

30. Fowler, B.A., Ed. *Biological and Environmental Effects of Arsenic* (Elsevier, New York, 1983).

31. Bailey, R.A., H.M. Clarke, J.P. Ferris, S. Krause and R.L. Strong. *Chemistry of the Environment* (Academic Press, 1978) pp. 396-402.

32. Cotton, F.A. and G. Wilkinson. *Advanced Inorganic Chemistry* (Interscience Publishers, A Division of John Wiley and Sons, 1972) pp.448-450.

33. Roy, G.L., A.L. Laferrer and J.O. Edwards. "A Comparative Study of Polyol Complexes of Arsenite, Borate, and Tellurate Ions," *J. Inorg. Nucl. Chem.* 4:106-114 (1957).

34. Ber. Der Deut. Chemi. Gesellschaft 52B:1316 (1919).

35. Emsley, J. and E. Hall, Ed. *The Chemistry of Phosphorous* (Halsted Press, A Division of John Wiley and Sons, 1976).

36. Grundl, T.J. "The Behavior of Redox Electrodes in Anaerobic Systems Containing Iron," PhD Thesis, Colorado School of Mines, Golden, CO (1987).

37. Bauman, E.W. "Determination of Parts per Billion Sulfide in Water with the Sulfide Selective Electrode," *Anal. Chem.* 46:1345-1349 (1974).

38. Hem, J.D. "Study and Interpretation of the Chemical Characteristics of Natural Water," U.S. Geol. Survey Water Supply Paper 2254, p. 263 (1985).

39. *Standard Methods for the Examination of Water and Wastes*, 14th ed. (Washington, D.C.: American Public Health Association, 1982) pp. 550-554.

40. Skougstad, M.W., M.J. Fishman, L.C. Freidman, D.E. Erdmann and S.S. Duncan. "Methods for Determination of Inorganic Substances in Water and Fluvial Sediments," Techniques for Water-Resources Investigations of the U.S. Geological Survey, Book 5, p. 626 (1979).

41. White, M.A., M.L. Peterson and R.B. Solbaur. "Measurements and Interpretation of Low Levels of Dissolved O_2 in Groundwater," *Ground. Monit. Rev.*, 28:584-590 (1990).

42. Grundl, T.J. and D.L. Macalady. "Electrode Measurements of Redox Potential in Anaerobic Aqueous Iron Systems," *J. Cont. Hydrol.*, 5:97-117 (1989).

43. Ramamoorthy, S., C. Guarhaschelli and D. Fecchio. "Equilibrium Studies of Cu^{++}-Nitrilotriacetic Acid with a Solid State Cupric Ion-Selective Electrode," *J. Inorg. Nucl. Chem.* 34:1651-1656 (1972).

44. Johnson, D.L. and M.Q. Pilson. "Spectrophotometric Determination of Arsenite, Arsenate, and Phosphate in Natural Waters," *Anal. Chim. Acta* 58:289-299 (1972).

45. Huyck, K.A., S.R. Daniel and D.L. Macalady. "Interaction of Arsenate and Phenols in Aqueous Media," paper presented at the 197th meeting of the American Chemical Society, Dallas, TX, April 14, 1989.

46. Thurman, E.M. *Organic Geochemistry of Natural Waters* (Martinus Nijhoff/Dr. W. Junk Publishers, Dordrecht, 1985).

47. Tester, D.J. and R.J. Harker. "Ground-water Pollution Investigations in the Great Ouse Basin. II. Solid Waste Disposal," *Wat. Pollut. Control,* pp. 309-328 (1982).

48. Reinhard, M., N.L. Goodman and J.F. Barker. "Occurence and Distribution of Organic Chemicals in Two Landfill Leachate Plumes," *Environ. Sci. Technol.* 18:953-961 (1984).

49. Chian, E.S.K. and F.B. DeWalle. "Evaluation of Leachate Treatment, Volume 1, Characterization of Leachate," EPA 600/2-77-186a (1977).

50. Norwood, D.L., L.C. Michael, S.D. Cooper, T.W. Pack, M.E. Montgomery, E.D. Pellizzari. "An Application of the 'Master Analytical Scheme' to Influent and Effluent Wastewaters," EPA 600/D-88-248 (1988).

51. Kempton, J.H. "Hetergeneous Electron Transfer Kinetics and the Measurement of Eh," M.S. Thesis, University of Colorado Boulder, CO (1987).

52. Sillen, L.G. and A.E, Martell Eds., *Stability Constants of Metal-Ion Complexes*, Special Publication No. 25 (The Chemical Society, 1971) p.149.

53. Washburn, E.W., Ed., *International Critical Tables* (McGraw Hill Book Co., Inc., New York, 1929) 6:73-110.

54. Bender, P. "Measurement of Dipole Moments," *J. Chem. Ed.* 4:179-181 (1946).

55. Apgar, M.A. and D. Langmuir. "Groundwater Pollution Potential of a Landfill Above the Watertable," *Groundwater* 9:76-96 (1971).

56. Baedecker, M.J. and M.A. Apgar. "Hydrogeochemical Studies at a Landfill in Delaware," in *Groundwater Contamination,* J.D. Bredehoeft, Panel Chairman (National Research Council Studies in Geophysics, 1984), pp. 127-138.

10

TRACE METALS IN AGRICULTURAL SOILS

Terry J. Logan and Samuel J. Traina
Department of Agronomy
The Ohio State University
Columbus, OH 43210

1.   INTRODUCTION

Trace metals have long been of interest to agriculture since the discovery of their essential roles in plant and animal nutrition. The behavior of trace nutrients, unlike the macronutrients such as nitrogen or phosphate, is delicately balanced in plant and animal metabolism, being both essential and toxic over a narrow concentration range. Thus, research on the uptake of trace nutrients has been concerned with deficiencies as well as excesses in soil-plant-animal systems. There has been the further concern for the environmental impact of industrial sources of metals, including non-nutrients and radioactive elements in the last twenty years. Concerns have been for the impact of metals on crop growth and economic yields (phytotoxicity), as well as for transmission of toxic trace metals through the animal and human food-chains, and for the more insidious effects of toxic metals on the ecosystem.

Trace metals enter the food-chain primarily by uptake from soil, and the behavior of trace metals in soil is the major factor regulating terrestrial metal contamination. In addition, interaction of soil with surface and ground water is a fundamental process controlling the solubility of trace metals in water and their impact on drinking water

0-87371-277-3/93/$0.00 + $.50

quality. It is unfortunate that only recently have aquatic scientists recognized the important role of surface soil in regulating metal solubility in water. Likewise, the preoccupation of agricultural scientists with crop and animal production has, until recently, minimized the impact that soil science might have had on our understanding of the important role of soil as both source and sink for potential pollutants such as the trace metals.

The literature on trace metal behavior in soil is extensive and it is not our intent to survey it here. Rather, we hope, in this paper, to review those properties of surface soil that are unique in their control of trace metal behavior, and to discuss some of the more recent approaches being used by soil chemists to study trace metal chemistry in soil. In particular, we discuss the successes and limitations of applying methodologies developed for pure chemical systems on the relatively homogeneous geological environments that are the domains of the groundwater chemists.

2.   TRACE METALS OF INTEREST IN AGRICULTURE

We have classified the trace metals of interest in agriculture according to their role as: plant and animal nutrients; plant and animal toxins; and human food-chain contaminants (Table 1). For each metal we also give reported ranges of concentrations in soil and surficial materials (Table 2) and the dominant chemical species (Table 3).

Metals are defined as those elements which readily give up electrons, i.e., they have low electronegativities. Sposito [1] defines a trace element as one that occurs at concentrations < 100 mg/kg in solids. Adriano [2] states that "It is a general consensus that an element is considered 'trace' in natural materials (i.e., lithosphere) when present at levels of less than 0.1%". Using Adriano's [2] compiled data on trace element concentrations in soils and surficial materials (Table 2), we see that Ba, Mn, and Sr would be excluded by Sposito's more restrictive definition. Adriano also includes as trace elements F, a nonmetal and, inexplicably, Ti which occurs at concentrations well above 1,000 mg/kg (0.1%).

TABLE 1. *Classification of Trace Metals as Plant and Animal Nutrients or Toxins.* Based in part on Adriano [2].
Toxicity considers the likelihood of uptake of metal.

Metal	Essential Plant	Beneficial Animal	Toxicity to Plant	Toxicity to Animal
Ag	No	No	No	Yes
As	No	Yes	Yes	Yes
B	Yes	No	Yes	
Ba	No	Possible	Low	Low
Be	No	No	Yes	Yes
Bi	No	No	Yes	Yes
Cd	No	No	Yes	Yes
Co	Yes	Yes	Low	Low
Cr	No	Yes	Yes	Yes (Cr^{6+})
Cu	Yes	Yes	Yes	Yes
Hg	No	No	No	Yes
Mn	Yes	Yes	Yes	Low
Mo	Yes	Yes	Yes	Yes
Ni	Possible	Yes	Yes	Yes
Pb	No	No	Yes	Yes
Sb	No	No	?	Yes
Se	Yes	Yes	Yes	Yes
Sn	No	Yes	?	Yes
V	Yes	Yes	Yes	Yes
W	No	No	?	?
Zn	Yes	Yes	Yes	Yes

We have considered As and Se in this paper although their electronegativities are high enough to exclude them as metals. However, their chemical species and behavior in soil are such that they are potential groundwater contaminants, and Se, in particular, has been shown to be a major water quality problem in the Central Valley of California [7].

TABLE 2. Median (M) or Average (A) Contents Reported for Elements in Soils and Other Surficial Materials.
From Adriano [2].

Element	Bowen [3] Median	Shacklette & Boerngen [4] Average	Vinogradov [5] Average	Rose et al. [6] Average	Rose et al. [6] Median
Ag	0.05	--	--	--	
As	6	7.2	5		7.5
B	20	33	10		29
Ba	500	580	--		300
Be	0.3	0.92	6	0.5 - 4	
Bi	0.2	--	--	--	
Cd	0.35	--	--	--	
Co	8	9.1	8		10
Cr	70	54	200		6.3
Cs	4	--	--	--	
Cu	30	25	20		15
F	200	430	200		300
Hg	0.06	0.09	--		0.056
Mn	1,000	550	850		320
Mo	1.2	0.97	2	2.5	
Ni	50	19	40		17
Pb	35	19	--		17
Sb	1	0.66	--	2	
Se	0.4	0.39	0.001		0.3
Sn	4	1.3	--	10	
Sr	250	240	300		67
Ti	5,000	2,900	4,600	--	
Tl	0.2	--	--	--	
V	90	80	100		57
W	1.5	--	--	--	
Zn	90	60	50		36

TABLE 3.   *Important Environmental Chemical Species of the Trace Metals.*

* Does not account for ion-pairs or complex-ion species.
\+ Considers degree of bioavailability.

Metal	Dominant Chemical Species*		Most Toxic Species+
	Soil	Water	
Ag	Ag^+	Ag^+	Ag^+
As	AsO_4^{3-}	AsO_4^{3-}; AsO_3^{3-}	AsO_4^{3-}
B	$B(OH)_3$	$B(OH)_3$	$B(OH)_3$
Ba	Ba^{2+}	Ba^{2+}	Ba^{2+}
Be	Be^{2+}; $Be_xO_y^{2x-2y}$	Be^{2+}	Be^{2+}
Bi	Bi^{3+}?	Bi^{3+}?	?
Cd	Cd^{2+}	Cd^{2+}	Cd^{2+}
Co	Co^{2+}	Co^{2+}	Co^{2+}
Cr	Cr^{3+}	Cr^{3+}; Cr^{6+}	Cr^{6+}
Cu	Cu^{2+}	Cu^{2+}-fulvate	Cu^{2+}
Hg	Hg^{2+}; CH_3Hg	$Hg(OH)_2^{\circ}$; $HgCl_2^{\circ}$	CH_3Hg
Mn	Mn^{4+}; Mn^{2+}	Mn^{2+}	Mn^{2+}
Mo	MoO_4^{2-}	MoO_4^{2-}	MoO_4^{2-}
Ni	Ni^{2+}	Ni^{2+}	Ni^{2+}
Pb	Pb^{2+}	$Pb(OH)^+$	Pb^{2+}
Sb	$Sb^{III}O_x$?	$Sb(OH)_6^-$?	?
Se	$HSeO_3^-$; SeO_4^{2-}	SeO_4^{2-}	SeO_4^{2-}
Sn	$Sn(OH)_6^{2-}$?	$Sn(OH)_6^{2-}$?	?
V	$V^{IV}O_x$?	?	?
W	WO_4^{2-}?	WO_4^{2-}?	?
Zn	Zn^{2+}	Zn^{2+}	Zn^{2+}

2.1   Trace Metal Toxicity

All elements are toxic when absorbed in excess of the organism's assimilative capacity. This is particularly true of the trace elements. The trace metals most often found to be phytotoxic are B, Cu, Mn, Ni and Zn. Toxicity to livestock has been observed for many of the trace metals, quite often associated with direct ingestion of metal-contaminated soil as well as metal taken up by the forage [8].

These include: As, Be, Cd, Co, Cr, Cu, Mo, Ni, Pb, Se, and V. Because of the minimal exposure to humans by direct ingestion and the impact of food processing on elimination of metals from vegetable and animal foods, the number of trace metals with potential to contaminate man is limited. While the list will vary with site-specific contamination problems, the most ubiquitous metals toxic to humans are Cd, Pb and Hg. To this list could be added Cr^{6+} for specific cases of Cr contamination of drinking water. Lead remains the most serious of the human toxic metals and the most insidious. Major exposure pathways are drinking water from Pb-contaminated pipes, inhalation of Pb-contaminated dust and Pb-based paint particles, and direct ingestion of Pb-contaminated dust, soil, and paint by children known to suffer from "pica" (a disorder in which the child ingests inordinate quantities of such substances), and from more normal hand-to-mouth activities by young children. Exposure to children is of particular concern because of widespread Pb contamination of urban areas by leaded gasoline and the correlation of high blood Pb with childhood learning disorders [5].

2.2   Toxicity and Trace Metal Speciation

There is a growing body of evidence that dissolved metal speciation is important in uptake and toxicity by plants and animals. An increasingly accepted paradigm is that the free metal species is the most bioactive and the most toxic [8]. Ion-pairs (e.g., $CdCl^+$), complex ions (e.g., Cu-fulvate), polymers (e.g., $Al_x(OH)_y^{(3x-y)+}$), and microparticulates (e.g., colloidal iron oxide or humic substances) reduce the free ion form of the metal and its potential for uptake and toxicity. Chaney [8] also points out the importance of other metals in regulating uptake and toxicity of specific metals by competing for metal-binding sites on biomolecules. He emphasizes the importance of the metal matrix in determining bioavailability. Metals bound in sewage sludge or manure are much less bioavailable than equivalent doses of metal salts [8,10].

The significance of free metal association with bioavailability is twofold: 1) in correctly assessing environmental exposure to metals, and 2) in devising intervention or remediation strategies for trace metal toxicity. In its proposed comprehensive regulations on sludge disposal [11], the U.S.EPA has used a risk assessment of exposure to sludge-borne trace metals (specifically As, Cd, Cr, Cu, Pb, Hg, Mo, Ni, Se and Zn) that makes no distinction between sludge-bound metals and metal salts. As a result, data on metal uptake by crops used in a number of the environmental exposure pathways included pot studies with metal salts instead of the more valid field studies with sludge [12]. This was done because the Agency would not accept the existence of a no-observed-effect plant uptake response from the sludge studies with metals such as Cu and Pb which have low bioavailability in the sludge matrix. Instead, studies with metal salts were used because they gave a positive uptake response.

The complexation of metals by dissolved ligands and the resultant decrease in plant uptake can be used to reduce the potential for phytotoxicity or food-chain contamination. Pavan (Dr. Marcus Pavan, IAPAR, Parana, Brazil, personal communication) has been studying the possibility of using gypsum ($CaSO_4 \cdot 2H_2O$) as an alternative to limestone as a means of reducing Al phytotoxicity. This is based on his earlier work [13] which showed that Al toxicity in coffee was better correlated with Al^{3+} activity than with Al_T nutrient solution concentration, and that $AlSO_4$ was a major complexed species.

2.3   Exposure Pathways for Soil-Derived Metals

Exposure to humans from chemicals originating in, or applied to, soil is shown schematically in Figure 1. For trace metals, the most important pathways are soil-plant-human, and soil-human (particularly so for young children and the extreme case of the pica child). Volatilization is a minor pathway, even for potentially volatile metals such as As, Hg and Se. Exposure through the soil-plant-animal-human and soil-animal-human pathways is also limited by the mechanisms of metal exclusion by both plants and animals, and the tendency for metals to accumulate in organs such as liver and kidney which are not readily consumed. Exposure through surface drinking water is limited by metal retention on sediments and the efficient removal of sediment in water treatment. Groundwater metal contamination from soil sources also

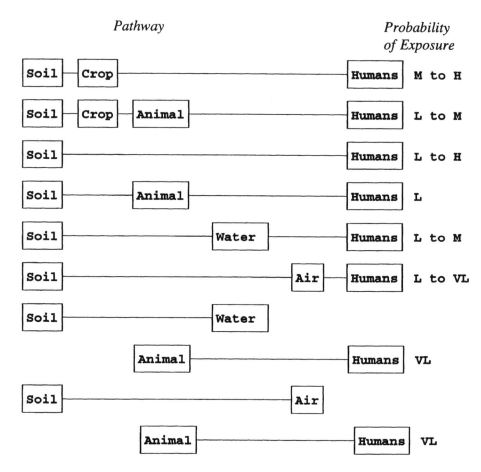

FIGURE 1. *Pathways for Human Exposure to Trace Metals from Soil.*
VL - very low; L - low; M - medium; H - High.

presents a low probability of human metal exposure because of the excellent attenuation of metals by soils, even in instances of high metal loadings [10]. Exceptions to this have been found for rapid-infiltration wastewater systems where attenuation was probably limited by rates of metal removal versus percolation rates [10], and below sludge disposal sites where metal accumulations were found to depths up to 3.5 meters [15].

3.   TRACE METAL SOURCES IN AGRICULTURE

Agricultural land is both a source of, and a sink for, chemicals. Anthropogenic sources are responsible for almost all cases of high trace metal concentrations in soil, with the exception of localized areas where soils reflect high-metal parent materials, such as the high-Cd shales of the California coastal region [16].   Other sources of metal in agricultural soils are the metal-containing pesticides, such as $CaAsO_4$ and $PbAsO_4$, Bordeaux mixture (a combination of $CuSO_4$ and $CaCO_3$), Paris Green (copper aceto-arsenite), $HgCl_2$, $ZnSO_4$, Maneb (1-17% Zn), Mancozeb (16% Mn, 2% Zn), methyl and phenyl mercuric salts, phenyl mercuric acetate, Zineb and Ziram (1-18% Zn), $CaAsO_3$ and Na_2AsO_3; micronutrient fertilizers; and phosphate fertilizers (Cd and Zn) [2]. Many of the pesticides have been applied to orchard crops and with repeated applications over many years have resulted in very high soil metal concentrations. Metal-containing pesticides have declined in use over the last two decades as a result of stronger environmental laws, but residual levels will remain high for years until erosion or mixing through plowing decrease surface concentrations. Many developing countries still use metal pesticides for specialty crops such as coffee.

Agricultural soils are far more likely to be metal sinks than sources. This can occur by atmospheric deposition from metal smelters, automobile emissions (Pb), land disposal of metal-containing wastes, and conversion of minelands or metal waste deposits to farmland. Loading from these various sources may be diffuse and low (1-5 times background) in the case of atmospheric depositions (except for areas within a few to several kilometers of smelters, depending on the height and emission rate from the stack; for a single large source, see Hutchinson and Whitby [17]), or more localized and concentrated in the case of disposal sites [18] or former minespoil materials [19].

In most cases of trace metal contamination of agricultural soils, the impact on crops and on drainage water quality is mitigated by the normal management used by the farmer. Practices such as periodic liming and maintenance of high surface organic matter levels, together with the generally better quality of agricultural soils - deeper, medium to fine textured, relatively high cation exchange capacities - tend, in most cases, to favor trace metal attenuation in the soil.

4.   AGRICULTURAL SURFACE SOIL AND ITS IMPACT ON TRACE METAL ATTENUATION

Sustained high crop production requires soils with certain essential chemical characteristics. These soils are sought by farmers in selecting land for crop production and are managed so as to optimize those characteristics. These same characteristics tend to attenuate trace metals by mineral adsorption, organic matter complexation, oxidation, cation exchange, and precipitation. Dissolved soil organic matter is also important in solution metal speciation. Table 4 summarizes surface soil characteristics important in trace metal attenuation and contrasts them with expected values in subsoil or groundwater substrata.

TABLE 4. Characteristics of Agricultural Surface Soil and Subsoil/Groundwater that Affect Trace Metal Attenuation.

Characteristic	*Surface Soil*	*Subsoil/Groundwater Media*
pH	Narrow (4-8)	Wide (3-9)
Soil Organic C	High (0.5-5 %)	Low (0-0.5 %)
Soluble Organic C	High (10-500 mg/L)	Low (<10 mg/L)
Cation Exchange Capacity	High (5-100 $cmol_c$/kg)	Low (<10 $cmol_c$/kg)
Mineralogy	Complex, more weathered	More homogeneous, less weathered
Soil Solution	Complex	More homogeneous
Redox	Oxic (E_h > 500 mV), except under transient conditions	Oxic in sandy media, anoxic if DOC is present
Microbial Activity	High, dominated by aerobes, heterotrophs	Medium (20-90 % of surface), dominated by heterotrophs, but chemoautotrophs can also occur [63]

Perhaps the most important determinant of trace metal solubility is pH. With the exception of the oxyanion metals Mo and Se, metal solubility increases with decreasing soil pH. As pH falls, most metal-containing minerals become less stable, and mineral and organic matter surface functional groups protonate with concurrent metal desorption. These processes are significant at pH's below 5.5. This is below the metal adsorption edge for most minerals [20-23], and below the pK_a for most soil organic matter functional groups [24]. Agricultural management by liming to maintain soil pHs in the range of 5.0-7.0 provides an important control on metal solubility.

Modern farmers have increasingly come to appreciate the importance of soil organic matter in the productive capacity of their soils. The shift from inversion tillage to conservation tillage - including the extreme case of no-till - has resulted in increases in the organic matter content of agricultural surface soils. This, together with managed high pH, increases the potential for trace metal attenuation. In fact, farmers on organic soils have to maintain lower soil pH (usually <5.5) and use trace nutrient fertilizers to overcome the strong binding of metals. The low metal bioavailability found for sewage sludges is partly attributable to metal complexation by sludge organic matter [25].

The role of soluble organic matter on metal attenuation in soils is not clear. First, there are very little data on DOC concentrations in soil and soil drainage. In our long-term study of tile drainage chemistry under plowed and no-till farming systems, we have monitored DOC concentrations in tile drainage of 0 to 2 mmole C L^{-1}. Studies of metal speciation in soil solution indicate that complexes with DOC account for a significant fraction of M_T for metal ions such as Fe^{3+} and Al^{3+}, and for Cu^{2+} [26-28]. For metals such as Cd^{2+}, DOC complexes are of minor importance. The positive impact of DOC-metal complexation is on reduction of metal phytotoxicity, while a potential negative effect is on increased metal mobility. The enhanced mobility of soil Fe by fulvic acids in spodosols is well known, but these complexes tend to precipitate in subsoil as a result of restricted drainage. The potential for small amounts of DOC-complexed trace metal to migrate to groundwater is not known but, for very insoluble metals such as Pb, any enhanced solubility could have a marked effect on metal leaching.

5.   CURRENT APPROACHES TO TRACE METAL CHEMISTRY IN SOILS

The conceptual tools applied to studies of trace metals in soil chemistry are the same as those used in other branches of geochemistry. The three principal processes which must be evaluated are: 1) solution speciation, 2) trace metal sorption, and 3) solid phase equilibria.

5.1   Solution Speciation

The chemical composition of the soil solution can be modeled with two chemically distinct approaches. These are herein referred to as the Specific Interaction Model (SIM) and the Ion Association Model (IAM) [29]. The SIM describes the composition of an aqueous electrolyte through the use of the total molalities of the stoichiometric components of neutral solutes. This approach makes use of mean ionic activities and mean ionic activity coefficients. The activity coefficients are typically calculated with the empirically based Pitzer equations [30,31] which generally reflect pairwise interactions between ions of opposite charge sign. Whereas the SIM provides a strictly rigorous thermodynamic description of electrolyte solution chemistry, its utility to soil chemistry is limited by its failure to account for the formation of individual molecular species. Yet an extensive body of previous research clearly indicates the primacy of specific dissolved molecular species in such diverse processes as metal uptake by organisms, adsorption and ion exchange reactions, and solute transport.

Numerous studies have elucidated the effects of ion speciation on metal uptake by plants. Greenhouse experiments (Figure 2) have shown the concentration of Cd in the leaves of Swiss chard (*Beta vulgaris* L. var. Cicla) to be strongly correlated (through a parametric fit) to the activity of Cd^{2+} in the soil saturation extract, and thus presumably in the soil solution, at the time of harvest [32]. No other measure of Cd, including total dissolved Cd, was found to be closely related to Cd uptake. The phytotoxicity of Al to plants grown in solution-culture experiments has been shown to be related to the specific molecular species of dissolved Al present in the solution, with the greatest toxicity attributed to the Al^{3+} species [33].

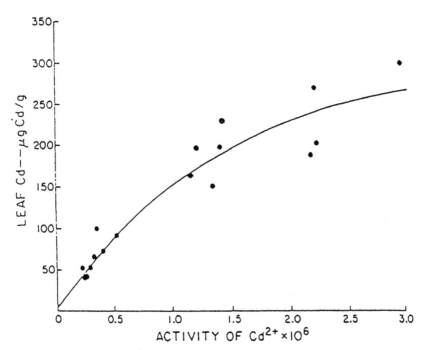

FIGURE 2. *Leaf Cd Concentration of Swiss Chard Plotted as a Function of the Activity of Cd²⁺ in the Soil Extract [32].*

The impact of specific molecular species on sorption processes is demonstrated in Figure 3. In this study, Cu sorption by a montmorillonitic soil, in a background electrolyte of $NaClO_4$, exhibited an S-shaped isotherm, indicated that, at low total Cu (Cu_T) concentrations dissolved, Cu was more stable than sorbed Cu [34]. This was attributed to the complexation of Cu^{2+} by dissolved organic ligands present in the soil solution, which inhibited Cu sorption. As the solution concentration of Cu_T increased, the Cu-complexation capacity of the dissolved organic ligand was exceeded to allow the sorption of either Cu^{2+} or the sorption/precipitation of the $CuL_x^{(2-x)+}$ complex, where L is the organic ligand. Washing of the soil materials with dilute $NaClO_4$ solutions, prior to Cu sorption, resulted in removal of dissolved organic matter as evidenced by the brown-colored extracts and increases

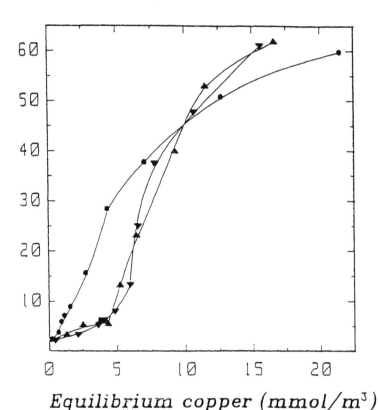

$$Equilibrium\ copper\ (mmol/m^3)$$

FIGURE 3.   *Sorption of Copper by Altamont Soil (▲), Altamont Soil*
               Separate Washed (■), and by Altamont Soil Separate to
               which 0.484 g of CONTECH Fulvic Acid per kg of
               Suspension was Added (▼) [34].

in measurable dissolved carbon. The subsequent sorption of Cu on
these "washed" soil samples produced L-shaped isotherms. That the
original S-shaped isotherms were due to the formation of dissolved
$CuL_x^{(2-x)+}$ species was confirmed by the addition of a commercial fulvic
acid to the "washed" soil samples which reproduced the original
S-shape. Similar results have also been reported for the effects of
organic ligands on Cd sorption by soils [35].

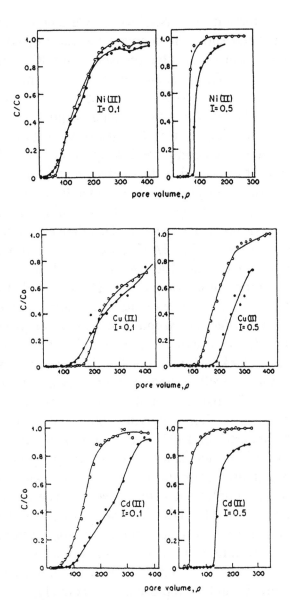

FIGURE 4. *Relative Concentrations, c/c₀ of Ni(II), Cu(II) or Cd(II) in Effluent after p Pore Volumes of either NaCl (○) or NaClO₄ (●) Solutions Passed through Columns of Hanford Sandy Loam Soil (Typic Xerorthent).* c = μmol/ml effluent; c₀ = 10 μmol/ml, where M = Ni(II), Cu(II), and Cd(II) [36].

The formation of Ni-chloro, Cu-chloro, and Cd-chloro complexes has been shown to increase the rates of Ni(II), Cu(II) and Cd(II) leaching, relative to rates measured in ClO_4^- solutions, in soil columns (Figure 4). This was attributed to preferential sorption of the M^{2+} species over MCl^+ or MCl_2° [36]. This hypothesis was supported by Hirsch *et al.* [37] who reported greater sorption of Cd by specimen montmorillonite in ClO_4^- than in Cl^- electrolyte solutions.

It is apparent that complete descriptions of metal bioavailability, the partitioning of dissolved metals between solid and solution phases in soils, and the transport of trace metals in soil solutions, all necessitate an explicit recognition and accounting of the formation of specific dissolved molecular species. This dictates the use of the extra-thermodynamic IAM. The IAM describes the composition of an electrolyte solution through the use of individual molecular species, comprised of cation-anion complexes. The thermodynamic properties of these complexes are treated with the extra-thermodynamic convention of *single ion activities* and *single ion activity coefficients*. This is typically done with the use of accepted, empirical equations, such as the extended Debye-Huckel equation or the Davies equation, and thus this approach is not as well vested in thermodynamics as is the SIM. Nevertheless, the inclusion of molecular species makes the IAM the method of choice for those studying ionic, chemical process in soil solutions. At present, the IAM is most commonly applied through the use of "equilibrium speciation" computer programs which solve simultaneous equations for multiple equilibria by explicitly allowing for the formation of complex molecular species. The computer programs GEOCHEM [38] and MINTEQ [39] are perhaps the most commonly used versions of the IAM in studies of soil systems. These two programs are very similar in their overall approach. Both describe the formation of a large number of solution complexes of the form $C_x A_y^{(y-x)+}$, where C and A represent, respectively, a metal or ligand species and x and y are stoichiometric reaction coefficients. Whereas GEOCHEM contains a larger number of organic ligands in its thermodynamic database, MINTEQ allows for the consideration of a larger number of possible solids. Both programs allow for the addition of new complex species and solids to their respective databases.

TABLE 5. ***Aqueous Half-lives for Selected Ion Association and Dissociation Reactions at 25°C [1].***

Reaction	Half-life (s)
$CO_2 + H_2O = H_2CO_3$	10
$Fe^{3+} + H_2O = FeOH^{2+} + H^+$	10^{-7}
$FeOH^{2+} + H^+ = Fe^{3+} + H_2O$	10^{-6}
$Mn^{2+} + SO_4^{2-} = MnSO_4^{\circ}$	10^{-5}
$MnSO_4^{\circ} = Mn^{2+} + SO_4^{2-}$	10^{-9}
$Ni^{2+} + C_2O_4^{2-} = NiC_2O_4^{\circ}$	1
$NiC_2O_4^{\circ} = Ni^{2+} + C_2O_4^{2-}$	10^{-1}
$Al^{3+} + F^- = AlF^{2+}$	10^3
$CO_2 + OH^- = HCO^{3-}$	10
$HCO_3^- = CO_2 + OH^-$	10^3
$Ca(H_2O)_6^{2+} + H_2O' = Ca^{2+}(H_2O)_5H_2O' + H_2O$	10^{-8}

GEOCHEM and MINTEQ both make provisions for ion exchange and surface complexation reactions.

Even though GEOCHEM and MINTEQ use different algorithms to calculate the aqueous concentration of a given molecular species, the main limitation in the use of these and other simultaneous chemical equilibrium computer models lies not in the specific numerical routines utilized in the computer codes, but rather in the quality of the complex-ion stability constants present in the thermodynamic databases, and in the common assumption of local chemical equilibrium. The former caveat requires a careful review and evaluation of published critical stability constants for aqueous complex-ions, prior to the use of a given computer model. In contrast, the latter problem requires a more fundamental evaluation of the nature of the specific chemical reactions under consideration. Sposito [1] has tabulated experimentally measured reaction orders and half-lives for a number of "complexation" reactions which are likely to occur in soil solutions (Table 5). The range in half-lives (10^{-9} to 10^3 s) would most certainly increase if one considers the complexation of a cationic species by a large macromolecule, such as fulvic acid, which might undergo time-dependent conformational changes in solution with changes in the associated cation content.

Thus, depending upon the rates of complexation, sorption/desorption, precipitation/dissolution, root exudation, and various biochemical processes, it is possible that true equilibrium between dissolved complex-ion species may not exist in soil solutions.

5.2   Trace Metal Sorption

The sorption of alkaline earth and alkali metals by soils has been traditionally studied with equilibrium thermodynamic cation exchange models. For a homovalent, binary cation exchange reaction this is described as:

$$v\ AX_u(s)\ +\ u\ B^{v+}(aq)\ =\ u\ BX_v(s)\ +\ v\ A^{u+}(aq) \tag{1}$$

where A^{u+} and B^{v+} are the exchangeable cations, X^{1-} represents one mole of charge in the exchanger phase and v and u are stoichiometric reaction coefficients. A complete description of the chemical reaction in Equation (1) is achieved by a calculation of the exchange equilibrium constant, which is of the form:

$$K_{ex}\ =\ \frac{(BX_v)^u\ (A^{u+})^v}{(AX_u)^v\ (B^{v+})^u} \tag{2}$$

where the parentheses denote activities. The determination of (A^{u+}) and (B^{v+}) can be made with direct analytical methods such as specific ion electrodes, or alternatively the IAM can be used to calculate individual ion activities from the total dissolved concentrations of the metals of interest. Unfortunately, it is not possible to directly measure the activities of the ions in the exchanger phase and one must thus use model expressions to calculate the exchanger-phase activities from total composition data. Ion exchange composition data is often expressed with the Vanselow selectivity coefficient (K_v), defined as:

$$K_v = \frac{[BX_v]^u\ (A^{u+})^v}{[AX_u]^v\ (B^{v+})^u} \qquad (3)$$

where the brackets denote concentration units, for the composition of the exchanger phase. The K_v can be formally linked to K_{ex} by the rational activity coefficients for the exchanger phase, f_A and f_B

$$K_{ex} = K_v\ \frac{f_B^u}{f_A^v} \qquad (4)$$

where

$$f_A = \frac{(AX_u)}{N_A}, \qquad f_B = \frac{(BX_v)}{N_B} \qquad (5)$$

and N_A and N_B are the respective mole fractions of AX_u and BX_v in the exchanger phase. Conversion to the equivalent fraction concentration scale for the exchanger-phase and incorporation of the Gibbs-Duhem equation provides a direct link between the rational activity coefficients and the exchanger phase composition:

$$v\ \ln f_a = E_B\ \ln K_v - \int_0^{E_B} \ln K_v\ dE_B \qquad (6)$$

$$u\ \ln f_B = -(1 - E_B)\ \ln K_v + \int_{E_B}^1 \ln K_v\ dE_B \qquad (7)$$

where E_B is the equivalent fraction of cation B in the exchanger phase [29].

The conceptual approach presented in Equations (1)-(7) has been reasonably successful in describing the binary exchange chemistry of alkali and alkaline-earth metals in soil materials, and these equations have recently been extended to ternary and quaternary ion exchange systems [40, 41]. Additionally, the formal relationships presented in Equations (1)-(7) have been used to model cation exchange reactions for many trace metals on specimen 1:1 and 2:1 layered aluminosilicate clays. However, little success has been realized in applying this approach to trace metal sorption in whole soils or even in isolated mineral fractions from soils. This is, in part, due to the requirement to obtain compositional data for the exchanger-phase over the range of 0 < E_B < 1. Clearly a value of $E_B = 1$ is unrealistic for any substance that would be considered a **trace** metal. An additional difficulty lies in the nature of trace metal sorption by soil surfaces. For the most part, alkali and alkaline earth metals form outer sphere surface complexes with the surface functional groups found in soils. This typically occurs on permanently-charged, cation-exchange sites present on 1:1 and 2:1 layered aluminosilicate clays. In contrast, many of the trace metals are thought to form inner sphere surface complexes with hydrous metal oxides and with functional groups present on soil organic matter. Both of these types of surfaces sites exhibit strong pH-dependent sorption.

The sorption of trace metals in soils is commonly modeled with the various forms of the Constant Capacitance Model and the Triple Layer Model [42]. Both of these models make specific postulates about the nature of the **surface complex** which forms between the sorbate and the sorbent. Whereas each model considers a unique set of possible adsorbed species and chemical reactions, in both models the inorganic surface hydroxyl is the primal functional group responsible for trace metal sorption [42]. Perhaps the most extensive use of Constant Capacitance Model in soil systems has been made by Goldberg and co-workers [43-47]. These investigators have successfully described the sorption of P, B, Se and As by soils and soil materials. Sposito *et al.* [48] used the previously published data to calibrate the Constant Capacitance Model for Se on a single soil. The intrinsic surface complexation constants (K_{int}) were then used to predict Se sorption envelopes for additional soils. As evidenced by the data in Figure 5, excellent agreement was obtained between the predicted and the observed selenite sorption data. The Triple Layer Model has been used to describe the sorption of Na^+, K^+, Cl^-, and NO_3^- by a Brazilian oxisol [49].

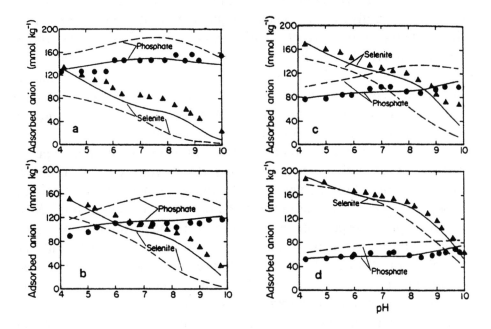

FIGURE 5. *Actual Competitive Adsorption of Orthophosphate (•) and Selenite (▲) on Goethite and Competitive Adsorption Predicted with the Constant Capacitance Model (lines).* Dashed lines are based on values for K_{int} from single ligand experiments. Solid lines are based on values for K_{int} from mixed ligand experiments [45].

The application of surface complexation models in pure mineral systems requires a determination of the adsorption site density. This is an arduous experimental task, and the values reported in the literature are somewhat dependent upon the experimental methods used. However, reasonable estimates of surface site density have been reported for a number of synthetic and specimen mineral systems. More difficult is the task of determining surface site density in soils and soil materials. Many experimental methods used to measure surface site densities on pure mineral systems are not applicable to soil systems, because of the heterogeneous nature of the sorbent phase. In practice, this difficulty is overcome by assuming that the maximum adsorption capacity for a given trace metal is equivalent to the surface density of the **surface sites which will react with that trace metal**. Unfortunately, this makes the surface site density an operational parameter which will vary with each trace metal. Furthermore, this definition requires that the investigator(s) measure the maximum adsorption capacity under conditions which will preclude precipitation, so as not to overestimate the surface site density.

Soils invariably contain a large number of different types of minerals, each with a different set of surface hydroxyls which, in turn, have a unique set of pK_a values. Additionally, other types of functional groups are present on the mineral and organic surfaces in soils. Since macroscopic adsorption experiments measure the partitioning of a given trace metal between the solid and solution phases, it is generally not possible to distinguish the contributions of each type of surface hydroxyl, nor is it possible to distinguish between adsorption of a trace metal to inorganic hydroxyls versus carboxyls or other surface functional groups. Thus, application of surface complexation models in soils requires the adoption of the concept of an "**average surface hydroxyl group**" [45]. In reality this approach is also taken in studies on pure mineral systems since a given mineral particle will have two or more different types of hydroxyls present on its surfaces.

The ubiquitous presence of mixed electrolyte solutions in natural soils necessitates a consideration of competitive and/or synergistic adsorption phenomena. Goldberg [50] and Goldberg and Traina [45] examined the utility of the Constant Capacitance Model to describe competitive adsorption reactions in pure mineral systems. A **qualitative** description of the competitive adsorption of Se and P on goethite was obtained by using values for the intrinsic surface complexation constants, K_{int}, obtained for each oxyanion in **noncompetitive** adsorption experiments [50]. Unfortunately, a **quantitative** description of the competitive adsorption reactions was only obtained when the surface complexation constants for each oxyanion were fit simultaneously to experimental data obtained in competitive adsorption experiments [45].

Since soil systems are typically far more complex than those considered by Goldberg [50] and Goldberg and Traina [45], it seems likely that accurate treatment of trace metal adsorption data, in solutions that realistically depict typical soil solutions, requires the collection and evaluation of true competitive sorption data. As indicated above, the formation of dissolved, complex ion-pairs can greatly influence the sorption of a trace metal in soil systems. Neal *et al.* [51] observed a significant increase in Se sorption by arid zone soils when $CaCl_2$ was present in the background electrolyte solution (Figure 5). However, due to the high levels of dissolved Ca present in these soils, precipitation of $CaSeO_3 \cdot 2H_2O(c)$ could not be excluded.

5.3   Solid Phase Equilibria

The formation of discrete solid-phase forms of trace metals in soils has generally been evaluated through the use of equilibrium solubility relationships [52]. This approach is based on the assumption that equilibrium has been reached between the solution phase and a thermodynamically stable or meta-stable solid phase which controls the aqueous activity of the trace metal of interest. Implicit in this approach are the assumptions that: 1) the aqueous activity of a given trace metal can be measured directly or be calculated from total solution composition data, 2) sufficient contact time has occurred between the solid and solution phases for the system to reach a steady-state condition, and 3) the solid phases of interest are present in the soil in their **thermodynamic standard states**. Commonly, solubility experiments resort to a determination of the total or near total, elemental composition of the aqueous phase, followed by a model calculation of solution speciation with the IAM. The extent to which this meets the first assumption hinges upon the completeness of the soil solution composition data. In particular, determination of the aqueous concentrations of all relevant **complexing ligands** is of critical importance since they can strongly influence the aqueous activity of the trace metal of interest. The ubiquitous presence of complex mixtures of organic ligands in many soil solutions makes such a comprehensive analysis extremely difficult. Alternatively, attempts have been made to employ specific ion electrodes to obtain more direct determinations of aqueous metal activities [53]. Unfortunately this method is sometimes faced with numerous analytical interferences [54,55]. Assumption 2) is often not addressed in solid phase equilibria studies in soils. Rather, experimental reaction times often seem to have been based on practical considerations instead of the attainment of some steady state condition. Pierzynski *et al.* [56] examined the influence of reaction time on the solubility of orthophosphate in eleven different soils and determined that attainment of steady-state conditions was highly sample dependent. Assumption 3) is perhaps the most difficult to address in soil systems. Because of the low concentrations typically present, it is difficult to directly ascertain the specific solid phase form of a given trace metal in soil. In a study of Cu equilibria in Cu-contaminated soil, McBride and Bouldin [57] found the soil solution to be undersaturated with respect to malachite and tenorite. Yet infrared analysis of green coatings present on dolomitic limestone collected from the soil indicated the presence of malachite. The apparent disagreement between solution and solid-phase chemistry may have been due to "surface enhanced" precipitation of the Cu-hydroxycarbonate on the surfaces of dolomite particles. Clearly such a solid would not be in its thermodynamic standard state. Another condition which would lead to the occurrence of nonstandard-state solids is the formation of mixed-solids or solid solutions. As is discussed below, ample direct experimental evidence exists to indicate the presence of mixed-solids in soils. If a mixed-solid forms an ideal solid solution, then the solid-phase activity of an individual metal in the mixed-solid can be calculated from its mole fraction in the solid-phase [29]. However, this requires a knowledge of the chemical composition of the mixed solid and such information is not provided by equilibrium solubility measurements made on systems, such as soils, which contain multiple solid-phases.

Obviously the application of equilibrium solubility measurements to characterize trace-metal bearing solid phases in soils, without **direct** observations of the solid-phase, can only lead to questionable conclusions. Fortunately, several emerging technologies have been recently applied to directly study the solid-phase forms of metals in soils and sediments. Using solid state ^{113}Cd NMR, Bank *et al.* [58]

identified four different types of solid-phase Cd: (1) Cd in a Cd-Ca-carbonate, (2) an oxo-Cd species present in a "clay-like" matrix, (3) a Cd-hydroxycarbonate, and (4) a Cd-hydroxycarbonate species in a "clay-like" matrix. Hinedi and Chang [59] studied solid phase P-chemistry in two municipal sewage sludges and in sludge-amended soil. Using ^{31}P magic angle spinning nuclear magnetic resonance, the presence of Ca-phosphate, carbonated apatite, pyrophosphate and Al-phosphate solids was established [59]. Equilibrium solubility measurements made on samples of sludge-amended soil, suggested that the solution-phase chemistry was controlled by carbonated-apatite [60]. Essington and Mattigod [61] combined equilibrium solubility measurements with density separation techniques, followed by energy dispersive x-ray analysis (EDAX) analytical electron microscopy to examine the solid-phase forms of La, Ce, and Nd present in municipal sewage sludge and sewage sludge-amended soils. Indirect, solubility measurements indicated that the stable solid phases were cerianite [CeO_2], lanthanite [$La_2(CO_3)_3 \cdot 8H_2O$], Nd-lanthanite [$Nd_2(CO_3)_3 \cdot 8H_2O$], La-Nd-sulfate [$(La,Nd)_2(SO_4)_3 \cdot 8H_2O$], yet direct EDAX measurements indicated more complex elemental associations. Lanthanum, Ce and Nd were found associated with Th and P, Ce with Si, Ba, and S, and La and Nd with S. Whereas these associations were often suggestive of the presence of standard-state solid phases, many of the EDAX spectra indicated the presence of additional ions suggesting that some mixed solids had formed (Figure 6).

Pierzynski *et al.* [56] combined equilibrium solubility measurements with EDAX-analytical scanning-transmission electron microscopy (STEM) measurements of particle size separates and density separates [62], to study solid-phase P chemistry in eleven soils. The equilibrium solubility measurements suggested the presence of hydroxy-Al-phosphates in acidic soils and the occurrence of basic Ca-phosphates in slightly acid to neutral soils [56], yet the EDAX-STEM analysis indicated that Al was present in virtually all P-rich particles from the samples with higher pH values. Additionally, in most of the soils studied, the presence of mixed P-bearing solids containing two or more dominant cations was the rule rather than the exception [56]. Clearly, careful consideration must be given to the

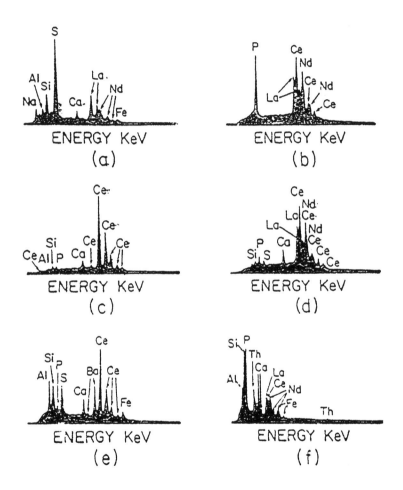

FIGURE 6.   ***Typical SEM/EDAX Spectra of Sewage Sludge and***
Sludge-amended Soil Lanthanide-bearing Particles.
Particles consistent with spectra (a) occurred only in the sludge
only. Those consistent with spectra (f) occurred in sludge-
amended soil only. Spectra (b) through (e) illustrate the
chemistry of particles observed in both sewage sludge and the
sludge-amended soil [61].

solid-phase compositional data which these direct methods provide [56,58,59,61], and great caution must be exercised when using equilibrium solubility measurements alone.

6. MAJOR KNOWLEDGE GAPS IN TRACE METAL SOIL CHEMISTRY

Study of soil chemistry is greatly hampered by the complexity of soil chemical constituents. Soils are not merely admixtures of mineral and organic solid phases, but complex products of the chemical interaction of these materials. In addition, the diverse and dynamic biological activity of soils makes application of chemical research methods virtually impossible. This is particularly true for the trace metals. Soil chemistry encompasses both solid and liquid phases and their chemical interactions, and there are major limitations to the study of all three.

6.1 Solid Phase Soil Trace Metal Chemistry

Given their low abundance in rocks and soil, trace metals likely do not occur as discrete mineral phases but as mineral coprecipitates and complexed with soil organic matter. Sposito and Mattigod [38] include as coprecipitates surface complexed metals, inclusions and solid solutions. Corey [62] has suggested that most minerals, especially soil minerals, are solid solutions of trace constituents. With the exception of solid solutions formed during precipitation of primary minerals, the composition of trace metals in various soil forms is likely to be highly variable. This makes it impossible to determine thermodynamic solubility constants for coprecipitated trace metals. Likewise, Corey [62] points out the additional problem that there is a tendency for incongruent dissolution of solid solutions. This inability to explicitly determine thermodynamic data for trace metal-containing soil minerals has led Lindsay [52] to use the concept of a general, but undefined, solid phase for trace metals. "Soil-Zn" or "soil-Cu" is presented in the form of a solubility constant based on soil solution data for a wide range of soils. The chemical impropriety of this approach is that the assumption that a distinct trace metal solid phase exists in soil is unsupported by direct evidence and neglects the possibility of adsorption or organic matter control of solubility. The current and more scientifically rigorous approach of combining direct spectroscopic evidence with solubility data [56,60] in proposing the existence of specific solid phases is likely to be inappropriate for trace metals because of their low soil abundance, which makes most direct observation methods insensitive. In the absence of direct analytical methods, empirical approaches, such as Lindsay's, will continue to be used to describe soil trace metal solubility. We do not discourage this approach, only the assertion that the solubility constant derived with this approach represents a unique solid phase rather than a mixed constant representing the complex interactions of adsorption, complexation and coprecipitation reactions in soil.

6.2   Solution Phase Soil Trace Metal Chemistry

Our knowledge of soil solution trace metal chemistry has improved markedly in the last decade. These improvements are attributable to the wide array of more sensitive analytical methods available today for both trace metals and soil solution ligands. We can also better quantify many of the metal complexes in soil solution. There are major limitations in analytical detection limits, however, that make it difficult to conduct experiments at realistic trace metal soil solution concentrations. Concentrations of metals such as Cd, Cu and Pb in soil solution are often $< 10^{-8}$ M, levels below the detection of commonly used tools like atomic absorption spectrophotometry. As a result, studies have been conducted at unrealistically high levels [63], or have used synthetic chelates to perturb equilibrium and increase M_T in solution [64]. The problem of the latter approach is that equilibrium may be so perturbed as to compromise interpretation of the results.

Speciation of soluble soil organic ligands is a difficult task. Not only is chromatographic separation and quantification of dissolved organic matter challenging, but the organic ligand composition of the soil solution is likely to be highly dynamic in surface soils as a result of biological activity. This aspect has received little attention. Even if the ligand composition could be determined, there are few thermodynamic formation constants for these ligands with many of the metals of interest. In addition, Chaney [8] has shown that plants and microbes can produce low levels of metal-binding organic compounds with very high selectivity for specific metals. These are poorly characterized as to their concentrations in the rhizosphere soil solution, and the mechanisms that trigger their synthesis are also unknown. These deficiencies in our knowledge led Sposito and Mattigod [38] to simulate soil solution organic ligand chemistry as a mixture of simple organic acids (e.g., citrate, acetate) in their model GEOCHEM. Not only does this inaccurately reflect the true solution composition, but we have found in a recent use of GEOCHEM that unique combinations of pH and DOC concentration caused unrealistic complexation of Al to be predicted [56].

There is clearly a need for better characterization of the major metal-binding ligands in soil solution, description of their temporal distribution under various soil conditions, and development of thermodynamic formation constants for trace metal-ligand complexes.

6.3   Trace Metal Surface Processes in Soils

Previous work on trace metal sorption by soil has centered on studies of metal selectivity by soil organic matter [27,28], ion exchange on clay minerals [65, 66], and pH-dependency of metal adsorption (adsorption edge) [20]. Recent work has examined the tendency of oxyanion metals (As, Cr and Se) to form inner or outer sphere complexes [51,67], the effect of metal adsorption on surface charge [68], and the use of the surface complexation model to describe the pH-dependency of oxyanion metal adsorption [44-47,51,67]. Many, if not most, of these studies have been conducted with pure minerals or humic materials, and many have employed initial metal concentrations that are orders of magnitude higher than normally found in soil solution.

Only a few studies have used direct spectroscopic observation to study trace metal sorption. Noteworthy are the studies of Cu^{2+}

sorption to oxides, organic matter and clay minerals using electron spin spectroscopy (ESR) [69]. Rapid developments in solid phase spectroscopy offer the potential for more direct observation of surface processes, but the low levels of trace metals realistic for soil systems will limit use of these techniques with trace metals.

7.   CONCLUSIONS

Trace metals are important in agriculture as nutrients and potential plant, animal and human toxins. While surface agricultural soils may occasionally serve as sources of high levels of trace metals, they are more likely to be sinks for anthropogenic metals. The physical, chemical and biological properties of soil that are considered optimum for crop production are also important in attenuation of trace metals, and agricultural soils pose little problem for surface or groundwater metal contamination. The most serious exception to this generalization is the contamination of irrigation return flow by Se (in the form of soluble SeO_4^{2-}) mobilized from evaporite minerals in the Central Valley of California.

Trace metal studies in soil have included identification and quantification of important soil solution complexes, and research on trace metal sorption to oxides, clay minerals and soil organic matter. Little work has been done on characterization of trace metal-containing metals in soil, and very few of the trace metal studies in general have been conducted with soil or soil separates. Likewise, many trace metal studies have had to employ unrealistically high metal concentrations to overcome poor sensitivities of existing analytical methods.

In the future, trace metal soil chemistry must concentrate on the use of more advanced spectroscopic techniques, available to pure chemistry for a decade or more, but increasingly being employed by soil chemists and geochemists. Chromatography should also be employed to better characterize soil solution organic ligands, and thermodynamic formation constants with the environmentally important metals need to be developed for these ligands as they are identified. This work should focus on the plant and microbe synthesized metal-binding compounds which have been shown [8] to have very high metal selectivity. Soil chemists need to develop, in the absence of direct analytical methods, a more geochemically acceptable approach to representation of equilibrium solubility data for trace metals in soils. As a general principle, we do not support the use of stability isotherms to represent equilibrium solution data unless there is supporting spectroscopic evidence for existence of soil trace metal solid phases.

Soil chemists have made important contributions to our knowledge of trace metal chemistry in the larger environment, and, on the other hand, have benefitted from advancements in analytical chemistry and geochemical modeling with more pristine systems. Our challenge is to stretch the limits of these tools as we study the more complex soil system, a particularly difficult task for the trace metals.

REFERENCES

1.   Sposito, G. *The Chemistry of Soils.* (New York: Oxford University Press, 1989).

2.   Adriano, D.C. *Trace Elements in the Terrestrial Environment.* (New York: Springer-Verlag, 1986).

3.   Bowen, H.J.M. *Environmental Chemistry of the Elements.* (New York, NY: Academic Press, 1979) 333 pp.

4.   Shacklette, H.T. and J.G. Boerngen. *Element Concentrations in Soils and Other Surficial Materials of the Conterminous United States.* U.S.G.S. Prof. Paper 1270. (Washington, DC: U.S. Govt. Printing Office, 1984).

5.   Vinogradov, A.P. *The Geochemistry of Rare and Dispersed Chemical Elements in Soils.* (New York, NY: Consultants Bureau Inc., 1959) 209 pp.

6. Rose, A.W., H.E. Hawkes and J.S. Webb. *Geochemistry in Mineral Exploration.* (London, UK: Academic Press, 1979) 658 pp.

7. Bureau, R.G. "Environmental Chemistry of Selenium." *Calif. Agric.* 39:16-18 (1985).

8. Chaney, R.L. "Metal Speciation, Agriculture and Food-Chains". In J.R. Kramer and H.E. Allen Eds. *Metal Speciation. Theory, Analysis and Application.* (Chelsea, MI: Lewis Publishers, Inc., 1988).

9. National Research Council. *Lead in The Human Environment.* (Washington, DC: National Academy of Sciences, 1980).

10. Logan, T.J. and R.L. Chaney. "Metals". In A.L. Page, T.L. Gleason, J.E. Smith, I.K. Iskander and L.E. Sommers Eds. *Utilization of Municipal Wastewater and Sludge on Land.* (Riverside, CA: University of California-Riverside, 1983).

11. U.S. EPA. "Standards for the Disposal of Sewage Sludge. U.S. EPA Proposed Rule 40 CFR Parts 257 and 503." *Federal Register* 54:5746-5902 (1989).

12. Peer Review Committee. "Peer Review of Standards for the Disposal of Sewage Sludge. U.S. EPA Proposed Rule 40 CFR Parts 257 and 503." W-170 Regional Research Committee, CSRS-USDA (1989).

13. Pavan, M.A. and F.T. Bigham. "Toxicity of Aluminum to Coffee Seedlings Grown in Nutrient Solution." *Soil Sci. Soc. Am. J.* 46:993-997 (1978).

14. Chang, A.C. and A.L. Page. 1983. "Fate of Trace Metals During Land Treatment of Municipal Wastewater." In A.L Page, T.L. Gleason, J.E. Smith, I.K. Iskander and L.E. Sommers Eds. *Utilization of Municipal Wastewater and Sludge on Land.* (Riverside, CA: Univiversity of California-Riverside, 1983).

15. Lund, L.J., A.L. Page and C.O. Nelson. "Movement of Heavy Metals Below Sewage Disposal Ponds." *J. Environ. Qual.* 5:330-334 (1976).

16. Lund, L.J., E.E. Betty, A.L. Page and R.A. Elliott. "Occurrence of naturally high cadmium levels in soils and its accumulation by vegetation." *J. Environ. Qual.* 10:551-556 (1981).

17. Hutchinson, T.C. and L.C. Whitby. "Heavy-metal pollution in the Sudbury mining and smelting region of Canada. 1. Soil and vegetation contamination by nickel, copper and other metals." *Environ. Conserv.* 1:123-132 (1974).

18. van Drill, W. and J.P.J. Nijssen. "Development of dredged material disposal sites: Implications for soil, flora and food quality." In W. Salomons and U. Forstner Eds. *Chemistry and Biology of Solid Waste: Dredged Material and Mine Tailings.* (New York, NY: Springer-Verlag, 1988).

19. Asami, T. "Soil pollution by metals from mining and smelting activities". In W. Salomons and U. Forstner Eds. *Chemistry and Biology of Solid Waste: Dredged Material and Mine Tailings.* (New York, NY: Springer-Verlag, 1988).

20. Kinniburgh, D.G., M.L. Jackson and J.K. Syers. "Adsorption of alkaline earth, transition and heavy metal cations by hydrous oxide gels of iron and aluminum." *Soil Sci. Soc. Am. J.* 40:796-799 (1976).

21.   Leckie, J.O. and R.O. James. "Control mechanisms for trace metals in natural waters." In A.J. Rubin Ed. *Aqueous-Environmental Chemistry of Metals*. (Ann Arbor, MI: Ann Arbor Science Pub., 1974).

22.   Bourg, A.C.M. "Metals in aquatic and terrestrial systems: sorption, speciation, and mobilization." In W. Salomons and U. Forstner Eds. *Chemistry and Biology of Solid Waste: Dredged Material and Mine Tailings*. (New York, NY: Springer-Verlag, 1988).

23.   Benjamin, M.M. and J.O. Leckie. "Multiple site adsorption of Cd, Cu, Zn and Pb on amorphous iron oxyhydroxide." *J. Colloid Interface Sci.* 79:209-211.

24.   Perdue, E.M. "Acidic functional groups of humic substances." In G.R. Aiken, D.M. McKnight, R.L. Wershaw and P. MacCarthy Eds. *Humic Substances in Soil, Sediment and Water*. (New York, NY: John Wiley and Sons, 1985).

25.   Logan, T.J. "Sludge metal bioavailability." In *Proc. Battelle Int. Symp. Solid/Liquid Separations*. (Columbus, OH: Battelle Press, 1990). (In press).

26.   Camerlynck, R. and L. Kiekens. "Speciation of heavy metals in soils based on charge separation." *Plant and Soil*. 68:331-339 (1982).

27.   Schnitzer, M. and S.I.M. Skinner. "Organo-metallic interactions in soils: 5. Stability constants of Cu^{2+}, Fe^{2+}, and Zn^{2+} fulvic acid complexes." *Soil Sci.* 102:361-365 (1966).

28.   Schnitzer, M. and S.I.M. Skinner. "Organo-metallic interactions in soils: 7. Stability constants of Pb^{2+}, Ni^{2+}, Mn^{2+}, Co^{2+}, and Mg^{2+} fulvic acid complexes." *Soil Sci.* 103:247-252 (1967).

29.   Sposito, G. *The Thermodynamics of Soil Solutions*. (New York: Oxford University Press, 1981).

30.   Pitzer, K.S. "Electrolyte theory-improvements since Debye and Huckel." *Acc Chem. Res.* 10:371-377 (1977).

31.   Pitzer, K.S. "Theory: Ion interaction approach". In R.M. Pytkowicz Ed. *Activity Coefficients in Electrolyte Solutions*. Vol I. CRC Press, Boca Raton, FL. (1979).

32.   Bingham, F.T., J.E. Strong, and G. Sposito. "Influence of chloride and salinity on cadmium uptake by swiss chard." *Soil Sci.* 135:160-165 (1983).

33.   Parker, D.R., T.B. Kinraide, and L.W. Zelazny. "Aluminum speciation and phytotoxicity in dilute hydroxy-aluminum solutions. *Soil Sci. Soc. Am. J.* 52:438-444 (1988).

34.   LeClaire, J.P. "Adsorption of copper and cadmium onto soils: Influence of organic matter." Ph.D. Thesis. Univ. of California, Riverside. (1985).

35.   Neal, R.H. and G. Sposito. "Effects of soluble organic matter and sewage sludge amendments on cadmium sorption by soils at low cadmium concentrations." *Soil Sci.* 142:164-172 (1986).

36.   Doner, H.E. "Chloride as a factor in mobilities of Ni(II), Cu(II), and Cd(II) in soil." *Soil Sci. Soc. Am. J.* 42:882-885 (1978).

37.   Hirsch, D., S. Nir, and A. Banin. "Prediction of cadmium complexation in solution and adsorption to montmorillonite." *Soil Sci. Soc. Am. J.* 53:716-721 (1989).

38.   Sposito, G. and S.V. Mattigod. *GEOCHEM: A Computer Program for the Calculation of Chemical Equilibria in Soil Solutions and other Natural Water Systems.* (Riverside, CA: University of California-Riverside, 1980) 92 pp.

39.   Felmy, A.R., D.C. Girvin and E.A. Jenne. *MINTEQ - A Computer Program for Calculating Aqueous Geochemical Equilibria.* Final Project Report. U.S. EPA Contract No. 68-03-3089. (Washington, DC: U.S. EPA, 1983).

40.   Sposito, G. and P. Fletcher. "Sodium-calcium-magnesium exchange reactions on a montmorillonitic soil: III. Calcium-magnesium selectivity coefficient." *Soil Sci. Soc. Am. J.* 49:1160-1163 (1985).

41.   Thellier, C. and G. Sposito. "Quaternary cation exchange on silver hill illite." *Soil Sci. Soc. Am. J.* 52:979-985 (1988).

42.   Sposito, G. *The Surface Chemistry of Soils.* (New York, NY: Oxford University Press, 1984).

43.   Goldberg, S. and G. Sposito. "A chemical model of phosphate adsorption by soils. II. Noncalcareous soils." *Soil Sci. Soc. Am. J.* 48:779-783 (1984).

44.   Goldberg, S. and R.A. Glaubig. "Boron adsorption on California soils." *Soil Sci. Soc. Am. J.* 50:1173-1176 (1986).

45.   Goldberg, S. and S.J. Traina. "Chemical modeling of anion competition on oxides using the constant capacitance model-mixed-ligand approach." *Soil Sci. Soc. Am. J.* 51:929-932 (1987).

46.   Goldberg, S. and R.A. Glaubig. "Anion sorption on a calcareous, montmorillonic soil-selenium." *Soil Sci. Soc. Am. J.* 52:954-958 (1988).

47. Goldberg, S. and R.A. Glaubig. "Anion sorption on a calcareous, montmorillonic soil-arsenic." *Soil Sci. Soc. Am. J.* 52:1297-1300 (1988).

48. Sposito, G., J.C.M. de Wit and R.H. Neal. "Selinite adsorption on alluvial soils: III. Chemical modeling." *Soil Sci. Soc. Am. J.* 52:947-950 (1988).

49. Charlet, L. and G. Sposito. "Monovalent ion adsorption by and oxisol." *Soil Sci. Soc. Am. J.* 51:1155-1160 (1987).

50. Goldberg, S. "Chemical modeling of anion competition on goethite using the constant capacitance model." *Soil Sci. Soc. Am. J.* 49:851-856 (1985).

51. Neal, R.H., G. Sposito, K.M. Holtzclaw and S.J. Traina. "Selenite adsorption on alluvial soils: II. Solution composition effects." *Soil Sci. Soc. Am. J.* 51:1165-1169 (1989).

52. Lindsay, W.L. *Chemical Equilibria in Soils.* (New York, NY: John Wiley and Sons, 1979). 449 pp.

53. Street, J.J., W.L. Lindsay, and B.R. Sabey. "Solubility and plant uptake of cadmium in soils amended with cadmium and sewage sludge." *J. Environ. Qual.* 6:72-77 (1977).

54. Minnich, M.M. and M.B. McBride. "Copper activity in soil solution: I. Measurement by ion-selective electrode and Donnan dialysis." *Soil Sci. Soc. Am. J.* 51:568-572 (1987).

55. Sikora, F.J. and F.J. Stevenson. "Interference of humic substances on chloride analysis using the chloride ion-selective electrode." *Soil Sci. Soc. Am. J.* 51:924-929 (1987).

56. Pierzynski, G.M., T.J. Logan, S.J. Traina and J.W. Bigham. "Phosphorus chemistry and mineralogy in excessively fertilized soils: equilibrium solubility." *Soil Sci. Soc. Am. J.* 54:1589-1595 (1990).

57. McBride, M.B. and D.R. Bouldin. "Long-term reactions of copper(II) in a contaminated calcareous soil." *Soil Sci. Soc. Am. J.* 48:56-59 (1984).

58. Bank, S., J.F. Bank, P.S. Marchetti, and P.D. Ellis. "Solid state cadmium-113 nuclear magnetic resonance study of cadmium speciation in environmentally contaminated sediments." *J. Environ. Qual.* 18:25-30 (1989).

59. Hinedi, Z.R. and A.C. Chang. "Phosphorous-31 magic angle spinning nuclear magnetic resonance of wastewater sludges and sludge-amended soil." *Soil Sci. Soc. Am. J.* 53:1053-1056 (1989).

60. Hinedi, Z.R. and A.C. Chang. "Solubility and phosphorus-31 magic angle spinning nuclear magnetic resonance of phosphorus in sludge-amended soils." *Soil Sci. Soc. Am. J.* 53:1057-1061 (1989).

61. Essington, M.E. and S.V. Mattigod. "Lanthanide solid phase speciation." *Soil Sci. Soc. Am. J.* 49:1387-1393 (1985).

62. Corey, R.B. "Adsorption vs precipitation." In M.A. Anderson and A.J. Rubin, Eds. "Phosphorus Chemistry and Mineralogy in Excessively Fertilized Soils:Quantitative Analysis of Phosphorus-Rich Particles," *Soil Sci. Soc. Am. J.* 54:1567-1583 (1990).

63. Hendrickson, L.L. and R.B. Corey. "Effect of equilibrium metal concentrations on apparent selectivity coefficients of soil complexes." *Soil Sci.* 131:163-171.

64. Corey, R.B., R. Fujii and L.L. Hendrickson. "Bioavailability of heavy metals in soil-sludge systems." *Proc. Fourth Annual Madison Conf. Applied Res. and Practice on Munic. and Ind. Waste.* (Madison, WI: Univ. Wisconsin-Extension, 1981) pp. 449-465.

65. Garcia-Miragaya, J. and A.L. Page. "Influence of ionic strength, and inorganic complex formation in the sorption of trace amounts of Cd by montmorillonite." *Soil Sci. Soc. Am. J.* 40:658-663 (1976).

66. Garcia-Miragaya, J. and A.L. Page. "Influence of exchangeable cation on the sorption of trace amounts of cadmium by montmorillonite." *Soil Sci. Soc. Am. J.* 41:718-721.

67. Zachara, J.M., C.C. Ainsworth, C.E. Cowan and C.T. Resch. "Adsorption of chromate by subsurface soil horizons." *Soil Sci. Soc. Am. J.* 53:418-428 (1989).

68. Bleam, W. and M.B. McBride. "Cluster formation versus isolated-site adsorption. A study of Mn(II) and Mg(II) adsorption on boehmite and goethite. *J. Colloid and Interface Sci.* 103:124-132 (1984).

69. McBride M.B. "Surface chemistry of soil minerals." In J.B. Dixon and S.B. Weed Eds. *Minerals in Soil Environments.* 2nd Ed. (Madison, WI: Soil Sci. Soc. Am., 1989) pp. 35-88.

11

URANIUM TRANSPORT IN THE SUB-SURFACE ENVIRONMENT KOONGARRA - A CASE STUDY

T. David Waite and T.E. Payne
Australian Nuclear Science and
  Technology Organisation (ANSTO)
Menai, New South Wales 2234, Australia

1.   INTRODUCTION

The safe disposal of radioactive wastes continues to pose problems in many countries with burial at selected sub-surface sites the preferred option in most cases. In the United States, the U.S. Nuclear Regulatory Commission has proposed a set of rules (10 CFR 60) specifying criteria for the disposal of radioactive wastes. Any chosen site must also comply with standards set by the U.S. Environmental Protection Agency. In this regard, draft USEPA standards have been formulated and apply to integrated releases over a period of 10,000 years.

Birchard and Alexander [1] point out that assessment of proposed repository sites over these time-scales poses three critical questions:

•   Can the performance of a system be predicted for thousands of years or more without a record of equal or greater length?

0-87371-277-3/93/$0.00 + $.50
©1993 by Lewis Publishers

- How can the interactions of man-made elements such as technetium, plutonium and neptunium be predicted in the repository?

- How can laboratory experiments on complex processes be scaled from the laboratory to repository dimensions (in both space and time)?

A partial answer to these questions may be obtained by studying carefully selected geochemical analogs of a waste repository. The most direct analog to the waste repository is the natural fission reactor which occurred at Oklo in Gabon, West Africa about 1.8×10^9 years ago. However, since the phenomenon occurred in remote geologic time, direct information on transport over the time scales of interest (10^3-10^6 years) is no longer available. Potentially more useful analogs are uranium ore deposits such as those of the Alligator Rivers Uranium Province in the Northern Territory of Australia (Figure 1) which have the advantages that [2,3]:

- there are a number of accessible deposits which may be compared to establish general features of elemental redistribution within and downgradient of the deposit;

- a number of the deposits intersect the ground surface. The uranium series nuclides are therefore distributed between the main metamorphosed sequences and the overlying weathered profile. It is therefore possible, in principle, to compare the elemental transport within crystalline schists, gneisses and silicified carbonates, and the iron-bearing clays and quartz into which they weather;

- weathering initiates mobilization of uranium and provides a rational basis for the definition of zero time - an essential component in constructing a mathematical model of transport;

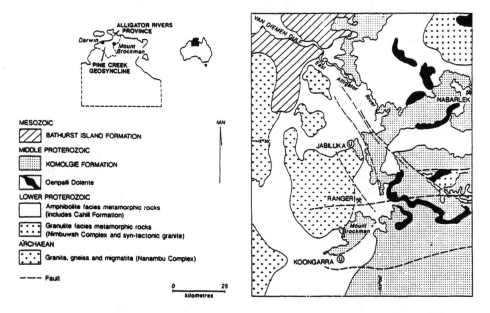

FIGURE 1. *Map of Regional Geology Showing the Location of the Koongarra Uranium Deposits (after [7]).*

- at Koongarra, the timescale of migration is around 1-3 X 10^6 years [4] - a relevant period for waste repository studies;

- groundwater intersects all deposits in this region and not only induces elemental redistribution within the ore zone but also leads to downgradient transport; and

- low levels of the transuranics ^{237}Np and ^{239}Pu and the fission products ^{129}I and ^{99}Tc, which are relevant to the calculation of the total dose commitment to the public, are present and may be incorporated into the analog.

In 1981, the Australian Nuclear Science and Technology Organisation (ANSTO) (then called the Australian Atomic Energy Commission) contracted with the United States Nuclear Regulatory Commission to undertake a systematic investigation of radionuclide migration around the Narbalek, Jabiluka, Koongarra and Ranger ore bodies in the Alligator Rivers Region. Support for this work was broadened under the auspices of the Organization for Economic Cooperation and Development (OECD) Nuclear Energy Agency in 1987 with funding for further investigations provided by ANSTO, the Japan Atomic Energy Research Institute, the Swedish Nuclear Power Inspectorate, the United Kingdom Department of the Environment, the United States Nuclear Regulatory Commission and the Power Reactor and Nuclear Fuel Development Corporation of Japan. In this continuing latter project, referred to as the Alligator Rivers Analog Project (ARAP), attention is focussed on the Koongarra ore body and associated downgradient uranium redistribution in the weathered zone [5]. The primary goal of the project is to assist in the long-term prediction of the rate of transport of radionuclides through the geosphere. Specific aims are to evaluate the significance of processes which may play a role in radionuclide transport and to establish a database which may be used to assist in the validation of radionuclide transport and geochemical modeling programs. The approach taken in this project involves the integration of mathematical and physical modeling with hydrological and geochemical field and laboratory investigations of the analog site.

2.   ORE MINERALOGY, MINERAL ALTERATION AND HYDROGEOLOGY AT KOONGARRA

2.1   Ore Mineralogy

The Koongarra uranium ore bodies lie 225 km east of Darwin and 25 km south of Jabiru in the Northern Territory of Australia within the Pine Creek Geosyncline. This geosyncline is a region of Lower Proterozoic sediments with interlayered tuff units resting on a granitic late Archaean basement - the Nanambu Complex (Figure 1) [6-9]. The Lower Proterozoic Cahill Formation flanking the Nanambu Complex consists of two members. The lower member is dominated by a thick basal dolomite. The uranium mineralization is associated with carbonaceous horizons within the overlying chloritised quartz-mica schist. The lower member passes transitionally upwards into the more psammitic upper member which is largely feldspathic schist and quartzite. A 150 Myr period of weathering and erosion followed metamorphism. The thick, essentially flat lying Middle Proterozoic Kombolgie Formation (primarily sandstone) was then deposited unconformably on the Archaean-Lower Proterozoic basement and metasediments. At Koongarra, subsequent reverse faulting has juxtaposed the lower Cahill formation schists and Kombolgie Formation sandstone. Indeed, the dominant structural feature at Koongarra is a reverse fault system that trends almost the entire SE side of the Mount Brockman outlier of the Kombolgie Formation.

There are two discrete ore bodies at Koongarra, separated by a 100 m wide barren zone (Figure 2). The main (No. 1) ore body (that of major interest here) has a strike length of 450 m and persists to 100 m depth. Within the primary body (unweathered zone), the ore consists of lenses containing coalescing veins of uraninite conformable within the steeply dipping host Cahill quartz-chlorite schists. Secondary uranium mineralization is present in the weathered schists, from below the surficial sand cover to the base of weathering at depths varying between 25 and 30 m. This mineralization has been derived from decomposition and leaching of the primary mineralized zone and forms a tongue like fan of ore grade material dispersed downslope for about 80 m to the SE (see Figure 3). Details of the discovery and evaluation of the ore bodies are given by Foy and Pedersen [10]. A large number of additional studies of these ore deposits have since been undertaken and reported [11-16].

The primary ore consists of uraninite veins and veinlets (1 to 10 mm thick) with groups of veinlets intimately intergrown with chlorite, which forms the matrix to the host breccias. Associated with the ore are minor volumes (to 5%) of sulfides which include galena and lesser chalcopyrite, bornite and pyrite. Chlorite, predominantly magnesium chlorite, is the principal gangue, and its intimate association with the uraninite indicates that the two minerals formed together. The No. 2 ore body is located entirely in the unweathered zone at depths greater than 50 meters, hence there is no secondary mineralization.

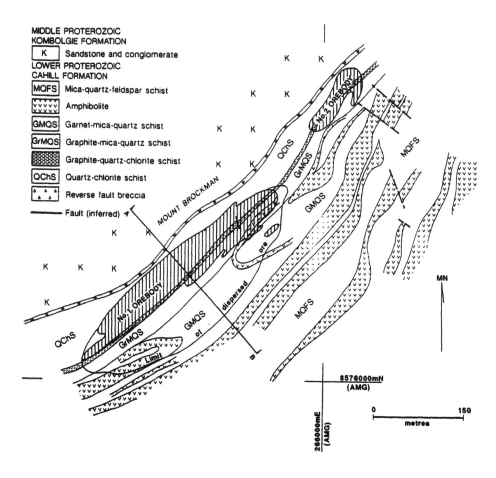

FIGURE 2. *Map of Local Geology Showing Location of the Koongarra No. 1 and 2 Ore Bodies (after [6]).* Because of surficial cover, the geological units and outline of the mineralization are projected to the surface from the base of weathering.

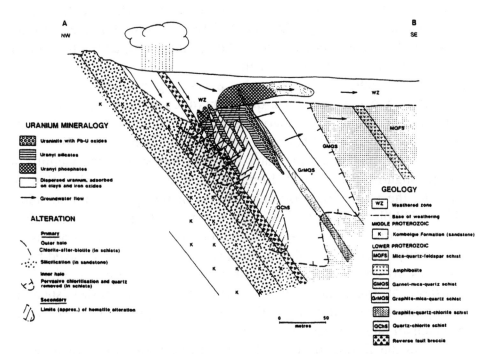

FIGURE 3. *Simplified Cross-section through the Koongarra No. 1 Ore bodies.* Geology, distribution of uranium minerals, hematite alteration and groundwater flow are shown (after [9]).

2.2   Mineral Alteration

Oxidation and alteration of uraninite within the primary ore zone have produced a variety of secondary minerals, particularly the uranyl silicates kasolite, sklodowskite and uranophane [9]. Uraninite veins, even veins over 1 cm wide, have been completely altered *in situ.* Some uraninite veins are concentrically sheathed outwards by uranium-lead oxides (vandendriesscheite, fourmarierite and then curite) and then sklodowskite.   Other veins are partially or totally replaced by intergrown uranyl silicates.

The secondary mineralization in the weathered schists above the No. 1 ore body is characterized by uranyl phosphates, particularly saleeite ($Mg(UO_2)_2(PO_4) \cdot 8H_2O$), metatorbernite ($Cu(UO_2)_2PO_4)_2 \cdot 8H_2O$) and renardite ($Pb(UO_2)_4(PO_4)_2(OH)_4 \cdot 7H_2O$). This phosphate-bearing zone constitutes weathered and leached primary ore, and is the former upward extension of the present primary ore lenses. The uranyl phosphates occur as veins, stringers, euhedral grains, intergrowths and rosettes amongst the clay-mica-quartz mixtures of the weathered schists. Further away from the primary ore zone, uranium is dispersed in the weathered schists and adsorbed onto clays and iron oxides [9].

The weathered zone clays and iron minerals in the vicinity of the Koongarra ore deposit form, at least in part, by alteration of chlorite, the major rock-forming mineral of the quartz-chlorite schist [17]. Dibble and Tiller [18] note that the authigenic, or secondary minerals, formed by water-rock interaction (i.e. the alteration products) are usually metastable with the attainment of equilibrium delayed by as much as 10^7 years. The relationship between mineral alteration and transport of transuranic elements (mediated by processes such as adsorption and desorption) is likely to be strongly influenced by the mechanism and kinetics of alteration.

In the Koongarra case, the extent of alteration is a function of depth. In the unweathered zone, chlorite is abundant but at lesser depths has been transformed to vermiculite through regularly interstratified chlorite/vermiculite. Finally, vermiculite is replaced by kaolinite and possibly smectite. Using average compositions of chlorite and vermiculite, the reaction may be expressed as follows [17]:

$$(Al_{1.11}Fe_{2.93}{}^{2+}Mg_{1.97})(Si_{2.87}Al_{1.13})O_{10}(OH)_8 + 0.24\ K^+ + 0.06\ Ti^{4+} \rightarrow$$
$$\text{CHLORITE}$$

$$K_{0.24}(Al_{0.34}Fe_{1.01}{}^{3+}Mg_{1.55}Ti_{0.06})(Si_{2.61}Al_{1.61})O_{10}(OH)_2 + 1.82\ FeO(OH).xH_2O$$
$$\text{VERMICULITE} \qquad\qquad\qquad\qquad \text{FERRIHYDRITE}$$

$$+ 0.42\ Mg^{2+} + 0.26\ Si^{4+} + 0.29\ Al^{3+} + (3 - 1.82\ x)\ H_2O$$

During the transformation, excess Fe most likely forms ferrihydrite around vermiculite. The K and Ti required to form vermiculite may be released from mica and anatase of the quartz-chlorite schist respectively. Excess Mg, Si and Al may be consumed in the formation of kaolinite and possibly smectite. Additional Fe, Mg, Si and Al are released during the decomposition of vermiculite and may be used again for the formation of ferrihydrite, kaolinite and smectite. The results of backscattered electron image (BEI) micrography and the similar crystal structures of chlorite and vermiculite suggest that cation diffusion may account for the formation of vermiculite from chlorite. Kaolinite is most likely formed by dissolution and reprecipitation [17].

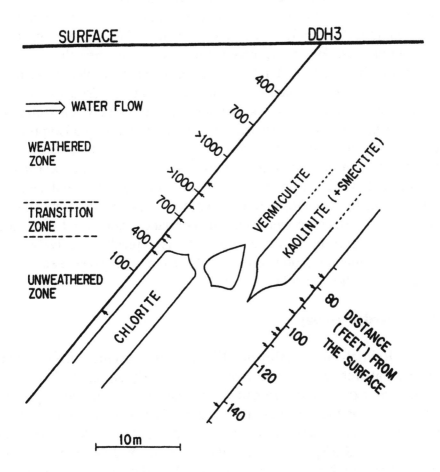

FIGURE 4.   *Comparison of Uranium Concentration and Chlorite, Vermiculite and Kaolinite - Smectite) Abundances along the Koongarra DDH3 Core.* The arrows indicate sampling locations for mineralogy determination by X-ray diffraction (after [17]).

Analysis of material from a core in the vicinity of the Koongarra deposit (Bore DDH3) indicates that the uranium concentrations are closely related to mineral abundances. The chlorite predominant zone corresponds to the zone of lower U concentrations, the vermiculite predominant zone to the zone of intermediate U concentrations, and the kaolinite predominant zone to the zone of highest U concentrations (Figure 4). Isobe and Murakami [17] suggest that the crystal chemistry of chlorite may account for these relationships. In any of the minerals, large cations like UO_2^{2+} cannot be accommodated in tetrahedral or octahedral sites. However, large cations can easily enter the interlayers of vermiculite hence, after the initiation of chlorite alteration, uranyl species are accommodated in the interlayers. Ferrihydrite, or other iron oxide phases, which occur as interstitial materials between vermiculite and chlorite slabs, may also scavenge uranyl species by adsorption. As for chlorite, the entry of large cations into the kaolinite structure is limited thus, in the kaolinite rich zone, uranium would be expected to be associated with iron oxides (ferrihydrite, goethite and hematite) formed via the transformation of chlorite and the decomposition of vermiculite. Smectite can accommodate large cations but the amount of smectite present in these substrates is typically relatively small. Similar "vermiculitization" processes have previously been examined by a range of authors [19-21].

2.3   Hydrogeology

Rainfall in the Alligator Rivers region occurs during a humid monsoonal wet season from November to early April. Relatively dry conditions exist for the remainder of the year. Rainfall intensity is variable during the wet season while June, July and August usually lack rain. A six year record of rainfall (ending with the 1976/77 water year) measured at Koongarra has an annual mean of 1798 mm with 95% falling between November and April. Over the same time period, the mean annual pan evaporation rate was 2580 mm [22]. Seasonal variations in water levels of up to 15 m reflect groundwater recharge in the wet season and discharge in the dry season.

Five hydrologic units compose the Koongarra site [23]. These are the Kombolgie Formation, unweathered Cahill Formation, weathered Cahill Formation, surficial sands and the Koongarra fault. Water-level and aquifer test data suggest the presence of two major aquifers. The "hanging wall" aquifer consists of (1) unweathered Cahill formation forming a partially confined fractured-rock aquifer, (2) weathered Cahill Formation acting as a laterally variable aquitard, and (3) the surficial sands which form an unconfined aquifer during the wet season. The vertical hydraulic conductivity of the weathered zone and the amount of leakage possible through it are not known at this stage. The second major aquifer is a "footwall" aquifer within the Kombolgie formation. The Koongarra fault forms the hydraulic boundary between the two aquifers. The hydraulic characteristics of the Kombolgie Formation and the Koongarra fault are poorly understood [23].

Water levels at Koongarra have been obtained from bores that are typically cased but not cemented through the surficial deposits and weathered zone and are open in the unweathered schist. As such, the water levels probably represent the hydraulic head for the unweathered Cahill Formation and the base of the weathered zone. The hydraulic gradient dips primarily to the south and varies slightly in magnitude with the season. Drawdown patterns, during aquifer tests by Norris [23], indicate a heterogeneous aquifer with pronounced north to east-northeast anisotropy. The combination of southerly hydraulic gradients and major transmissivity axes aligned more from northeast to southwest results in southerly to southwesterly groundwater flux vectors.

3.   SOLUTION AND MINERAL EQUILIBRIA OF KOONGARRA GROUNDWATERS

As a result of both the investigation of the Koongarra deposit as an analog for a waste repository and the interest by mining companies in defining the extent of the ore bodies, the Koongarra region is rich in boreholes of varying type and quality.

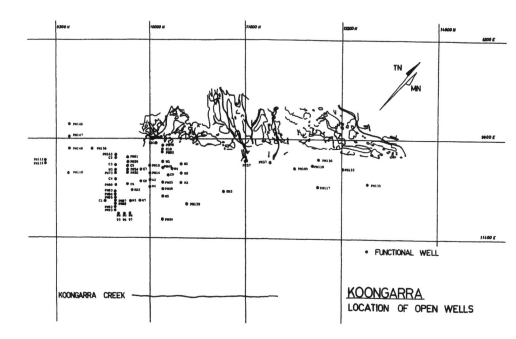

FIGURE 5.   *Location of Open Wells in the Vicinity of the Koongarra Uranium Ore Deposit Used in the Alligator Rivers Analogue Project.*

The location of currently open boreholes is shown in Figure 5. Many of these boreholes (the PH and KD series) were drilled and cased in the early 1970's while 22 new holes (the W, M and C series) were drilled and cased in late 1988. Most of the PH series of boreholes are cased through the weathered zone and extend into the unweathered zone. As of May 1988, water has been extracted from these bores using two collection systems. The first uses a variable cycle time bladder pump drawing from a 2 m region defined by upper and lower packers. The second system uses a submersible pump and a single bottom packer to block water from deeper aquifers in the system. Of the new set of wells, the W series were drilled to 25 m using a cable tool and back-filled to encase two PVC tubes slotted at 13-15 m and

23-25 m (i.e. approximately in the middle and base of the weathered zone respectively). Rotary percussion drilling was used for the deeper M and C series which were sunk to a depth of either 40 or 50 m. These recent wells have now settled and samples are being collected for analysis for a range of chemical parameters at frequent intervals.

3.1   General Observations on Groundwater Chemistry

Analytical results for waters pumped from a number of bores in the vicinity of the Koongarra ore deposit in May and November 1988 are summarized in Tables 1 and 2. Groundwaters taken from boreholes in the deposit (PH14, PH15, PH49, PH55, PH56, PH58) and directly south of the deposit (PH80, PH88) are characterized by high concentrations of magnesium resulting from the intense chlorite alteration that accompanies uranium migration within these deposits (as discussed above). Bicarbonate is the major anion for these waters. PH14, PH15 and PH49 pass through the primary and secondary uranium mineralization zones and, not surprisingly, waters from these bores exhibit high concentrations of dissolved uranium. PH56 is on the southwestern extremity of the deposit and U-238 concentrations are somewhat lower. PH55 and PH58 are cased through the region exhibiting the uranium "dispersion fan" and groundwaters drawn from these wells using packers at the bottom of the weathered zone and in the consolidated (unweathered) rock exhibit relatively low concentrations of uranium.

The decrease in groundwater uranium concentrations on moving away from the mineralized zone is best seen in the recent results from the W-series of bores (Table 3). The location of these boreholes is shown in Figure 5. W6 is in the Kombolgie sandstone near KD1, upgradient of the deposit. The remaining boreholes, except W3, provide samples from within and downgradient of the No. 1 ore body. The order of increasing distance from the fault, W1, W4, W2, W5 and W7 corresponds to the order of decreasing standing water level in these boreholes (Figure 6a). (Note that these bores do not lie on the same flow line).

TABLE 1. Results for Selected Parameters for Groundwaters obtained from Bores in the Vicinity of the Koongarra Uranium Deposit in May, 1988 (from Duerden and Payne [25]).

Bore No.	Cased Depth (m)	Sampled Depth (m)	pH[1]	Eh mV	Ca[2] mg/L	Mg mg/L	Na mg/L	K mg/L	HCO3 mg/L	Cl mg/L	SO4 mg/L	F mg/L	Fe mg/L	Mn mg/L	Si mg/L	Al mg/L	Pb[3] µg/L	V[3] µg/L	Zn[3] µg/L	Co[3] µg/L	PO4[4] µg/L	U[5] µg/L	234U/238U[6]
PH14	20	<26	6.42	+162	5.0	20.3	1.32	0.62	128	5.8	<1	0.10	1.11	0.42	10.9	<0.02	0.4	1.4	59	1.8	180	53.7	0.73±0.03
PH15	24	<30	6.66	+115	3.6	21.5	1.01	0.59	137	2.7	3	0.15	0.42	0.35	8.2	0.02	1.2	2.2	1042	0.31	560	140	1.00±0.02
PH49	16	<28	6.44	+155	3.1	20.1	1.32	0.53	130	4.9	<1	0.15	0.82	0.18	9.9	0.02	ND	ND	ND	ND	310	ND	ND
		28-30	6.74	+130	3.9	26.4	1.35	0.68	154	3.7	2	0.17	0.78	0.154	8.9	0.24	1.4	2.3	28	1.5	310	272	1.06±0.02
		44-46	ND[7]	ND	7.9	28.7	1.36	0.81	ND	2.8	3	0.27	0.46	0.032	9.8	0.41	ND	ND	ND	0.12	650	308	1.20±0.02
PH55	20	26-28	6.57	+110	4.4	17.9	3.16	1.05	127	3.7	3	0.12	5.6	0.14	14.0	<0.02	0.3	1.2	51	0.12	650	0.415	0.97±0.08
		40-42	6.53	+125	3.7	18.3	1.57	1.65	124	3.8	<1	0.20	0.70	0.097	12.1	0.12	ND	1.8	ND	ND	ND	1.03	1.08±0.05
PH56	22	26-28	6.50	+145	3.6	16.0	1.15	0.47	108	4.7	<1	0.18	0.87	0.040	10.1	0.04	0.4	1.8	6	0.14	385	12.3	0.82±0.02
		43-45	6.95	+120	4.4	16.7	1.17	0.53	113	3.1	<1	0.13	0.53	0.035	9.4	0.15	ND	ND	ND	ND	ND	11.7	1.10±0.02
PH58	23	25-27	6.57	+100	4.3	18.0	2.30	1.43	125	3.9	2	0.24	0.52	0.11	14.4	0.03	0.7	2.7	175	0.14	260	1.65	0.91±0.06
		38-40	6.60	+145	3.7	17.7	2.22	1.45	114	3.8	2	0.10	0.28	0.09	15.8	0.03	ND	ND	ND	ND	ND	4.06	1.41±0.10
PH61	33	43-45	7.03	+160	8.8	18.0	1.12	1.00	151	2.6	1	0.20	0.25	0.054	7.2	0.02	ND	ND	ND	ND	145	5.64	1.15±0.05
PH80	17	<20	6.31	+160	1.3	16.2	1.15	0.37	96	4.3	<1	0.09	0.35	0.23	10.9	0.03	1.0	2.3	<4	0.20	230	13.5	0.74±0.01
PH88	21	28-30	6.54	+127	1.4	14.1	1.52	0.56	88	3.6	1	0.12	0.32	0.020	14.5	<0.02	ND	ND	ND	ND	185	2.18	0.93±0.03
		38-40	6.62	+90	0.95	13.0	1.62	0.63	81	3.8	8	0.10	1.05	0.067	14.1	0.02	ND	ND	ND	ND	185	3.69	0.78±0.02
PH94	24	26-28	7.17	+250	1.4	5.7	1.99	1.00	38	2.2	8	0.09	0.035	0.008	12.1	<0.02	ND	ND	ND	ND	ND	0.175	1.38±0.08
		40-42	6.77	+125	3.8	6.1	3.53	2.00	56	3.8	2	1.2	0.165	0.023	10.4	0.02	ND	ND	ND	ND	185	0.584	1.01±0.02
PH96	16	<24	6.55	+120	2.1	7.6	2.18	1.37	56	3.5	2	0.15	0.41	0.055	13.6	0.02	ND	ND	ND	ND	210	0.258	1.06±0.04
KD1	70[6]	40-42	5.95	+185	3.1	3.6	1.09	0.37	69	ND	ND	ND	18.9	0.19	6.1	0.02	ND	ND	ND	ND	<5	4.73	1.12±0.01
		80-82	5.82	+235	1.9	2.4	0.99	0.18	43	2.6	<1	0.05	0.097	0.009	6.5	<0.02	ND	ND	ND	ND	30	3.67	1.11±0.04

[1] pH, Eh and HCO3 determined by field measurement
[2] All samples filtered through 1.0 µm membranes in the field and subsequently analyzed for cations by ICP-AES (unless indicated otherwise)
[3] Analyses for these cations by ICP/MS
[4] Dissolved acid-hydrolysable phosphate concentrations determined by standard ascorbic acid colorimetric procedure
[5] Uranium-238 concentrations determined by α-spectrometry (with close correspondence to concentrations determined by ICP/MS)
[6] KD1 casing slotted in 40-42 m range
[7] ND = not determined
[8] 234U/238U activity ratio

TABLE 2. Results for Selected Parameters for Groundwaters Obtained from Bores in the Vicinity of the Koongarra Uranium Deposit in November, 1988 (from Duerden [26] and Payne [27]).

Bore No.	Cased Depth (m)	Sampled Depth (m)	pH[1]	Eh mV	Ca[2] mg/L	Mg mg/L	Na mg/L	K mg/L	HCO₃ mg/L	Cl mg/L	SO₄ mg/L	F mg/L	Fe mg/L
PH14	20	<26	6.80	+70	4.7	25	1.2	0.85	157	9.1	0.32	0.24	3.8
PH15	24	<30	7.05	+120	2.7	24	1.2	0.83	149	6.0	0.04	0.37	0.53
PH49	16	<28	ND	ND	ND	ND	ND	ND	ND	ND	ND	ND	ND
		28-30	6.84	+130	5.2	25	1.2	0.75	159	8.5	0.29	0.34	1.1
		44-46	7.01	+280	8.5	24	1.3	0.91	180	2.0	0.11	0.38	0.45
PH55	20	26-28	6.55	+105	2.0	19	1.4	1.0	117	6.2	0.20	0.24	2.3
		40-42	6.60	+75	2.7	19	1.4	1.1	120	5.5	0.24	0.24	3.5
PH56	22	26-28	6.72	+120	3.5	18	2.3	0.64	116	7.3	0.24	0.23	0.93
		43-45	6.93	+135	6.8	19	1.3	0.85	120	5.7	0.18	0.29	0.71
PH58	23	25-27	6.70	+115	2.2	19	2.2	1.3	118	6.1	0.16	0.21	2.4
		38-40	6.59	+135	2.1	18	2.2	1.3	114	6.2	0.10	0.20	1.1
PH61	33	43-45	7.30	+80	11.0	21	1.5	1.1	156	5.1	0.22	0.36	0.59
PH80	17	<20	6.68	+130	0.90	17	1.1	0.39	99	6.6	0.43	0.17	1.3
PH88	21	28-30	6.92	+120	0.65	15	1.6	0.72	87	5.5	0.08	0.21	1.1
		38-40	6.82	+90	0.71	15	1.6	0.74	92	3.2	<0.02	0.22	1.2
PH94	24	26-28	7.06	+225	0.64	6.0	14	0.84	38	2.8	0.08	0.14	0.06
		40-42	6.94	+225	0.70	6.1	2.1	1.0	38	3.4	0.84	0.11	0.05
PH96	16	<24	7.14	+145	2.3	9.5	2.1	1.6	66	4.6	0.37	0.21	0.78
KD1	70[6]	40-42	6.07	+140	2.3	4.8	1.0	0.41	81	3.1	<0.02	0.12	22
		70-72	6.18	+220	4.6	8.9	8.5	0.22	67	3.4	0.20	0.29	1.6

[1] pH, Eh and HCO₃ determined by field measurement
[2] All samples filtered through 0.45 µm membranes in the field and subsequently analyzed for cations by ICP-AES (unless indicated otherwise)
[3] Analyses for these cations by ICP/MS
[4] Dissolved acid-hydrolysable phosphate concentrations determined by standard ascorbic acid colorimetric procedure
[5] Uranium-238 concentrations determined by α-spectrometry (with close correspondence to concentrations determined by ICP/MS)
[6] KD1 casing slotted in 40-42 m range
[7] ND = not determined
[8] Preliminary copper concentrations determined by anodic stripping voltammetry
[9] Total organic carbon (TOC) concentrations determined by TiO₂-catalyzed photolytic oxidation method
[10] ²³⁴U/²³⁸U activity ratio

Bore No.	Cased Depth (m)	Sampled Depth (m)	Mn mg/L	Si mg/L	Al mg/L	Pb[3] µg/L	V[3] µg/L	Zn[3] µg/L	Co[3] µg/L	Cu[8] µg/L	TOC[9]	PO₄[4] µg/L	U[5] µg/L	²³⁴U/²³⁸U[10]
PH14	20	<26	0.23	10.3	<0.01	3.3	0.8	15	1.0	0.6	0.97	130	17.8	0.88±0.05
PH15	24	<30	0.090	8.1	<0.01	1.3	1.6	640	0.35	0.7	1.74	295	30.6	0.99±0.03
PH49	16	<28	ND	ND	ND	ND	ND	ND	ND	ND	ND	ND	ND	ND
		28-30	0.14	9.2	<0.01	1.0	1.0	<4	1.8	0.5	1.51	90	102	1.02±0.03
		44-46	0.067	8.7	<0.01	0.3	1.1	<4	0.85	0.3	1.24	60	131	1.08±0.02
PH55	20	26-28	0.059	14.7	<0.01	0.4	1.0	<4	0.15	ND	1.30	475	0.62	1.16±0.10
		40-42	0.096	14.5	<0.01	0.7	0.6	<4	0.12	ND	1.38	445	1.01	1.17±0.01
PH56	22	26-28	0.036	10.0	0.01	0.5	1.0	<4	0.15	ND	1.09	215	7.73	0.86±0.03
		43-45	0.032	8.8	0.01	0.6	0.7	44	0.08	ND	1.33	175	4.35	1.03±0.04
PH58	23	25-27	0.086	15.2	<0.01	1.3	0.9	<4	0.11	1.0	2.16	685	2.02	ND
		38-40	0.075	15.7	<0.01	0.7	1.4	<4	<0.06	0.6	1.24	665	1.01	1.41±0.06
PH61	33	43-45	0.067	6.5	<0.01	0.6	0.8	<4	0.21	ND	1.34	130	2.16	1.02±0.06
PH80	17	<20	0.28	10.7	<0.01	0.9	0.7	<4	0.17	1.4	0.97	305	4.55	0.82±0.03
PH88	21	28-30	0.073	14.9	<0.01	0.7	2.3	7	0.10	ND	1.66	105	2.94	0.81±0.02
		38-40	0.071	14.9	<0.01	0.9	1.5	4	0.13	ND	0.94	155	2.70	0.78±0.05
PH94	24	26-28	<0.005	12.1	<0.01	ND	4.1	<4	<0.06	0.4	0.87	285	0.23	1.26±0.29
		40-42	0.006	12.1	<0.01	1.1	2.2	<4	<0.06	<0.1	1.58	200	0.26	1.26±0.24
PH96	16	<24	0.055	15.7	<0.01	2.1	3.1	6	0.16	ND	0.76	210	0.30	0.91±0.06
KD1	70[6]	40-42	0.12	5.3	<0.01	1.0	0.6	<4	0.16	1.4	1.42	10	0.49	1.00±0.04
		70-72	<0.005	8.2	<0.01	1.1	0.8	<4	0.10	1.8	ND	5	0.49	1.21±0.09

[1] pH, Eh and HCO₃ determined by field measurement
[2] All samples filtered through 0.45 µm membranes in the field and subsequently analyzed for cations by ICP-AES (unless indicated otherwise)
[3] Analyses for these cations by ICP/MS
[4] Dissolved acid-hydrolysable phosphate concentrations determined by standard ascorbic acid colorimetric procedure
[5] Uranium-238 concentrations determined by α-spectrometry (with close correspondence to concentrations determined by ICP/MS)
[6] KD1 casing slotted in 40-42 m range
[7] ND = not determined
[8] Preliminary copper concentrations determined by anodic stripping voltammetry
[9] Total organic carbon (TOC) concentrations determined by TiO₂-catalyzed photolytic oxidation method
[10] ²³⁴U/²³⁸U activity ratio

TABLE 3.   *Field and Laboratory Results for Selected Parameters*
              for Groundwaters Extracted from the W-series of Bores
              in October, 1989. Magnesium samples taken in June,
              1989. [1]- Residual influence of bore construction may
              contribute to elevated pH.

Bore No.	Sample Depth (m)	pH	Eh mV	Mg mg/L	HCO_3 mg/L	PO_4 µg/L	U µg/L	$^{234}U/^{238}U$
W1	13-15	6.55	+220	2.7	98	<5	318±0.09	0.87±0.03
	23-25	7.09	+255	9.6	120	415	65.9±1.9	0.78±0.02
W2	13-15	8.94[1]	+135	0.5	119	<5	16.1±0.6	0.65±0.02
	23-25	6.48	+235	15.4	98	110	79.4±2.7	0.66±0.01
W3	13-15	8.35[1]	+120	3.8	41	25	5.50±0.17	0.91±0.03
	23-25	7.08	+100	14.4	89	180	1.58±0.07	0.97±0.05
W4	13-15	5.85	+330	8.8	66	30	412±10	0.81±0.01
	23-25	6.80	+140	23.2	165	90	431±11	1.21±0.01
W5	13-15	8.25[1]	+145	4.3	101	590	0.42±0.04	0.91±0.11
	23-25	6.80	+185	10.6	82	545	1.89±0.07	0.81±0.02
W7	13-15	8.77[1]	+215	0.2	27	80	0.2±0.4	0.8±0.2
	23-25	7.15	+220	3.2	52	450	0.2±0.4	1.0±0.2

The uranium concentration in groundwaters from both depths
sampled in these bores is highest in W4 (in the centre of the deposit)
and decreases successively in bores with lower standing water levels
(Figure 6b). A combination of dilution and/or sorption processes
presumably account for this decrease in uranium concentration on
moving downgradient.

Groundwaters from the region directly east of the ore body
(Bores PH94 and PH96) are chemically rather different to waters from
the mineralized zone and exhibit significantly lower cation dominance
by magnesium [24]. The chemical characteristics of these waters
suggest a component of recent recharge though a direct flow path from
the Kombolgie is also possible [25-27].

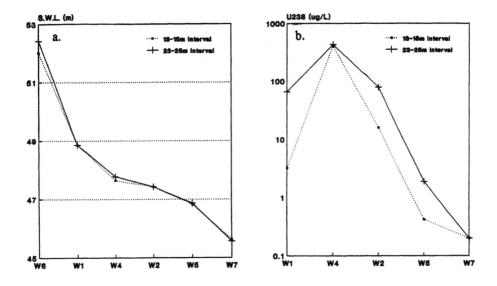

FIGURE 6. *(a) Standing Water Levels. (b) Uranium Concentrations for Groundwaters from the W-series of Bores.*

Bore KD1 intersects the Kombolgie "footwall" aquifer upfield of the uranium ore deposit. Groundwaters from this bore are of low ionic strength with magnesium concentrations markedly lower than in the ore zone. The pH of KD1 waters is slightly acidic (5.5-6.0) and notably lower than that of downfield waters which typically exhibit pH's in the 6.5-7.0 range.

Some insight into the likely age, source and rate of flow of these waters can be obtained from the results of tritium, carbon-14 and chlorine-36 analyses given in Table 4 [28]. While the errors on the percentage of modern carbon (% MC) are quite large (±10%), it appears that only one of the groundwater samples (PH88) can be considered modern though the results for PH94 are somewhat ambiguous. The ^{14}C ages for the bulk of the samples are in the range of 1000-3000 years with a tendency to decrease on moving downgradient (possibly as a result of gradual mixing with modern waters).

TABLE 4.   *Carbon-14, Tritium and Chlorine-36 Results for*
                Groundwaters from Selected Bores in the Vicinity of the
                Koongarra Uranium Deposit [25, 28]. M C - modern
                carbon, estimated from samples obtained in May, 1988. Tritium
                concentrations for samples obtained in May, 1988. ^{36}Cl/Cl ratios
                from samples obtained in November, 1988. ND - not determined.

Bore No.	Sampling Depth (m)	Percent M C	$\delta^{13}C$ o/oo	^{14}C Age (yrs)	Tritium Activity (TU)	$^{36}Cl/Cl$ X 10^{-1}
PH14	< 16	69.3±8.3	-20.89	3040	0.3±0.2	ND
PH15	< 30	69.0±7.7	-19.62	3070	0.3±0.2	203±15
PH49	28-30	ND	ND	ND	0.4±0.2	231±21
	44-46	ND	ND	ND	0.3±0.2	ND
PH55	26-28	84.8±9.7	-20.64	1360	0.2±0.2	177±17
	40-42	88.4±9.9	-20.28	1020	0.2±0.2	243±22
PH58	25-27	90.0±10.6	-19.83	790	0.2±0.2	ND
	38-40	ND	ND	ND	0.3±0.2	ND
PH61	43-45	ND	ND	ND	0.0±0.2	103±7
PH80	< 20	84.5±9.6	-20.79	1390	0.4±0.2	193±21
PH88	28-30	105.0±12.0	-20.41	Modern	0.8±0.2	556±42
	38-40	101.9±12.2	-19.86	Modern	1.2±0.2	496±38
PH94	26-28	93.5±11.6	-21.08	550	ND	ND
	40-42	ND	ND	ND	0.2±0.2	ND
KD1	40-42	ND	ND	ND	0.0±0.2	ND
	80-82	82.9±9.4	-20.65	1550	0.4±0.2	ND

$\delta^{13}C$ results for these samples (Table 3) indicate that matrix
effects (particularly carbonate dissolution) are relatively unimportant
since the $\delta^{13}C$ value for normal rainfall is on the order of -25 o/oo.
The tritium results support these ages with significant levels of tritium
only observed in groundwaters from PH88 and PH94 [29]. While the
possibility of neutron capture reactions complicates the interpretation of
chlorine-36 results, the relatively high ^{36}Cl/Cl ratios for waters from
PH88 are suggestive of some bomb-pulse influence [30].
        Analyses on ultrafiltrates of these waters [31] indicate that the
bulk of the iron and manganese observed in these waters is in true
solution. Under the redox conditions exhibited by these waters, this

dissolved iron and manganese is expected to be in the reduced form. Organic complexation of iron and other trace elements in these low organic content waters will be minor. The low levels of organic materials found in these waters (TOC's of 0.8-2.2 mg/L) are to be expected as the soils are intensely leached under these monsoonal conditions. The ultrafiltrate analyses also indicate that dissolved aluminum concentrations are very low with the occasional high values in the 1.0 μm filtered samples due to the presence of colloidal clays or oxides. Very little uranium was found to be associated with particulate (including colloidal) phases [31].

Relatively high concentrations of phosphate are observed in some instances (particularly in PH55 and PH58 - the bores passing through the dispersion fan). The presence of phosphate is of particular interest because of its role in secondary uranium mineralization and because of its possible impact on solution phase uranium speciation (and hence transport). These issues are discussed in more detail below. The dissolution of apatite is considered to be the major source of phosphate in these waters.

The major ion composition of the groundwaters sampled in May 1988 and November 1988 show close correspondence with a possible small general increase in pH. However, significant differences in uranium and phosphate concentrations were observed with the concentrations for both elements typically substantially lower in November than in May (PH58 being an exception). The concentrations of these elements may be influenced by the seasonal change in water table that occurs (low in November, high in May) but more substantial records are required in order to make definitive comment.

The major ion chemistry of waters taken from the same borehole but at different depths (bottom of weathered zone and in the unweathered zone) are generally similar. Again, the most distinct differences are evident for uranium and phosphate and presumably reflect the mineral content of the strata being sampled. The change in uranium content of waters with depth is variable but phosphate concentrations are typically somewhat lower in the consolidated (unweathered) rock than in waters extracted at the base of the weathered zone. More distinct differences in uranium and phosphate concentrations with depth are evident in the recently acquired data for the weathered zone from the W-series of bores (Table 3). In most cases (W5 is an exception), groundwaters extracted from the 13-15m depth exhibit much lower phosphate concentrations than waters from the base of the weathered zone (23-25m). It is also typical that higher concentrations of dissolved uranium are observed in groundwaters extracted through sampling ports at the base of the weathered zone than at the 13-15m depth though very high dissolved uranium concentrations are observed at intermediate depths in some instances (e.g. W4). The $^{234}U/^{238}U$ activity ratios for groundwaters extracted from the W-series of bores are also given in Table 4 and provide valuable insight into the likely source and mixing characteristics of these waters. These activity ratios are discussed in detail below.

3.2   Solubility Relationships

Sverjensky [32-34] has investigated the state of saturation of Koongarra groundwaters with respect to predominant minerals in the region using the equilibrium speciation code EQ3NR [35].

TABLE 5.   *State of Saturation, as Indicated by the Magnitude of Log (Q/K) ({ion activity product} / {equilibrium constant}).* Data are for waters from bores near the Koongarra uranium deposit [32 - 34].

Mineral	Log (Q/K)		
	PH49 (28-30 m)	*PH55 (26-28 m)*	*PH58 (25-27 m)*
Chalcedony	+ 0.13	+ 0.33	+ 0.34
Dolomite	- 1.3	- 2.1	- 2.1
Chlorite	- 10.4	- 13.3	- 13.3
Illite	- 1.6	- 1.6	- 1.5
Apatite	- 1.7	+ 0.1	- 1.7
Uraninite	- 22.8	- 27.3	- 25.9
Uranophane	- 8.8	- 17.8	- 15.9
Saleeite	- 5.0	- 12.2	- 10.3
Torbernite	- 9.4	- 16.5	- 14.1
Carnotite	- 5.3	- 14.6	- 10.8

The results for waters from PH49, PH55 and PH58 are given in Table 5. Values of Log (Q/K) refer to the logarithm of the ratio of the ionic activity product for dissolution of the mineral (Q) to the corresponding thermodynamic equilibrium constant (K). Values of Log (Q/K) near zero indicate that the solution is near saturation with respect to that mineral (as found for chalcedony in all three waters and apatite for PH55). Negative values of Log (Q/K) indicate that the solution is capable of dissolving that mineral.

These results indicate that the Koongarra groundwaters close to the ore body are undersaturated or near saturation with respect to hydroxyapatite. It is also apparent that the waters are significantly undersaturated with respect to both primary and secondary uranium minerals and will thus be actively dissolving these minerals.

3.3   Uranium Speciation in Koongarra Groundwaters

In the absence of significant concentrations of carbonate and phosphate, the chemistry of the uranyl ion in aqueous solution is controlled by a series of pH dependent hydroxy complexes (Figure 7a). However, the presence of carbonate markedly influences the solution speciation of U(VI). Using the chemical equilibrium code MINTEQA2, Version 2.01 (an updated version of MINTEQA1) [36] and constants validated by Tripathi [37] (Table 6), it is seen that for a solution in equilibrium with the atmosphere ($P_{CO_2} = 10^{-3.5}$ atm) and a total uranium concentration of 10^{-6}M (approximately 250 ppb), the carbonate containing species $UO_2CO_3^{\circ}$, $(UO_2)_2CO_3(OH)_3^{-}$, $UO_2(CO_3)_2^{2-}$ and $UO_2(CO_3)_3^{4-}$ are of varying significance for pH > 5.5 (Figure 7b). Not surprisingly, the species distribution is highly dependent on carbonate concentration. Thus, if the total carbonate concentration is maintained at 2 millimolar, the hydroxy uranyl species are not particularly important (Figure 7c).

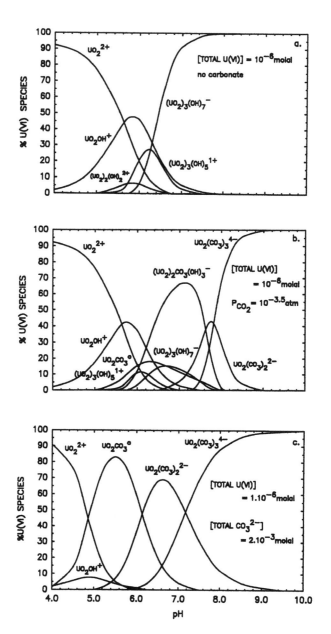

FIGURE 7. *Speciation of $10^{-6}M$ U(VI) as a Function of pH.* a) in the absence of carbonate, b) in equilibrium with the atmosphere ($PCO_2 = 10^{-3.5}$ atm), c) in solutions containing 2 mM total carbonate.

TABLE 6.   *Values of Stability Constants for Uranyl, Hydroxy, Carbonate and Phosphate Complexes Used in MINTEQA2 Calculations.* Values are the same as in [37]. β - overall stability constant for complex; *β - Overall stability constant using protonated ligand.

Complex	Type	Value (log_{10})
UO_2OH^+	*β	-5.30
$(UO_2)_2(OH)_2^{2+}$	*β	-5.68
$(UO_2)_3(OH)_4^{2+}$	*β	-11.68
$(UO_2)_3(OH)_5^+$	*β	-15.82
$(UO_2)_3(OH)_7^-$	*β	-28.34
$(UO_2)_4(OH)_7^+$	*β	-21.90
$UO_2CO_3^\circ$	β	9.65
$UO_2(CO_3)_2^{2-}$	β	17.08
$UO_2(CO_3)_4^{4-}$	β	21.70
$UO_2CO_3(OH)_3^-$	β	-1.18
UO_2F^+	β	6.08
$UO_2F_2^\circ$	β	8.84
$UO_2F_3^-$	β	11.24
$UO_2F_4^{2-}$	β	12.43
$UO_2H_2PO_4^-$	β	23.2
$UO_2H_3PO_4^{2+}$	β	22.9
$UO_2(H_2PO_4)_2^\circ$	β	45.24
$UO_2(H_2PO_4)H_3PO_4^+$	β	46.0

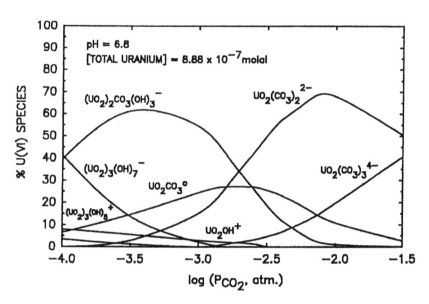

FIGURE 8.   *Results of MINTEQA2 Calculations Showing the Dependency of U(VI) Species Distribution on CO₂ Partial Pressure.*

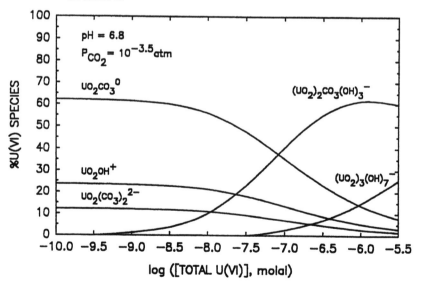

FIGURE 9.   *Results of MINTEQA2 Calculations Showing the Dependency of U(VI) Species Distribution on Total Uranium Concentration at $PCO_2 = 10^{-3.5}$ atm.*

The dependency of uranium speciation on carbon dioxide partial pressure is also marked (Figure 8) - an issue of major importance for these (and most other) groundwaters where typically $P_{CO_2} > 10^{-2}$ atm but major fluctuations (particularly in the shallower groundwaters) might be expected [38]. Significant variations in uranyl species dominance with change in total uranium concentration are to be expected when multi-uranyl species dominate (Figure 9).

It is generally considered that phosphate is a strong complexing agent for uranyl ion and, if present in sufficient quantity, would be expected to outcompete carbonate for U(VI). Indeed, the results of MINTEQA2 calculations for Koongarra groundwaters from a number of bores in the vicinity of the deposit are shown in Table 7 and indicate that the proportion of uranyl ion complexed by phosphate increases from less than 5% at KD1 (upgradient of the ore body) to 98.9% at PH55 which is in the dispersion fan immediately downgradient of the primary ore zone. It should be noted, however, that considerable confusion still surrounds the role of phosphate in controlling solution speciation of uranium (and concomitantly, influencing sorption behavior). Literature values (see Table 6) for the formation of the $UO_2(HPO_4)_2^{2-}$ species [39] have been used in these calculations however the results of Tripathi [37] and Markovic and Pavkovic [40] cast doubt on the existence of this species. Interestingly, Tripathi [37] has shown that there is very little difference between the extent of uranium adsorption in phosphate-free and phosphate-bearing systems, and that almost complete removal of uranium is possible by "adsorption" when $UO_2(HPO_4)_2^{2-}$ is calculated to be the dominant uranyl species in solution. A recent critique of the thermodynamic data for uranium [41] strongly supports Tripathi's view that the $UO_2(HPO_4)_2^{2-}$ species cannot be of significance. This issue is of considerable relevance to the Koongarra system and is being investigated further.

TABLE 7.   Uranium (VI) Speciation Computed Using MINTEQA2 for Groundwater Samples from Selected Bores in the Koongarra Region with and without the Inclusion of the Uranyl Phosphate Species $UO_x(HPO_4)_2^{2-}$. Stability Constants as Reported in Table 6 Used in all Computations. A log K of 42.99 Assumed for $UO_x(HPO_4)_2^{2-}$. (After Dongarra and Langmuir [39]) if Considered to be Present.

Bore No.	Depth (meters)	Total Mg (mg/L)	Total U (µg/L UO_2)	Total P (µg/L PO_4)	Alkalinity (mg/L CO_3^{2-})	pH	%U(VI) Species(>1%)				%U(VI) Species(>1%) ($UO_2(HPO_4)_2^{2-}$ excluded)			
							$UO_x(HPO_4)_2^{2-}$	$UO_2CO_3^o$	$UO_2(CO_3)_2^{2-}$	$UO_2(CO_3)_3^{4-}$	$UO_2CO_3^o$	$UO_2(CO_3)_2^{2-}$	$UO_2(CO_3)_3^{4-}$	$UO_2HPO_4^+$
PH49	28-30	26.4	240	310	75.7	6.80	76.4	2.3	20.2		9.5	85.5	4.4	
PH55	26-28	17.9	0.47	650	62.2	6.57	98.9				18.4	78.3	1.8	1.2
PH61	43.5-45.5	18.0	4.6	130	74.0	7.03	54.0	2.7	40.0	3.2	5.9	87.0	7.0	
PH94	26-28	5.7	0.20	290	18.6	7.17	98.8				16.9	79.7	1.7	1.1
PH14		20.3	61	160	62.7	6.42	78.6	5.2	15.8		24.2	73.8	1.2	
PH88	28-30	14.1	2.5	110	43.1	6.54	83.7	4.3	11.6		26.5	71.5		
KD1°	70.5-72.5	8.9	0.55	10	32.8	6.18	6.3	49.1	43.0		52.4	45.9		
KD1†	70.5-72.5	8.9	0.55	5	32.8	6.18	1.7	51.6	45.2		52.5	46.0		

° Phosphate at detection limit (10 µg/L PO_4)
† Phosphate at half detection limit

4.   URANIUM SERIES ISOTOPE DISEQUILIBRIA IN SOLID AND SOLUTION PHASES

4.1   Background to Isotope Disequilibrium

Uranium-238, the chief constituent of natural uranium (99.27% abundance) decays via alpha particle emission with a half-life of 4.468 X 10^9 years to the relatively short-lived ^{234}Th ($t_{1/2}$ = 24.1 days). Thorium-234 then decays via β emission to ^{234}Pa which in turn decays by β emission (with half-life of 6.7 hours) to the longer-lived ^{234}U ($t_{1/2}$ = 2.48 X 10^5 years). This beta stable, alpha-emitter then decays to ^{230}Th. Thorium-230, with a half-life of 7.52 X 10^4 years, is relatively stable decaying to ^{226}Ra ($t_{1/2}$ = 1602 y) and finally to ^{206}Pb via a number of other relatively short-lived intermediates (including ^{222}Rn and ^{210}Pb).

In the uranium-238 decay chain, the half-life of the parent nuclide is very long compared to the half-lives of all other members. Under this condition, all members of the decay chain acquire the same activity (i.e. the total number of disintegrations per unit time) as the parent, a phenomenon known as secular equilibrium [42]. Secular equilibrium is typically observed in primary uranium minerals such as uraninite which typically remain unaltered for over 10^9 years in which time it would have sustained negligible loss or gain of daughter nuclides. However, the uranium-238 decay series comprises a diverse array of elements each with its own chemical behavior. Alteration of the primary uranium mineral by, for example, chemical weathering, may result in conditions favourable for the selective removal of certain elements of the decay chain and resulting in what is termed radioactive disequilibrium. Solution and precipitation are by far the most important sorting processes leading to disequilibrium and the locales of disequilibrium production are primarily at liquid/solid phase boundaries. For example, leaching of uranium may occur during weathering leaving an excess of nuclides of the more insoluble elements such as thorium and protactinium.

The degree of disequilibrium between the longer-lived members of the ^{238}U series, namely ^{238}U, ^{234}U, and ^{230}Th, has been of particular interest in the study of groundwaters since analysis of activity ratios between these nuclides may provide valuable insight into the factors controlling uranium migration in the sub-surface environment [43,44]. Activity ratios in both solid and liquid phases, and the interrelationship between the two have been investigated extensively in the Alligator Rivers Analogue Project and are discussed briefly below.

4.2   Solid Phase ^{238}U-Series Disequilibria

Uranium and thorium concentrations, and ^{234}U/^{238}U and ^{230}Th/^{234}U activity ratios have been determined on "amorphous" and "crystalline" solid phases at Koongarra. In this case, amorphous solids are operationally defined as those solubilized by the Tamms acid oxalate (TAO) procedure in which the soil sample is shaken in the dark for four hours with 40 mL/g sample of 0.2 M ammonium oxalate and buffered to pH 3.0 with oxalic acid. TAO removes amorphous minerals of iron (including ferrihydrite), aluminum and silicon as well as metal-organic complexes and adsorbed metals [45]. The TAO extractable fraction is nominally equated with the fraction of solid readily accessible to groundwaters though it is recognized that it gives only a crude estimate of this fraction with a rather indistinct cut-off between amorphous and crystalline phases. For example, lepidocrocite, magnetite, maghemite, goethite, nontronite, trioctahedral chlorite, hydrous mica, biotite and interlayered smectites (in this approximate order of susceptibility) are partially dissolved by TAO [45].

Most of the uranium and thorium concentrations and ^{234}U/^{238}U and ^{230}Th/^{234}U activity ratios on TAO extract and residual phases have been determined for solids from a transect of the ore body running from diamond drill hole DDH52 to percussion hole PH89 perpendicular to the major fault and along the dispersion fan for about 400 m. On average, approximately 45% of the uranium in these samples was extractable by TAO though the mass of material in this "amorphous" phase was only on the order of 5%.

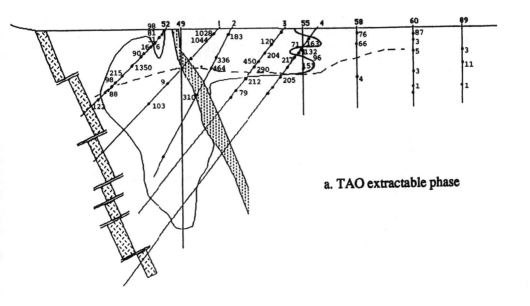

a. TAO extractable phase

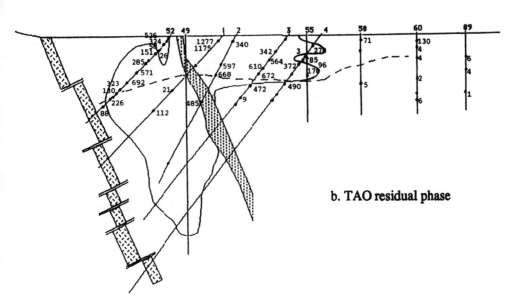

b. TAO residual phase

FIGURE 10. *Uranium Concentrations (µg/L)* (a) in Tamm's acid oxalate
(TAO) extractable and (b) TAO residual phases for solid samples
obtained from DDH52 - PH89 transect (from [46]).

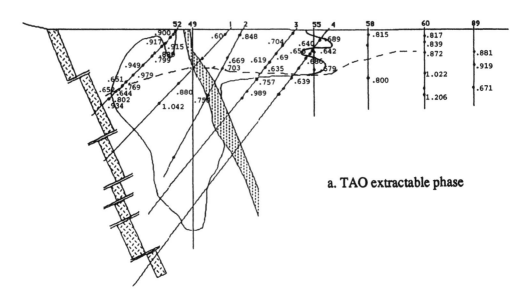

a. TAO extractable phase

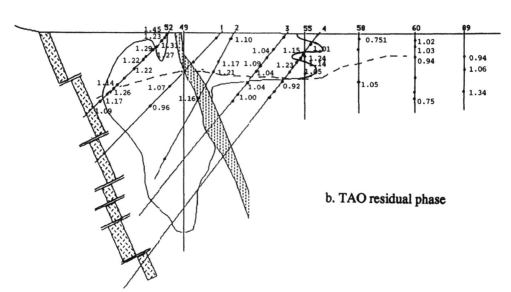

b. TAO residual phase

FIGURE 11. $^{234}U/^{238}U$ *Activity Ratios* (a) in TAO extractable and (b) TAO residual phases for solid samples obtained from DDH52 - PH89 transect (from [46]).

The distribution of uranium concentration in the transect in both the TAO extractable and TAO residual phases is shown in Figures 10a and b. The uranium concentrations in both phases follow similar patterns. Edghill [46] notes that the drop in uranium concentration near the end of the dispersion fan appears to be more abrupt in the residual phases than in the extractable phases. Downgradient of the dispersion fan, the uranium concentrations are relatively high in the top 5-10 m in both the extracted and residual phases suggesting that rapid uranium transport occurs in (or at the base of) the surface sands, presumably particularly during the wet season when this material is saturated.

The $^{234}U/^{238}U$ activity ratios for solids from the transect are shown in Figures 11a and b for the TAO extract and residual phases respectively. The activity ratios are observed to be usually less than unity in the TAO extracted "amorphous" phase and greater than unity in the residual. These ratios are suggestive of a preferential mobility of ^{238}U over ^{234}U. One explanation proposed for this phenomenon is α-recoil of the daughter ^{234}Th nucleus into inaccessible phases upon decay of ^{238}U [47,48]. Because the more crystalline (inaccessible) phase makes up the bulk of the solid, the ^{234}Th nucleus, expelled in a random direction during α-decay, is most likely to enter this portion of the solid. The subsequent decay of ^{234}Th to ^{234}U renders the ^{234}U inaccessible to groundwater. Nightingale [49] found that most of the uranium that was determined to be inaccessible to groundwaters was trapped within crystalline iron oxide aggregates and suggested that the high levels of iron oxides, and the implantation of alpha recoil nuclei in or around these phases, accounted for the observed uranium isotope fractionation in the Koongarra case. It is also possible that the fractionation is induced because of the preferential sorption of thorium over uranium or similarly because of the preferential entrapment of Th in aging iron oxides.

In weathered samples, the highest $^{234}U/^{238}U$ ratios for TAO extracts were found near the top of DDH52 and in the barren zones downgradient of the dispersion fan. Unweathered samples also gave high ratios. Lowest ratios for TAO extracts are found near the edge of the dispersion fan. Uranium-234/uranium-238 ratios in residuals follows a similar pattern with the lowest values around DDH3 to DDH4 and PH55, and in some areas beyond the dispersion fan. The ratios were most different between the extractable and residual phases around DDH4, PH55 and the bottom of DDH3. They were nearest in holes downgradient of the dispersion fan. This suggests that the zones at the tip of the dispersion fan, and at the base of weathering, are the sites where the most recent deposition of ^{238}U may have occurred [46].

Thorium-230/uranium-234 activity ratios have also been determined for solids from the same transect and are generally below unity in the TAO extract with higher values in the residual phases [46,49]. Particularly low values (around 0.4 and less) occur near the surface at DDH1, DDH2, PH58 and PH60. Low values also occur near the base of weathering along the dispersion fan. These low ratios are indicative of sites of most recent deposition of uranium. High ^{230}Th/^{234}U ratios have been obtained for TAO extracts of samples taken from near the surface in DDH52 and are suggestive of leaching of uranium in this location.

Plotting ^{234}U/^{238}U versus ^{230}Th/^{238}U (in the style of Osmond et al. [44]) provides a particularly useful representation of zones of uranium leaching and deposition. This is done in Figure 12a for the TAO extract data with the location of the allocated groupings identified in Figure 12b. Zone 1 appears to represent an area of rapid accumulation of uranium with considerable preferential partitioning of uranium-234 into the crystalline phase and uranium-238 into the amorphous phase. Zone 2 appears to be similarly a uranium deposition area although possibly not as intensely accumulating as Zone 1. Zone 3 (at the base of surficial sands) appears to be an area of intense uranium accumulation but exhibits less tendency than Zones 1 and 2 to fractionate ^{234}U and ^{238}U. Edghill [46] suggests that this may be because of the different nature of the solid substrates (i.e. surficial sand versus weathered schist) and/or because of the different flow regimes operating in these zones (i.e. rapid and seasonal versus slow and continuous). Zone 4, if anything, can be considered to be a slightly leaching region. Strong leaching of uranium is a characteristic of Zone 5. Nightingale [49] suggests that an upward extension of the ore body in the vicinity of DDH52-DDH1 had been eroded and the uranium leached away, leading to very high ^{230}Th/^{234}U ratios in this area.

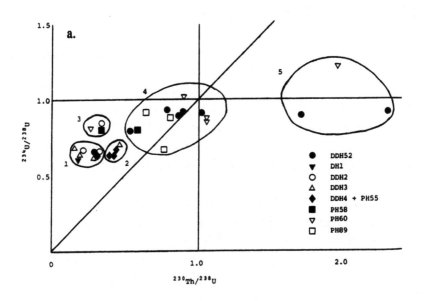

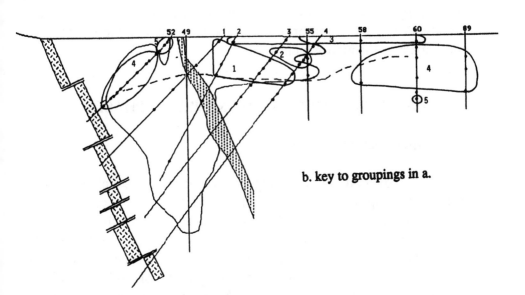

b. key to groupings in a.

FIGURE 12. $^{234}U/^{238}U$ *versus* $^{230}Th/^{238}U$ *for the TAO Extractable Phase of Solids from the DDH52 - PH89 Transect.* (a) is the plot, and the locations of groupings are shown in (b) [46].

4.3   Solution Phase ^{238}U-Series Disequilibria

The ^{234}U/^{238}U activity ratios for groundwaters obtained from the PH-series of bores in May and November 1988 are given in Tables 1 and 2 and from the W-series of bores at 13-15 m and 25-27 m (October 1989) in Table 4. The most striking observation on these results is the lower ratio found for the samples taken from shallower depths compared to waters extracted at depth from boreholes passing into the main ore zone (i.e. PH49,55,56,58) which exhibit ^{234}U/^{238}U ratios greater than unity.

Low ^{234}U/^{238}U ratios (below unity) are characteristic of weathered zone groundwaters at Koongarra and are in accord with the hypothesized mechanism of ^{234}U entrapment by α-recoil. In this respect, Koongarra groundwaters are atypical in that a direct α-recoil into groundwaters leading to substantially elevated ^{234}U/^{238}U ratios in solution phase is commonly found [44,50]. The activity ratio values for groundwaters from the weathered zone in this study are remarkably consistent with the activity ratios of amorphous phase weathered zone solids (Figure 11a) adding weight to the argument that these groundwaters are in equilibrium with the accessible components of the solid substrate through which it passes.

5.   URANIUM SORPTION TO MINERAL PHASES

5.1   Introductory Comments

As indicated above, it appears reasonable to conclude that the groundwaters in the vicinity of the Koongarra ore deposit are in equilibrium with the accessible solid phases (particularly the iron oxides) through which they pass. Thus, the chemistry of the groundwaters will reflect, in some way, the properties of these accessible solid phases. Given the complexity of both the solution chemistry of U(VI) and the nature of the solid substrates, description of the interplay between factors controlling uranium (VI) partitioning between solution and solid phases is (to say the least) non-trivial.

Indeed, while the interaction of uranyl species with a variety of solid substrates has been frequently studied in the laboratory [51-56], only a few authors have attempted to present a coherent mathematical description of sorption results obtained over a range of solution conditions. The most commonly used method for describing the interaction of solution phase species with solid substrates is the "surface complexation" approach [57-59] which has been used by both Tripathi [37] and Hsi and Langmuir [60] in modeling the adsorption of uranyl species to iron oxides. The approaches used by these authors have been critically reviewed and have provided a valuable basis for the preliminary modeling of the partitioning of uranium onto solid substrates from the Koongarra region.

5.2   Surface Complexation Modeling of Uranium Sorption

Payne and Waite [61] report on the results of uranium sorption studies to solid substrates from the weathered zones in the vicinity of the Koongarra and geologically similar Ranger uranium ore bodies. In these studies, a simulated groundwater with a composition similar to natural Koongarra groundwaters, and containing an artificial uranium isotope (^{236}U), was contacted with solid phases that contained natural uranium. Subsequently, the levels of all uranium isotopes in the liquid phase were measured by α-spectrometry, enabling both uranium adsorption and desorption to be studied.

The Koongarra sample (from drill core DDH52 at 15 meters) was a highly weathered schist, containing 31% quartz and with kaolinite the dominant clay mineral. Significant amounts of goethite and hematite and traces of aluminum hydroxides were present. The surface area, measured using ethylene glycol monoethyl ether (EGME), was $29(\pm 4)$ m^2 g^{-1}. The total uranium content was 392 μg g^{-1}, of which the readily leachable component as determined using TAO was 90 μg g^{-1}. Approximately 1.0 mg g^{-1} of amorphous iron was solubilized in the TAO extraction. The sample from the Ranger deposit (from drill core S1/146 at 4 meters) contained 10% quartz, with smectite and kaolinite the main clay minerals. Mica, goethite, anatase and hematite were present in small amounts. The uranium content was 83 μg g^{-1}, of which

20 µg g^{-1} was TAO-extractable. The EGME surface area was found to be 145($\pm$10) m^{2}g^{-1}, with the higher surface area in this case most likely due to the presence of smectites. For this sample, 1.9 mg of amorphous iron oxide per gram of substrate was removed in the TAO extraction.

TABLE 8.   *Hydrolysis Constants for Amorphous Ferric Oxyhydroxide Surface Groups and Intrinsic Constants for Major Ion Surface Complexation Reactions Used in Sorption Modeling.* * - Mg-FeO$^-$ sorption constant assumed the same as for Ca-FeO$^-$ value.

Reactions	Intrinsic Constants (log K)	Reference
FeOH$_2^+$ = FeOH + H$^+$	- 5.1	[64]
FeOH = FeO$^-$ + H$^+$	- 10.7	[64]
FeOH + Na$^+$ = (FeO$^-$-Na$^+$)$^\circ$ + H$^+$	- 9.0	[60]
FeOH + Cl$^-$ + H$^+$ = (FeOH$_2^+$-Cl$^-$)$^\circ$	7.0	[60]
FeOH + CO$_3^{2-}$ + 2H$^+$ = (FeOH$_2^+$-HCO$_3^-$)$^\circ$	20.7	[62]
FeOH + CO$_3^{2-}$ + 2H$^+$ = (FeOH-H$_2$CO$_3^-$)$^\circ$	20.0	[62]
FeOH + SO$_4^{2-}$ + H$^+$ = (FeOH$_2^+$-SO$_4^{2-}$)$^-$	11.6	[62]
FeOH + SO$_4^{2-}$ + 2H$^+$ = (FeOH$_2^+$-HSO$_4$)$^\circ$	17.3	[62]
FeOH + Ca^{2+} = (FeO$^-$-Ca^{2+}) + H$^+$	- 6.3	[62]
FeOH + Mg^{2+} = (FeO$^-$-Mg^{2+}) + H$^+$	- 6.3	*

The results of uranium sorption studies on the chosen Ranger and Koongarra substrates are shown in Figures 13a and b. It should be recognized that the total uranium concentration participating in sorption reactions in these laboratory studies is comprised of both the added ^{236}U to the groundwater and the "accessible" portion of ^{238}U present in the solid phase. However, the ^{236}U in these studies represents a minor component of the total mass of uranium involved in the sorption reactions and is essentially being used as a tracer.

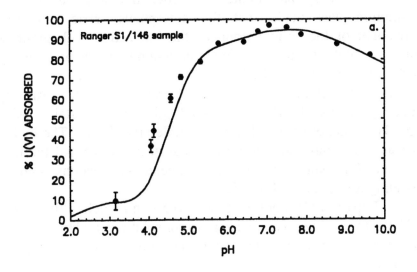

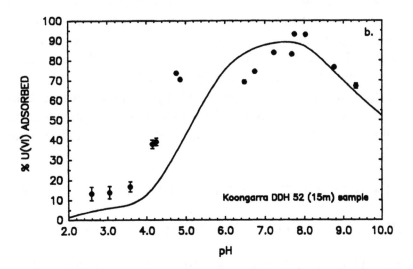

FIGURE 13. *Percentage of ^{236}U Adsorbed to Substrate.* (a) Ranger bore S1/146 (4 m depth), (b) Koongarra bore DDH52 (15 m depth). The solid lines represent best fits of model, assuming formation of surface complexes given in Table 9 (after [61]).

In initially attempting to model the sorption results, the assumption was made that the amorphous ferric oxyhydroxide component of the solid dominates the interaction with groundwater solutes (including uranium). Such an assumption has previously been used with some success by Zachara *et al.* [62] and Loux *et al.* [63] in modeling trace element adsorption to complex substrates. All sorption modeling reported by Payne and Waite [61] was undertaken using the triple layer model of the electrical double layer [64]. The hydrolysis constants for amorphous ferric oxide surface groups and the intrinsic constants for major ion surface complexation reactions used in the modeling are given in Table 8.

Tripathi [37] reported good success in modeling the adsorption of U(VI) to goethite over a wide range of solution conditions (including the presence of carbonate) using the single surface complexation reaction:

$$FeOH + UO_2^{2+} + 3H_2O = FeOH_2^+UO_2(OH)_3^- + 3H_2O$$

with intrinsic constant $p^*K^{int} = 7.0$. However, use of this equation and associated constant in modeling uranium sorption to the Ranger sample resulted in gross underprediction in the extent of adsorption. Hsi and Langmuir [60] found that the presence of carbonate resulted in significantly more uranium adsorption to iron oxides than could be accounted for by the formation of uranyl hydroxy surface complexes alone, and successfully modeled the experimental results assuming the formation of uranyl hydroxy and uranyl di- and tri-carbonate complexes; i.e.

$$FeOH + UO_2^{2+} + H_2O = [FeO^- \text{-} UO_2OH^+] + 2H^+ \qquad p^*K^{int} = 8.0$$

$$FeOH + UO_2^{2+} + 2CO_3^{2-} + H^+ = [FeOH_2^+ \text{-} UO_2(CO_3)_2^{2-}] \quad p^*K^{int} = -30.0$$

$$FeOH + UO_2^{2+} + 3CO_3^{2-} + H^+ = [FeOH_2^+ \text{-} UO_2(CO_3)_3^{4-}] \quad p^*K^{int} = -38.5$$

Application of these equations and constants by Payne and Waite [61] again resulted in significant deviation from experiment.   Major deviations were apparent in the pH 4.8-6.4 region where the monocarbonate species $UO_2CO_3^0$ dominates the solution speciation of uranium under the conditions used. On this basis, Payne and Waite [61] added a uranyl monocarbonate surface complex to the suite of surface complexes used by Hsi and Langmuir [60] and obtained good agreement between model and experiment (see Figure 13a). Best fit constants found by these investigators for the significant uranyl surface complexes and the distribution of these species on the ferric oxide surface as a function of pH are summarized in Table 9. It should be noted, however, that such an approach to sorption modeling is essentially a curve fitting exercise with enough variables added to ensure an adequate fit. Care should thus be exercised in concluding that the chosen suite of surface complexes and associated constants represent reality.

TABLE 9. *Distribution of Proposed Uranyl Surface Species on Ferric Oxyhydroxide Surface as a Function of pH.*
log K (FeO⁻-UO₂OH⁺) = -8.2; log K (FeOH₂⁺-UO₂CO₃⁰) = 23.3;
log K (FeOH₂⁺-UO₂(CO₃)₂²⁻) = 31.2; log K (FeOH₂⁺-UO₂(CO₃)₃⁴⁻) = 34.7.

pH	Percentage of sorbed U(VI) present in indicated species				Total % adsorbed
	$FeO^--UO_2OH^+$	$FeOH_2^+-UO_2CO_3^0$	$FeOH_2^+-UO_2(CO_3)_2^{2-}$	$FeOH_2^+-UO_2(CO_3)_3^{4-}$	
2	2.0				2.0
3	8.8				8.8
4	19.1				19.0
5	3.2	66.7	1.1		71.0
6		49.2	38.9		88.1
7			93.8		93.8
8			78.9	15.2	94.1
9			28.8	58.3	87.1
10			7.3	70.3	77.6

While the adsorption data for the Koongarra sample appear somewhat scattered, a reasonable description of these data is also obtained by using the constants found to fit the Ranger data (see Figure 13b). Such a fit is satisfying given that the Koongarra sample contains significantly more TAO-extractable ^{238}U than the Ranger sample and contains a smaller proportion of amorphous iron oxide. A consistent deviation between the model results and the experimental data is still observed for both substrates in the region of $UO_2CO_3°$ predominance indicating that the description of the interaction of uranium (VI) with the solid substrate in this pH region requires further attention. More comprehensive studies of U(VI) adsorption to both simple and complex substrates are underway in an attempt to clarify further the role of carbonate (and other potential ligands including silicate and phosphate) in uranium sorption processes. In addition, the assumption that one phase - the amorphous ferric oxide - is dominating adsorption of uranium to these complex substrates still appears reasonable given the modeling results obtained, but the potential modifying effects of the other phases present is open to investigation.

6.   URANIUM TRANSPORT MODELING

6.1   Previous Modeling Approaches

A variety of models of the distribution of uranium in solid and liquid phases in the vicinity of the Koongarra uranium ore deposit have been developed through the Alligator Rivers Analogue Project. Of particular interest is a one dimensional transport model developed by Lever [65] and extended by Lever and others [66-68].

The simplest form of this transport model includes one dimensional transport by groundwater flowing at constant velocity, u, and linear, equilibrium sorption to one phase described by a constant retardation factor, R (where $R = 1 + \rho K(1 - \varphi)/\varphi$ for medium density ρ and porosity φ). As discussed earlier, different isotopes of the same element can have different distributions between the various rock phases as a result of isotope fractionation mechanisms. This has been modeled by Lever [65] by assigning different retardation factors to the different isotopes. Ignoring the possible effects of the kinetics of leaching and deposition, the distribution in mineralogical phases, non-linear sorption and hydrodynamic dispersion, the concentration (C_i) in the groundwater of the ith member of the chain (with decay constant λ) may be expressed as:

$$R_i \frac{\partial C_i}{\partial t} + u \frac{\partial C_i}{\partial x} + R_i \lambda_i C_i = R_{i-1} C_{i-1}$$

With the neglect of dispersion, the initial distribution of ^{238}U, $C^0_{238}(x)$, migrates with a retarded groundwater velocity, thus:

$$C_{238}(x,t) = C^0_{238} \left(x - \frac{ut}{R_{238}} \right) e^{-\lambda_{238} t}$$

The corresponding analytical solution for C_{234} is:

$$C_{234}(x,t) = e^{-\lambda_{234} t} C^0_{234} \left(\frac{x - ut}{R_{234}} \right) + e^{-\lambda_{238} t} \frac{R_{238} \lambda_{238}}{R_{234}}$$

$$\cdot \int_0^t e^{-(\lambda_{234} - \lambda_{238}) t'} C^0_{238} \left(x - \frac{ut}{R_{234}} - ut' \left[\frac{1}{R_{234}} - \frac{1}{R_{238}} \right] \right) dt'$$

As shown by Lever [65] an expression may also be derived for the activity ratio of successive decaying nuclides:

$$\frac{a_i}{a_{i-1}} = 1 - \frac{u \left(\dfrac{R_{238}}{R_i} - 1 \right)}{R_{238} \lambda_i} \frac{1}{a_i} \frac{da_{238}}{dx}$$

This simplistic model has now been extended to include i) partitioning of nuclides between amorphous and crystalline phases, ii) the possibility of ^{230}Th mobility in the amorphous phase, and iii) longitudinal dispersion. The governing equations for this extended model may be written in the following form:

$$R_i \frac{\partial C_i}{\partial t} + u \frac{\partial C_i}{\partial x} + R_i \lambda_i C_i = D \frac{\partial^2 C_i}{\partial x^2} - \tau (K_i C_i - S_i)$$

$$+ (1 - B_1) R_{i-1} \lambda_{i-1} C_{i-1} + B_2 \lambda_{i-1} S_{i-1}$$

$$\frac{\partial S_i}{\partial t} + \lambda_i S_i = \rho(K_i C_i - S_i) + (1 - B_2) \lambda_{i-1} S_{i-1} + B_1 R_{i-1} \lambda_{i-1} C_{i-1}$$

where i - decay series member
       C_i - concentration of i in groundwater
       S_i - concentration of i in crystalline phase
       τ - rate of phase exchange
       K_i - distribution coefficient
       R_i - retardation coefficient
       B_1 - recoil factor (amorphous-crystalline)
       B_2 - recoil factor (crystalline-amorphous)
       λ_i - decay constant
       u - groundwater velocity
       D - longitudinal dispersion coefficient

In order to test this model, a region near the bottom of the weathered zone and away from the fault where the flow direction is relatively well-defined has been selected (Figure 14). In this region, lateral flow dominates over downward flow with the boundary between the weathered and unweathered rock acting as a horizontal barrier to groundwater flow.

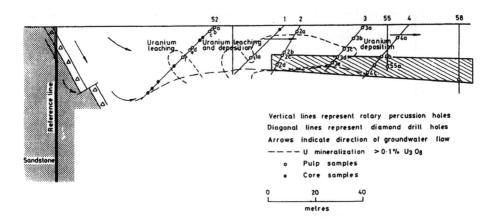

FIGURE 14. *Region Selected for Calibration and Validation of One Dimensional, Two Phase Transport Model (Shaded Area at Base of Weathered Zone).*

Experimental and theoretical activity ratios for this region are presented in an "Osmond and Cowart" type diagram in Figure 15 indicating that the essential isotope fractionation information can be approximately reproduced. The type of results that can be obtained for isotope concentrations as a function of distance from the source are shown in Figure 16 but the constants used to obtain these plots are, in many instances, based on too little information and on an extremely simplified representation of the chemical processes influencing sorption and fractionation of nuclides.

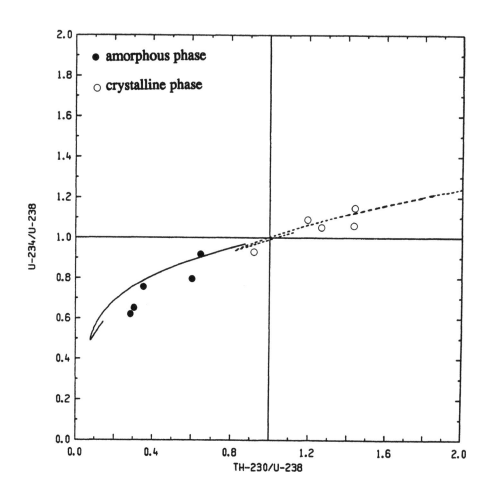

FIGURE 15. *Comparison of Measured and Calculated* $^{234}U/^{238}U$ *versus* $^{230}Th/^{238}U$ *Activity Ratios for Amorphous and Crystalline* **Solids.** Region selected is shown in Figure 14. Activity ratios estimated using 1D flow of groundwater interacting with a two phase solid (adapted from [68]). Crystalline phase - dashed line; amorphous phase - solid line.

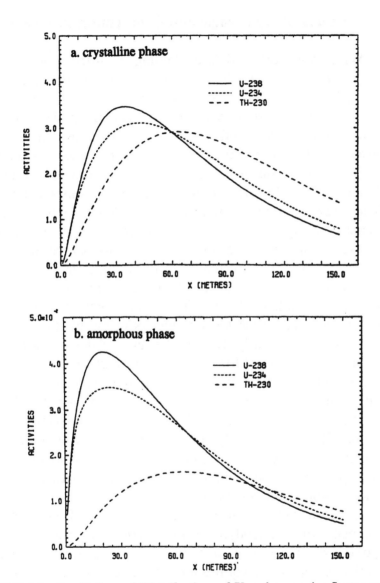

FIGURE 16. *Prediction of Distribution of Uranium-series Isotopes for (a) Crystalline and (b) Amorphous Solids Downfield of the Koongarra Ore Deposit.* The 1D two phase transport model of Lever and coworkers was used [68].

6.2   Coupling of Hydrologic and Geochemical Processes

While transport models containing a "distribution coefficient" approach to description of the partitioning of solute between solid and solution phase have been widely used [69-71], such an approach has been strongly criticized because of the lack of applicability of any determined K_d value to other situations involving different water chemistry or different substrate properties and because of the lack of insight into mechanism of interaction provided through use of such a coefficient [72].   In the Koongarra case, a constant distribution coefficient approach may provide a reasonable description of uranium partitioning between solid and solution phase over a small region (such as that used in the modeling study described above), but the results of such a study would be expected to have little relevance to transport in other regions exhibiting different water chemistry and substrate characteristics. Both solid and solution phase properties in some parts of the aquifer, such as the top few meters of the weathered zone, are expected to be particularly dynamic because of the seasonal variation in rate and extent of infiltration and the height of the water table. Adequate modeling of solute transport through such a complex region, other than simple ("lumped") multi-component mixing models such as that described by Airey *et al.* [73] and applied to Koongarra by Short *et al.* [74,75] may be nearly impossible.   However, more detailed description of hydrologic and geochemical aspects of uranium transport in this region and closer coupling of the major processes influencing migration would seem beneficial.

In recent years, considerable progress has been made in the development of thermodynamically based computer codes that incorporate chemical reactions (including adsorption), under the assumption of chemical equilibrium (or in a few cases, very simple kinetics of individual processes), into the solution of multicomponent solute transport equations.   This has been accomplished either by iteration between solutions of the transport equations and solutions of typically nonlinear, algebraic equations of chemical equilibrium (the "two-step" method) [76-79], or by simultaneous solution of both transport and chemical equations (the "one-step" or "direct" method)

[80-85]. As described briefly below, both approaches have previously been used to simulate the movement of uranium in groundwaters.

Walsh *et al.* [86] describe a two-step model (PHASEQL/FLOW) in which chemical equilibrium between an aqueous solution flowing in a linear, one-dimensional porous medium is assumed and mineral and aqueous-phase compositions are solved as a function of both space and time. The dissolution and precipitation of minerals, oxidation-reduction reactions and adsorption are included in this model. PHASEQL/FLOW has been applied to the simulation of uranium "roll front" deposition assuming the participation of 60 aqueous species and 26 minerals. They postulate [87,88] that uranium roll front deposits are formed as a result of deposition of primary uranous oxides on migration of originally oxidized uranyl-containing solution through porous, reduced sandstone (typically containing pyrite and organic carbon as reductants) have been strengthened as a result of the successful prediction of typical ore grades, deposit ages and sequencing of co-deposited minerals using this modeling approach.

The "direct" model THCC (developed from the CHEMTRN model of Miller and Benson) [83] is particularly well equipped to handle varying temperature and has been applied by Carnahan [89] to the estimation of uranium transport in a reducing environment following the mixing of solutions of different pH and temperature both initially at equilibrium with uraninite. In this case, the movement of the uranium front (made up of the dominant U(IV) species, $U(OH)_5^-(aq)$) is closely coupled to the flow of heat. Carnahan [90] has also used THCC to examine the effects of uranium thermodynamic database variations on uranium migration. Specifically, the sensitivity of uranium transport to formation constants for the species UO_2^+, $U(OH)_4^\circ$, $U(OH)_5^-$, and $(UO_2)_2CO_3(OH)_3^-$ were examined as were the effects of excluding UO_2^+ and $U(OH)_5^-$, and including the species $UO_2(CO_3)_3^{5-}$ and $U(CO_3)_5^{6-}$.

Liu and Narasimhan [91] describe the development of the multidimensional (2D or 3D), multiple species chemical transport model, DYNAMIX, and have applied the model to prediction of 1D uranium transport on mixing a static, alkaline, reducing solution containing uranium with an oxidizing, slightly acidic mobile phase also

containing uranium. A relatively simple solution composition of
eighteen aqueous species and two mineral phases were considered in
this study. Different methods of incorporating redox processes in the
modeling of chemical speciation were also considered by Liu and
Narasimhan [91,92]. Both an "effective internal redox approach" in
which the valence state of the system is conserved and an "external
approach" in which the electron activity is considered as a component
have been considered by these authors. Significant differences in
results were obtained for the two different approaches to modeling
redox processes and Liu and Narasimhan concluded that the effective
internal redox potential approach was more appropriate than the external
approach [92].

The models developed by Walsh *et al.* [86], Carnahan [89] and
Liu and Narasimhan [91,92] and applied to the prediction of uranium
transport provide some insight into the "state-of-the-art" of coupled
codes development. Of these codes, DYNAMIX embodies most of the
features that would be required to model uranium transport in the
weathered zone in the vicinity of the Koongarra ore deposit. However,
the seasonal change in state of saturation of the top few meters of the
weathered zone and the apparent heterogeneity in substrate
characteristics are issues that render detailed modeling of this region
extremely difficult.

7.   **SUMMARY**

In this chapter, aspects of the Alligator Rivers Analogue Project,
a large, internationally funded project being conducted in northern
Australia, have been reviewed. The focus of this project is the
examination of factors influencing radionuclide transport away from a
uranium ore body and use of the obtained database in validation of
appropriate transport models. The ore body is considered a useful
analog of a radioactive waste repository.

Oxidation of the primary ore, uraninite, has produced a range of
uranyl silicate minerals with further mineralization to uranyl phosphates
in the weathered zone. Further away from the primary ore zone,

uranium is dispersed in weathered schists and, for the most part, is adsorbed onto clays and iron oxides. The migration of uranium appears to be closely related to the alteration (vermiculitization) of chlorite, the major rock-forming mineral in the primary ore zone. Magnesium, the dominant cation in waters flowing through the deposit, is also released in the vermiculitization process. Bicarbonate is the dominant anion in this region.

The region in the vicinity of the ore body is hydrologically complex with a "footwall" aquifer interacting with a "hanging wall" aquifer through a deep fault which is slightly upstream of the ore body. Large changes in water table occur seasonally as a result of a monsoonal climate and dilution of waters in surface layers by infiltration is reflected markedly in seasonal variations in uranium and phosphate in the groundwaters. Groundwaters in the region appear to be significantly undersaturated with respect to primary and secondary uranium minerals.

Analysis of $^{234}U/^{238}U$ activity ratios in both solid and solution phases indicate that the groundwaters are essentially in equilibrium with uranium bound in accessible regions of the solids. There is good evidence that most of the accessible uranium is associated with amorphous iron oxides with the remainder associated with more crystalline iron oxide phases. Indeed, surface complexation modeling of uranium assuming that it is all adsorbed to the amorphous iron component of the solids gives good agreement with experiment.

To date, simple one-dimensional, two phase models have been applied to interpretation of the distribution of isotopes in solid and solution phases though considerable problems exist in procurement of parameter values by methods other than trial and error. Part of the problem in this regard lies in the difficulty in describing the processes controlling migration and thus in the physical meaning of the parameters required in the model. While more appropriate (elaborate, mechanistic) hydrologic and geochemical models (and the coupling of these) are desirable, the complexity of some portions of the aquifer (such as that in the surface layers of the weathered zone where water level fluctuations are most severe) may render application of such models, at best, difficult.

REFERENCES

1.   Birchard, G.F. and D.H. Alexander. "Natural Analogues - A Way to Increase Confidence in Predictions of Long-Term Performance of Radioactive Waste Disposal," in *Proceedings of Materials Research Society Symposia on Scientific Basis for Nuclear Waste Management VI*, D.G. Brookins, Ed. (San Francisco: Materials Research Society, 1983), Vol. 15 p. 323-329.

2.   Airey, P.L., D. Roman, C. Golian, S. Short, T. Nightingale, R.T. Lowson, and G.E. Calf. "Radionuclide Migration Around Uranium Ore Bodies - Analogues of Radioactive Waste Repositories," USNRC Report NUREG/CR-391 (1984).

3.   Airey, P.L. "Radionuclide Migration Around Uranium Ore Bodies in the Alligator Rivers Region of the Northern Territory of Australia - Analogue of Radioactive Waste Repositories - A Review," *Chem. Geol.* 55:255-268 (1986).

4.   Airey, P.L., C. Golian and D.A. Lever. "An Approach to the Mathematical Modelling of the Uranium Series Redistribution within Ore Bodies," Australian Atomic Energy Commission Report AAEC/C49 (1986).

5.   Hardy, C.J. and P. Duerden. "Alligator Rivers Analogue Project," Progress Report, 1 September 1987 - 30 April 1988, Australian Nuclear Science and Technology Organisation, Lucas Heights, New South Wales (1988).

6.   Pedersen, C.P. "The Geology of the Koongarra Deposits," in Koongarra Project Draft Environmental Impact Statement, Appendix 5 (Melbourne: Noranda Australia Limited, 1978).

7.   Needham, R.S. and P.G. Stuart-Smith. "Geology of the Alligator Rivers Uranium Field," in *Uranium in the Pine Creek Geosyncline,* J. Ferguson and A.B. Goleby, Eds. (Vienna: International Atomic Energy Agency, 1980), pp.233-257.

8.   Needham, R.S. "Alligator River, N.T.: 1:250,000 Geological Series," Bur. Miner. Resour. Geol. Geophys. Aust., Explanatory Notes SD 53-1 (1984).

9.   Snelling, A.A. "Koongarra Uranium Deposits," in *The Geology of the Mineral Deposits of Australia and Papua New Guinea,* F.E. Hughes, Ed. (Melbourne: The Australian Institute of Mining and Metallurgy, 1989), Monograph 14.

10.   Foy, M.F. and C.P. Pedersen. "Koongarra Uranium Deposit," in *Economic Geology of Australia and New Zealand,* Vol.1 Metals, C.L. Knight, Ed. (Melbourne: The Australasian Institute of Mining and Metallurgy, 1975), pp.317-321.

11.   Snelling, A.A. and B.L. Dickson. "Uranium/Daughter Disequilibrium in the Koongarra Uranium Deposit, Australia," *Miner. Deposita* 14:109-118 (1975).

12.   Dickson, B.L. and A.A. Snelling. "Movements of Uranium and Daughter Isotopes in the Koongarra Uranium Deposit," in *Uranium in the Pine Creek Geosyncline,* J. Ferguson and A.B. Goleby, Eds. (Vienna: International Atomic Energy Agency, 1980).

13.   Snelling, A.A. "A Geochemical Study of the Koongarra Uranium Deposit," PhD Thesis, University of Sydney, Sydney, Australia (1980).

14.   Johnston, D.J. "Structural Evolution of the Pine Creek Inlier and Mineralisation Therein", PhD Thesis, Monash University, Melbourne, Australia (1984).

15.   Wilde, A.R., V.J. Wall and M.S. Bloom. "Wall-Rock Alteration
        Associated with Unconformity-Related Uranium Deposits,
        Northern Territory, Australia: Implications for Uranium
        Transport and Depositional Mechanisms," Extended Abstracts,
        Conference on Concentration Mechanisms of Uranium in
        Geologic Environments, Nancy, France (1985), pp. 231-239.

16.   Wilde, A.R. "On the Origin of Unconformity-Related Uranium
        Deposits," PhD Thesis, Monash University, Melbourne,
        Australia (1988).

17.   Isobe, H. and T. Murakami. "Alteration of Chlorite in the
        Koongarra Uranium Deposit and its Implication for Uranium
        Migration," Alligator Rivers Analogue Project Progress Report,
        1 June 1989 - 31 August 1989, P. Duerden, Ed. (Lucas Heights:
        Australian Nuclear Science and Technology Organisation, 1989),
        pp.57-70.

18.   Dibble, Jr., W.E. and W.A. Tiller. "Kinetic Model of Zeolite
        Paragenesis in Tuffaceous Sediments," Clays and Clay Minerals
        29:323 (1981).

19.   Ross, G.J., C. Wang, A.I. Ozken and H.W. Rees. "Weathering
        of Chlorite and Mica in a New Brunswick Podzol Developed on
        Till Derived From Chlorite-Mica Schist," Geoderma 27:255
        (1982).

20.   Buurman, P., E.L. Meijer and J.H. van Wijck. "Weathering of
        Chlorite and Vermiculite in Ultramafic Rocks of Cabo Ortegal,
        Northwestern Spain," Clays and Clay Minerals 36:263 (1988).

21.   Proust, D., J.-P. Eymery and D. Beaufort. "Supergene
        Vermiculitization of a Magnesium Chlorite: Iron and
        Magnesium Removal Processes," Clays and Clay Minerals
        34:572 (1986).

22. Noranda Australia Ltd. Koongarra Project Draft Environmental Impact Statement (Melbourne: Noranda Australia Limited, 1978).

23. Norris, J.R. "Preliminary Hydraulic Characterization of a Fractured Schist Aquifer at the Kooongarra Uranium Deposit, Northern Territory, Australia," Alligator Rivers Analogue Project Progress Report, 1 March 1989 - 31 May 1989, P. Duerden, Ed. (Lucas Heights: Australian Nuclear Science and Technology Organisation, 1989).

24. Giblin, A.M. and A.A. Snelling. "Application of Hydrogeochemistry to Uranium Exploration in the Pine Creek Geosyncline, Northern Territory, Australia," *J. Geochem. Explorat.* 19:33-55 (1983).

25. Duerden, P. and T.E. Payne. "Results Obtained for Samples Obtained on May 1988 Field Visit," Alligator Rivers Analogue Project Progress Report, 1 May 1988 - 31 August 1988, P. Duerden, Ed. (Lucas Heights: Australian Nuclear Science and Technology Organisation, 1988), 194-236.

26. Payne, T.E. "Water Sampling Program - November 1988 Field Visit," Alligator Rivers Analogue Project Progress Report, 1 September 1988 - 30 November 1988, P. Duerden, Ed. (Lucas Heights: Australian Nuclear Science and Technology Organisation, 1988), pp.193-208.

27. Duerden, P., Ed. Alligator Rivers Analogue Project Progress Report, 1 June 1989 - 31 August, 1989, (Lucas Heights: Australian Nuclear Science and Technology Organisation, 1989), pp.273-298.

28. Duerden, P., Ed. Alligator Rivers Analogue Project Progress Report, 1 December 1988 - 28 February 1989, (Lucas Heights: Australian Nuclear Science and Technology Organisation, 1989), pp.120-130.

29. Calf, G.E. "Tritium Activity in Australian Rainwater 1962-1986," (Lucas Heights: Australian Nuclear Science and Technology Organisation Report ANSTO/E680, 1988).

30. Phillips, F.M., K.N. Trotman, H.W. Bentley, S.N. Davis and D. Elmore. "Chlorine-36 from Atmospheric Nuclear Weapons Testing as a Hydrologic Tracer in the Zone of Aeration in Arid Climates," in *Proceedings of the International Symposium on Recent Investigations in the Zone of Aeration (RIZA)*, P. Udluft, B. Merkel and K.-H. Prosl, Eds, Munich (1988).

31. Ivanovich, M., P. Duerden, T. Payne, T. Nightingale, G. Longworth, M.A. Wilkins, S.E. Hasler, R.B. Edghill., D.J. Cockayne and B.G. Davey. "Natural Analogue Study of the Distribution of Uranium Series Radionuclides Between the Colloid and Solute Phases in the Hydrogeological System of the Koongarra Uranium Deposit, Australia," United Kingdom Atomic Energy Authority, Harwell Laboratory, Report No. AERE-R-12975 (1988).

32. Sverjensky, D.A. "Geochemical Modelling of the Koongarra Uranium Deposit - Alligator Rivers Analogue Project," Alligator Rivers Analogue Project Progress Report, 1 May 1988 - 31 August 1988, P. Duerden, Ed. (Lucas Heights: Australian Nuclear Science and Technology Organisation, 1988), pp. 21-64.

33. Sverjensky, D.A. "Geochemical Modelling of the Koongarra Uranium Deposit - Alligator Rivers Analogue Project," Alligator Rivers Analogue Project Progress Report, 1 September 1988 - 30 November 1988, P. Duerden, Ed. (Lucas Heights: Australian Nuclear Science and Technology Organisation, 1988), pp. 57-70.

34.   Sverjensky, D.A. "Geochemical Modelling of the Koongarra Uranium Deposit - Alligator Rivers Analogue Project," Alligator Rivers Analogue Project Progress report, 1 December 1988 - 28 February 1989, P. Duerden, Ed. (Lucas Heights: Australian Nuclear Science and Technology Organisation, 1989), pp. 9-38.

35.   Wolery, T.J. "EQ3NR: A Computer Program for Geochemical Aqueous Speciation-Solubility Calculations, User's Guide and Documentation," Lawrence Livermore Report UCRL-53414 (1983).

36.   Brown, D.S., and J.D. Allison. "MINTEQA1, Equilibrium Metal Speciation Model: A User's Manual," U.S. Environment Protection Agency Report No. EPA/600/3-87/012 (1987).

37.   Tripathi, V.S. "Uranium Transport Modelling: Geochemical Data and Sub-models," PhD Thesis, Stanford University, Stanford, CA (1983).

38.   Payne, T.E. and T.D. Waite. "Modelling of Radionuclide Sorption Processes in the Weathered Zone in the Vicinity of the Koongarra Ore Body," Alligator Rivers Analogue Project Progress Report, 1 March 1989 - 31 May 1989, P. Duerden, Ed. (Lucas Heights: Australian Nuclear Science and Technology Organisation, 1989), pp. 157-183.

39.   Dongarra, G. and D. Langmuir. "The Stability of UO_2OH^+ and $UO_2(PO_4)_2^{2-}$ Complexes at 25°C," *Geochim. Cosmochim. Acta* 44:1747-1751 (1980).

40.   Markovic, M. and N. Pavkovic. "Solubility and Equilibrium Constants of Uranyl (2+) in Phosphate Solutions," *Inorg. Chem.* 22:978-982 (1983).

41.   Grenthe, I., R.J. Lemire, A.B. Muller, C. Nguyen-Trung and H. Wanner. "Chemical Thermodynamics of Uranium," Draft Report of the Nuclear Energy Agency Thermochemical Data Base Project (Saclay: OECD Nuclear Energy Agency, 1989), pp. 240-243.

42.   Ivanovich, M. "The Phenomenon of Radioactivity," in *Uranium Series Disequilibria: Application to Environmental Problems in the Earth Sciences,* M. Ivanovich and R.S. Harmon, Eds. (Oxford: Clarendon Press, 1982), pp. 1-19.

43.   Osmond, J.K. and J.B. Cowart. "Natural Uranium and Thorium Series Disequilibrium: New Approaches to Geochemical Problems," *Nucl. Sci. Appl. (B)* 1:303-355 (1982).

44.   Osmond, J.K., J.B. Cowart and M. Ivanovich. "Uranium Isotopic Disequilibrium in Ground Water as an Indicator of Anomalies," *Int. J. Appl. Radiat. Isot.* 34:283-308 (1983).

45.   Borgaard, O.K. "Phase Identification by Selective Dissolution Techniques," in *Iron in Soils and Clay Minerals,* J.W. Stuchi, B.A. Goodman and U. Schwertmann, Eds., NATO ASI Series, Series C, Vol. 217 (Dordrecht: D. Reidel Publ. Co. 1988) pp. 83-98.

46.   Edghill, R. "The Distribution of Uranium and Thorium Between Phases in Weathered Core from Koongarra", Alligator Rivers Analogue Project Progress Report, 1 June, 1989 - 31 August, 1989, P. Duerden, Ed. (Lucas Heights: Australian Nuclear Science and Technology Organisation, 1989), pp.167-194.

47.   Rosholt, J.N. "Isotopic Composition of Uranium and Thorium in Crystalline Rocks," *J. Geophys. Res.* 88:7315-7330 (1983).

48. Lowson, R.T., S.A. Short, B.G. Davey and D.J. Gray. "$^{234}U/^{238}U$ and$^{230}Th/^{234}U$ Activity Ratios in Mineral Phases of a Lateritic Zone," *Geochim. Cosmochim.* Acta 50:1697-1702 (1986).

49. Nightingale, T. "Mobilisation and Redistribution of Radionuclides During Weathering of a Uranium Ore Body," MSc Thesis, University of Sydney, Sydney (1988).

50. Petit, J-C, Y. Langevin and J.-C. Dran. "$^{234}U/^{238}U$ Isotopic Disequilibrium in Nature: Theoretical Reassessment of the Various Proposed Models," *Bull. Mineral.* 108:745-753 (1985).

51. Ames, L.L., J.E. McGarrah and B.A. Walker. "Sorption of Trace Constituents from Aqueous Solutions onto Secondary Minerals. 1. Uranium," *Clays and Clay Minerals* 31:321-334 (1983).

52. Ames, L.L., J.E. McGarrah and B.A. Walker. "Sorption of Uranium and Radium by Biotite, Muscovite, and Phlogopite," *Clays and Clay Minerals* 31:343-351 (1983).

53. Van der Weijden, C.H., M. van Leeuwen and A.F. Peters. "The Adsorption of U(VI) onto Precipitating Amorphous Ferric Hydroxide," *Uranium* 2:53-58 (1985).

54. Ho, C.H. and D.C. Doern. "The Sorption of Uranyl Species on a Hematite Sol," *Can. J. Chem* 63:1100-1104 (1985).

55. Ho, C.H. and D.C. Miller. "Adsorption of Uranyl Species from Bicarbonate Solution onto Hematite Particles," *J. Colloid Interface Sci.* 110:165-171 (1986).

56. Sagert, N.H., C.H. Ho and N.H. Miller. "The Adsorption of Uranium (VI) Onto a Magnetite Sol," *J. Colloid Interface Sci.* 130:283-287 (1989).

57.	James, R.O. and T.W. Healy. "Adsorption of Hydrolyzable Metal Ions at the Oxide-Water Interface," *J. Colloid Interface Sci.* 40:65-81 (1972).

58.	Stumm, W., H. Hohl and F. Dalang. "Interaction of Metal Ions with Hydrous Oxide Surfaces," *Croat. Chem. Acta* 48:491-504 (1976).

59.	Schindler, P.W. "Surface Complexes at Oxide-Water Interfaces," in *Adsorption of Inorganics at Solid-Liquid Interfaces,* M.A. Anderson and A.J. Rubin, Eds. (Ann Arbor, MI: Ann Arbor Science, 1980).

60.	Hsi, C-K, and D. Langmuir. "Adsorption of Uranyl Onto Ferric Oxyhydroxides: Application of the Surface Complexation Site-Binding Model," *Geochim. Cosmochim. Acta* 49:1931-1941 (1985).

61.	Payne, T.E. and T.D. Waite. "Surface Complexation Modelling of Uranium Sorption Data Obtained by Isotope Exchange Techniques," *Radiochim. Acta* 52-3:487-493 (1991).

62.	Zachara, J.M., C.C. Ainsworth, C.E. Cowan and C.T. Resch. "Adsorption of Chromate by Subsurface Soil Horizons," *Soil Sci. Soc. Amer. J.* 53:418-428 (1989).

63.	Loux, N.T., D.S. Brown, C.R. Chafin, J.D. Allison and S.M. Hassan. "Chemical Speciation and Competitive Cationic Partitioning on a Sandy Aquifer Material," *Chem. Speciation Bioavail.* 1:111-125 (1989).

64.	Davis, J.A., R.O. James and J.O. Leckie. "Surface Ionization and Complexation at the Oxide/Water Interface," *J. Colloid Interface Sci.* 63:480-499 (1978).

65. Lever, D.A. "Modelling Radionuclide Transport at Koongarra," Australian Atomic Energy Commission Report No. AAEC/C55 (1986), pp. 179-196.

66. Golian, C., M. Ivanovich, D.A. Lever and G. Longworth. "Modelling Radionuclide Migration in the Koongarra Ore Deposit," Alligator Rivers Analogue Project Progress Report, 1 May 1988 - 31 August 1988, P. Duerden, Ed. (Lucas Heights: Australian Nuclear Science and Technology Organisation, 1988), pp. 65-68.

67. Golian, C., D.A. Lever, G. Longworth and M. Ivanovich. "Application of One-Dimensional Migration Models to Data From a Selected Region of the Secondary Dispersion Fan of the Koongarra Ore Deposit," Alligator Rivers Analogue Project Progress Report, 1 September 1988 - 30 November 1988, P. Duerden, Ed. (Lucas Heights: Australian Nuclear Science and Technology Organisation, 1988), pp. 97-108.

68. Golian, C. "Uranium Mobility Through Porous Media Which Contains Two Iron Mineral Phases: Modifications and Extensions to the Two-Phase Transport Model," Alligator Rivers Analogue Project Progress Report, 1 March 1989 - 31 May 1989, P. Duerden, Ed. (Lucas Heights: Australian Nuclear Science and Technology Organisation, 1989), pp. 21-34.

69. Till, J.E. and H.R. Meyer. "Radiological Assessment," NUREG/UCR-3332, ORNL-5968 (1983).

70. Anderson, M.P. "Using Models to Simulate the Movement of Contaminants Through Groundwater Flow Systems," *CRC Crit. Rev. Environ. Control* 9:97 (1979).

71. Valocchi, A.J. "Describing the Transport of Ion-Exchanging Contaminants Using an Effective K_d Approach," *Wat. Resour. Res.* 20:499 (1984).

72.   Cederberg, G.A., R.L. Street and J.O. Leckie. "A Groundwater
        Mass Transport and Equilibrium Chemistry Model for
        Multicomponent Systems," *Wat. Resour. Res.* 21:1095-1114
        (1985).

73.   Airey P.L., Ed. "Radionuclide Migration Around Uranium Ore
        Bodies - Analogue of Radioactive Waste Repositories," Annual
        Report 1981-1982, Australian Atomic Energy Commission
        Report No. AAEC/C29 (1982).

74.   Short, S.A. "Chemical Transport of Uranium and Thorium in the
        Alligator Rivers Uranium Province, Northern Territory,
        Australia," PhD Thesis, University of Wollongong, Wollongong
        (1988).

75.   Short, S.A., R.T. Lowson and J. Ellis. "$^{234}U/^{238}U$ and $^{230}Th/^{234}U$
        Activity Ratios in the Colloidal Phases of Aquifers in Lateritic
        Weathered Zones," *Geochim. Cosmochim. Acta* 52:2555-2563
        (1988).

76.   Grove, D.A. and W.W. Wood. "Prediction and Field Verification
        of Subsurface-Water Quality Changes During Artificial
        Recharge, Lubbock, Texas," *Ground Water* 17:250-257 (1979).

77.   Jennings, A.A., D.J. Kirkner and T.L. Theis. "Multicomponent
        Equilibrium Chemistry in Groundwater Quality Models," *Wat.
        Resour. Res.* 18:1089 (1982).

78.   Haworth, A., S.M. Sharland, P.W. Tasker and C.J. Tweed. "A
        Guide to the Coupled Chemical Equilibria Code CHEQMATE,"
        NIREX Radioactive Waste Disposal Report No. NSS-R.113
        (Oxfordshire: Harwell Laboratory, 1988).

79. Hostetler C.J., R.L. Erikson, J.S. Fruchter and C.T. Kincaid. "FASTCHEM Package, Volume 1: Overview and Application to a Chemical Transport Problem," EPRI Report No. EA-5870 (1989).

80. Rubin, J. and R.V. James. "Dispersion-Affected Transport of Reacting Solutes in Saturated Porous Media: Galerkin Method Applied to Equilibrium-Controlled Exchange in Unidirectional Steady Water-Flow," *Wat. Resour. Res.* 9:1332-1356 (1973).

81. Valocchi, A.J., R.L. Street and P.V. Roberts. "Transport of Ion Exchanging Solutes in Ground water: Chromatographic Theory and Field Simulation," *Wat. Resour. Res.* 17:1517-1527 (1981).

82. Miller, C.W. "Toward a Comprehensive Model of Chemical Transport in Porous Media," *Mat. Res. Soc. Symp. Proc.* 15:481-488 (1983).

83. Miller, C.W. and L.V. Benson. "Simulation of Solute Transport in a Chemically Reactive Heterogeneous System: Model Development and Application," *Wat. Resour. Res.* 19:381-391 (1983).

84. Kirkner, D.J., A.A. Jennings and T.L. Theis. "Multisolute Mass Transport with Chemical Interaction Kinetics," *J. Hydrol.* 76:107-117 (1985).

85. Liew, S.K. and D. Read. "Development of the CHEMTARD Coupled Process Simulator for Use in Radiological Assessment," United Kingdom Department of the Environment Report No. DOE/RW/88.05 1 (1988).

86. Walsh, M.P., S.L. Bryant, R.S. Schechter and L.W. Lake. "Precipitation and Dissolution of Solids Attending Flow Through Porous Media," *AICHE Journal* 30:317-328 (1984).

87.   Hostetler, P.B. and R.M. Garrels. "Transportation and Precipitation of Uranium and Vanadium at Low Temperatures, with Special Reference to Sandstone Type Uranium Deposits," *Econ. Geol.* 57:137 (1962).

88.   Adler, H.H. "Concepts of Uranium-Ore Formation in Reducing Environments in Sandstones and Other Sediments," in *Proceedings of Symposium on the Formation of Uranium Ore Deposits* (Vienna: International Atomic Energy Agency, 1974).

89.   Carnahan, C.L. "Simulation of Chemically Reactive Solute Transport Under Conditions of Changing Temperature," in *Coupled Processes Associated with Nuclear Waste Repositories,* C-F Tsang, Ed. (London: Academic Press, Inc., 1987), pp. 249-257.

90.   Carnahan, C.L. "Some Effects of Data Base Variations on Numerical Simulation of Uranium Migration," *Radiochim. Acta* 44/45:349-354 (1988).

91.   Liu, C.W. and T.N. Narasimhan. "Redox-Controlled Multiple-Species Reactive Chemical Transport, 1. Model Development," *Wat. Resour. Res.* 25:869-882 (1989).

92.   Liu, C.W., and T.N. Narasimhan. "Redox-Controlled Multiple-Species reactive Chemical Transport, 2. Verification and Application," *Wat. Resour. Res.* 25:883-910 (1989).

12

METAL SPECIATION AND MOBILITY AS INFLUENCED BY LANDFILL DISPOSAL PRACTICES

Frederick G. Pohland
Department of Civil Engineering
University of Pittsburgh
Pittsburgh, PA 15261

Wendall H. Cross and Joseph P. Gould
School of Civil Engineering
Georgia Institute of Technology
Atlanta GA 30332

1.   INTRODUCTION

Many industrial activities generate complex solid residues or sludges which contain toxic heavy metals. These metal sludges require treatment and disposal often entailing considerable effort and expense to ensure protection against adverse environmental impacts. Earlier landfill practices governed by less stringent controls, resulted in the co-disposal of such sludges with municipal solid wastes (MSW). Today, such potentially toxic inorganic compounds may still be co-disposed with domestic refuse as a consequence of household and unregulated small generator inputs. Depending on the technology employed, such landfills may or may not be significant generator sources of environmental contaminants. Therefore, research studies were initiated to evaluate the potential environmental impact of co-disposal of heavy metal sludges with municipal solid waste (MSW) and to investigate the mechanisms controlling mobilization and attenuation. The data and discussions presented herein focus on heavy metal behavior and were part of a broader study on the fate of both organic and inorganic pollutants co-disposed with municipal refuse [2].

2.   MATERIALS AND METHODS

To accommodate the research objectives, five simulated landfill columns were constructed with the operational features illustrated in Figure 1. Each column employed two 0.9-m diameter steel sections with a total height of 3.0 meters. The columns were each operated to permit leachate collection and recycle after they were loaded and sealed gas-tight.

A control column was loaded with shredded MSW only, whereas the remaining four test columns received MSW as well as selected organic priority pollutants. (While the leachate concentrations of the organic priority pollutants were monitored routinely as part of the broader research study, they were consistently very low and there was little evidence that they impacted either the general behavior of the columns or that of the added heavy metals.) Three of these latter columns also received incremental loadings of heavy metals contained within alkaline metal finishing waste treatment sludge, supplemented with additional lead and mercury (HgO and PbO). Prior to addition to the columns, the metal sludge was blended with measured quantities of sawdust to enhance uniformity and contact. The metal loading levels for the simulated landfill columns are presented in Table 1; each column also contained 267 kg of dry municipal solid waste.

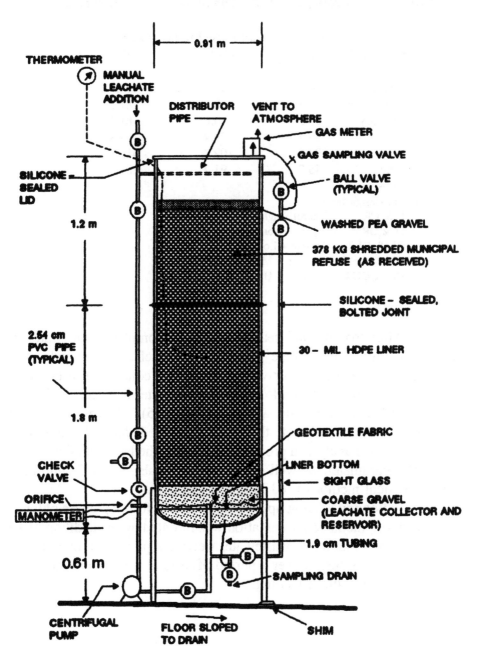

FIGURE 1. *Leachate Recycle Column (not to scale).*

TABLE 1.   *Loading Levels for Simulated Landfill Columns.*

	Column Loading (g/267 kg dry refuse)		
Heavy Metal	*Low*	*Medium*	*High*
Cadmium	26.1	52.3	104.6
Chromium	45.3	90.6	181.2
Mercury	20.3	40.6	81.2
Nickel	39.7	79.4	158.8
Lead	104.9	109.8	419.8
Zinc	45.7	91.4	182.8

Column Designations:

C & CR	Control Single Pass and Recycle
O & OR	Organic Priority Pollutants Only
OL & OLR	Organic + Low Metal Loading
OM & OMR	Organics + Moderate Metal Loading
OH & OHR	Organics + High Metal Loading

After loading and sealing had been completed, tap water was added to initiate immediate production of leachate for recycle and/or analysis. Thereafter, water was added to the columns at a rate of six liters per week and until sufficient leachate had accumulated to accommodate recycle and analysis (Day 135). Once leachate was generated from each of the five simulated landfill columns, a comprehensive analytical program was established for both the common indicator parameters as well as the heavy metals. Table 2 summarizes the analytical methods used in this assessment.

Following column operation in the acid forming phase of landfill stabilization, methane fermentation was initiated by the addition of seed sludge from a municipal wastewater treatment plant anaerobic digester. This addition occurred over a period of several weeks beginning about Day 600 after column loading.

TABLE 2. *Analytical Methods Used During Simulated Landfill Studies.* Analytical methods followed Standard Methods [1].

Parameter	Method	Instrument
Lead	Flame/Graphite Furnace Atomic Absorption	Perkin Elmer 303
Mercury	Flameless Atomic Absorption	Perkin Elmer 703
Chloride/Sulfate	Ion Chromatography	Dionex 2000i/SP
Sulfide, ORP, pH	Electrode methods	

3. RESULTS AND DISCUSSION

3.1 Theoretical Considerations

The potential for transport and/or attenuation of heavy metals within the simulated landfills involved a complex array of physical and chemical mechanisms, as determined by the following:

3.1.1 pH

The leachate pH directly influenced metal behavior, since metal solubilities generally increase with increasing hydrogen ion concentration. In addition, the hydrogen ion concentration indirectly influenced metal solubility by controlling dissociation acids, possibly to yield a precipitant anion, and influenced the intensity of reduction-oxidation (redox) reactions.

3.1.2 Precipitation

Many inorganic and organic anions can precipitate metals as sparingly soluble salts. Some anions, such as sulfide and hydroxide, have broad precipitating capabilities, while many others, such as sulfate and chloride, will form sparingly soluble salts with only a limited number of metals.

3.1.3   Complexation

The formation of complexes between heavy metals and ligands tends to increase metal solubility, although there are conditions under which the opposite may be expected [3,4]. Many organic and inorganic ligands capable of forming complexes with toxic heavy metals can be expected to be generated as a landfill proceeds through the phases of stabilization.

3.1.4   Reduction/Oxidation

It is well known that electrochemical processes can influence metal speciation and behavior both directly by modifying the nature of the metal itself, and indirectly by conversion of other species within the landfill environment. Thus, for example, selenium can be removed from landfills by reduction to the neutral element or conversion to the selenide ion which will be readily precipitated by ferrous ions. Likewise, reduction of sulfate is a significant process in controlling metal solubility in landfills by providing an abundant source of sulfide, a potent precipitant for many heavy metals.

3.1.5   Solid/Solute Interactions

The opportunity for interaction between dissolved species and refuse solids provides opportunities for changes affecting metal mobility. These include adsorption, ion exchange, interaction with solid phase ligands, and metal retention in interstitial water.

3.1.6   Source Interactions

Alkaline metal sludge forms a micro-environment within the landfill which will be substantially less acid than the leachate or refuse mass, especially during the early acid forming phase of landfill evolution. Therefore, mobility of dissolved metals can be expected to be significantly retarded as the local pH is increased by the reservoir of alkalinity contained in the sludge. In addition, interactions between the metal sludge and leachate anions, such as sulfate and sulfide, will tend to exchange hydroxide in the sludge with these anions, thereby forming less soluble salts. This process of replacement or encapsulation of the hydroxide solids by sulfide or other species has the effect of hindering the release of metals from the sludge solids.

3.2   Parametric Variations

The average results for the leachate heavy metals and important control parameters (pH, ORP, chloride, sulfate, sulfide) during the two major phases of the column operation are presented in Table 3. These results were averaged during periods of stable operations as characterized by an absence of dramatic parametric changes during the two phases including the acid formation phase which was maintained intentionally from about Day 100 through Day 600, and the methane fermentation phase which commenced at approximately Day 700 of column operation. The relatively brief transitional periods (Days 0 to 100 and Days 600 to 700) were not included in compiling the averages shown, since steady acid formation or methane fermentation, respectively, had not yet been established. Accordingly, the following was observed during these phases:

3.2.1 pH

By Day 100 of column operation, leachate pH stabilized in the range of approximately 5.2 to 5.6, a range maintained by the buffering effect of several low molecular weight volatile organic acids (acetic through hexanoic). It was only with the onset of active methane fermentation and the consequent depletion of these acids that the leachate pH increased to the more neutral range.

TABLE 3.   *Average Values of Selected Leachate Parameters During*
                Principal Operational Phases of Simulated Landfill Studies.
                (AFP): acid formation phase; (MFP): methane formation
                phase; BDL: below detection limit.

Parameter	(AFP) Heavy Metal Loadings			(MFP) Heavy Metal Loadings		
	Low	Med.	High	Low	Med.	High
pH	5.4	5.4	5.4	5.4	5.4	5.4
ORP, mV	-315	-305	-310	-230	-220	-215
Sulfate, mg/L	1500	1600	1650	140	110	110
Chloride, mg/L	1600	1500	1700	1400	1100	1200
Sulfide, mg/L	--	BDL	--	0.3	0.2	0.1
Cadmium, mg/L	10	15	50	1	2	6
Chromium, mg/L	2.0	2.6	5.0	BDL	BDL	BDL
Mercury, µg/L	17.2	14.3	15.4	15.3	12.8	12.2
Lead, mg/L	6.8	1.8	3.7	BDL	BDL	BDL
Zinc, mg/L	160	2240	725	25	55	65

3.2.2 Redox Potential

The measured leachate redox potential values were consistently
negative. Although substantially less negative than values considered
necessary for anaerobic stabilization reactions, the trend in reducing
conditions was maintained through the period of investigation.
Therefore this tendency suggests that appropriate reducing mechanisms
were operative.

3.2.3 Sulfate and Chloride

As indicated previously, sulfate and chloride will play a major
role in the behavior of the co-disposed heavy metals. Chloride can be
an important complexing agent for certain of the metals, while sulfate
is a significant precipitant for lead and, as the precursor of sulfide, is
vital in the attenuation of several of the heavy metals. However,
leachate chloride is relatively nonreactive and can be viewed as a

conservative substance, or tracer, as evidenced by an essentially constant concentration until approximately Day 700, followed by a slight concentration decrease as some dilution resulting from the addition of anaerobic sludge supernatant as seed took effect. In contrast, after a prolonged period of steady concentrations to approximately Day 700 of column operation, sulfate decreased precipitously to values approaching zero. This decrease paralleled the onset of sulfide formation coincident with the initiation of methanogenesis.

3.2.4 Sulfide

Leachate sulfide concentrations were consistently at or below the detection limit (≈ 0.1 mg/L) until Day 900, which were consistent with the behavior of sulfate. A 200- to 300-day lag in appearance of sulfide after the rapid decline in leachate sulfate concentrations can be ascribed to the high demand for sulfide as a precipitant, the most important likely being that imposed by a concomitant abundance of ferrous iron.

3.2.5 Chromium

Chromium, in many respects the chemically simplest of the heavy metals added to the columns, existed as Cr^{3+} at the redox potential values recorded and was subjected to precipitation only by hydroxide despite the presence of other significant leachate anions. Therefore, the chromic ion is unusual in that it does not form a sparingly soluble sulfide. On the other hand, the hydroxide species is a very sparingly soluble salt (Table 4) and, even at pH values as low as the 5.2 to 5.6 observed during the major portion of the acid formation phase of landfill stabilization, precipitation as the hydroxide could be expected to control the behavior of chromium.

TABLE 4. *Log$_{10}$ Values of Equilibrium Constants for Simulated Landfill Studies.*

Solubility Products		Formation Constants (ß)	
PbS	-26.6	PbSO$_4$$^\circ$	2.8
PbSO$_4$	-7.8	Pb(SO$_4$)$_2$$^{2-}$	3.6
Pb(OH)$_2$	-15.5	PbCl$^+$	1.3
PbCl$_2$	-4.7	PbCl$_2$	1.7
PbOHCl	-13.7	PbCl$_3^-$	1.9
Pb$_2$Cl(OH)$_3$	-36.1	PbCl$_4$$^{2-}$	1.5
HgS	-52.0	HgCl$_2$$^\circ$	13.9
Hg(OH)$_2$	-25.7	HgCl$_3^-$	14.8
CdS	-27.2	HgCl$_4$$^{2-}$	15.8
NiS	-23.8	H$_2$S: K$_1$	-7.0
ZnS	-24.8	K$_2$	-13.6

The behavior of leachate chromium concentrations during the two principal phases of operation revealed behavior totally consistent with hydroxide control. During the period of lowest leachate pH ($\leq$5.2), leachate chromium concentrations were subject to substantial variations and attained high values. However, by Day 500, chromium concentrations in the leachates from all columns had decreased to less than 1.0 mg/L, after which no significant chromium release was observed. The rapid decrease in leachate chromium levels preceded the onset of measurable sulfide formation at about Day 700, and can only be ascribed to precipitation by hydroxide. The transition to the methane fermentation phase, with its corresponding increase in pH, had the effect of further decreasing chromium solubility.

3.2.6 Nickel, Cadmium and Zinc

These three elements are similar in chemical behavior and, therefore, can be examined together. Because these elements existed only in the +2 oxidation state under the reducing conditions established within the column environments, they are subject to precipitation only by hydroxide at higher pH values or by sulfide at substantially lower pH values. Since the metal concentrations observed during the acid formation phase were much lower than would have been expected in the pH range characteristic of that phase, it is tempting to invoke sulfide as the controlling factor. Unfortunately, no evidence was obtained for the formation of sulfide during this phase. (As indicated in Table 4, it should be noted that the solubilities of the sulfides of these metals are so low that sulfide control was possible even at concentrations well below the detection limit of the electrode method used and at the pH conditions characteristic of the acid phase leachates.) In the absence of sulfide control, other mechanisms which might have acted to attenuate nickel, cadmium, and zinc, include sorption, ion exchange and precipitation with or in the alkaline micro-environments of the metal sludge solids.

The impact of the onset of methane fermentation, with the coincident development of sulfate reduction/sulfide production, was clearly evident. The average leachate concentrations of nickel, cadmium and zinc decreased rapidly, simultaneously with the onset of methane fermentation. These decreases reflected the effect of substantial precipitation of these metals as sparingly soluble sulfide salts and, possibly, the impact of sludge mass encapsulation by sulfide and consequent diminution of primary metal mobility at its source.

3.2.7 Lead

Changes in leachate lead concentrations during the acid formation and methane fermentation phases suggest a transition from one chemical control mechanism to another in response to the onset of methanogenesis in the simulated landfill columns. Examination of the data in Table 3 indicates that the lead concentrations during the acid phase were on the order of 30 μM. If sulfide was assumed to be the primary control on lead solubility during this phase, the total sulfide concentration in the leachate ($C_{T,S} = [H_2S] + [HS^-] + [S^{2-}]$) would be about 10^{-12} M in order to yield the observed lead concentration in the leachate at a pH of 5.5 as based on lead solubility equilibria (Table 4). This sulfide concentration is orders of magnitude lower than the limit of detection of the analytical method employed and, thus, could not be quantified.   However, examination of other solubility and complex equilibria for lead in equilibrium with sulfate and chloride suggests that there were alternate possibilities for solubility control during the acid formation phase in these systems.

A predominance area diagram for lead in equilibrium with chloride and sulfate (Figure 2) provides evidence that, in the absence of sulfide control, the major solid lead species to be expected in these systems was lead sulfate. Based on complex equilibria (Table 4), the distribution of soluble lead species in the leachate was computed and the total soluble lead concentration estimated on the order of 4 to 6 mg/L. These values agree reasonably well with the observed leachate lead concentrations. Again, the inability to measure sulfide at levels at which it would have been able to control lead solubility in these systems (Figure 3) necessitates a degree of caution in making unequivocal statements regarding sulfate control of lead solubility during the acid formation phase. However, the results of this analysis are highly suggestive of the presence of sulfate control during this period of landfill evolution.

In contrast, the response of lead to the onset of active sulfate reduction/sulfide production was dramatic. By Day 750, leachate lead concentrations had decreased below the analytical detection limit for all columns. Therefore, in the presence of measurable levels of sulfide, even at pH levels below 6.0, lead was essentially immobilized as the sulfide species.

3.2.8   Mercury

The changes in leachate mercury concentrations appear to have been dictated not primarily by solubility or complex equilibria (Table 4), but by its unusual redox chemistry.   Whether present as chloromercuric complexes ($HgCl_2$, $HgCl_3^-$, $HgCl_4^{2-}$) or as solid mercuric sulfide, HgS, the reducing conditions in the columns rendered reduction of mercury to the metal highly probable.

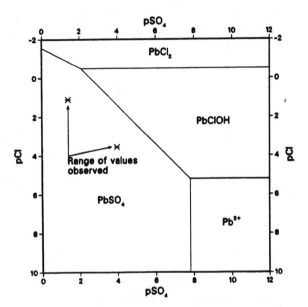

FIGURE 2. *Predominance Diagram for the System Pb^{2+} - SO_4^{2-} - Cl^- - OH^-.*

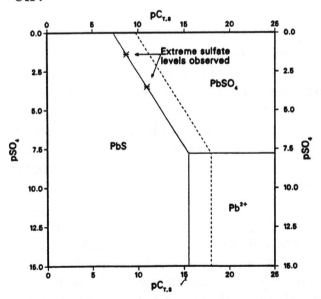

FIGURE 3. *Predominance Diagram for the System Pb^{2+} - SO_4^{2-} - Total Sulfide $(C_{T,S})$ (—— pH = 5.5; ---pH = 7.5).*

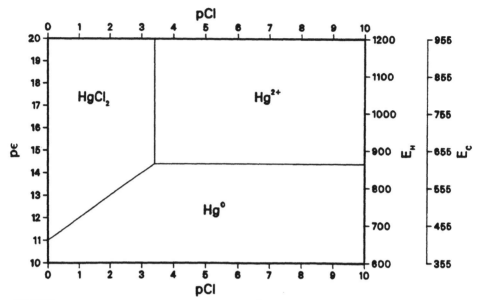

FIGURE 4.   *pC - pε Diagram for Hg²⁺ - Cl⁻ System.*

Figure 4 is a pC-pε diagram for mercury with respect to chloride. With column redox potential values sufficiently reduced to permit conversion of the mercury to the metallic form, leachate mercury levels stabilized rapidly early in column operation to values in the range of 10 to 30 µg/L. This concentration range is consistent with that reported for solubility of metallic mercury in water of 20 to 40 µg/L [5]. Moreover, mercury concentrations showed virtually no response to the onset of sulfide generation, in spite of the extremely low solubility of mercuric sulfide (Table 4). Hence, it would appear that sulfide had limited control over the behavior of mercury in these columns once the indicated reducing conditions had been established.

The possibility that mercury is subject to reduction to the metal in landfills is significant in that it presents the possibility of transport by a vapor phase mechanism due to the high volatility of metallic mercury. Hughes [6] cited this as an important mechanism for the transport of metallic mercury from landfills. Moreover, landfill conditions which include low pH and redox potential, high concentrations of organic compounds, and substantial biological activity are precisely the conditions under which alkylation of mercury has been observed [6]. While it has not been possible to detect alkylmercuric compounds in these investigations, it is reasonable to expect that they may have been produced and were present. Moreover, the unusual toxicological properties of these compounds may make them of particular significance in assessing the health impact of co-disposal of mercury containing wastes with municipal solid waste.

3.3   Sludge Loading Influences

3.3.1   Encapsulation

A process of major potential significance with respect to the mobility of heavy metals co-disposed in landfills is the replacement of hydroxide at the sludge mass surface by other anions such as carbonate, sulfate or sulfide. This process of encapsulation will be enhanced by the alkaline nature of the sludge solids and will permit reactions involving species such as carbonate which are not likely to be in great abundance in acid phase leachate. Similarly, during the methane fermentation phase, displacement of previously deposited anions by sulfide would be expected to occur. In both cases, the consequent reduction of solubility at the surficial layers of the sludge will act to limit further mobilization of metals from the source sludge.

3.3.2   Leachate Neutralization in the Sludge Layer

Mobility of metals, especially during the acid formation phase, is strongly enhanced by low pH. It is probable, however, that contact between the alkaline metal sludge and the leachate will result in some localized neutralization of the leachate pH and associated changes in the chemical composition of the leachate. This process is shown schematically in Figure 5.

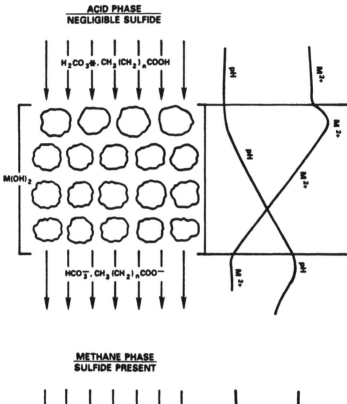

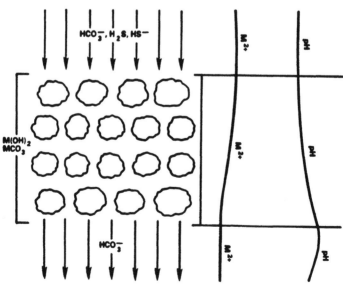

FIGURE 5.   *Impact of Sludge Mass on Attenuation of Heavy Metals.*

Of major interest, following an early increase in leachate metal levels due to their initial dissolution from the sludge by the acidic leachate, the metal concentrations will probably decrease in response to sludge-engendered neutralization. This process would act to retard further metal mobility in a fashion similar to adsorption on a chromatographic column.

Once the methane phase has been initiated, the impact of the physical distribution of the sludge within the columns on leachate composition will become much less significant, involving comparatively minor changes in the concentrations of dissolved species. The active sulfide formation characteristic of this phase will provide additional protection against further metal release into the leachate that will more than nullify any loss of retardation potential in the sludge layer. Thus, this sludge layer retardation of metal mobility will function during the acid forming phase when it is most needed, and cease to be as significant at a time when more powerful attenuation processes become available.

Assuming that sludge encapsulation and sludge mass neutralization operate as suggested, the practical implications are considerable. To derive optimal benefit from these processes, separate, discrete deposits of sludge, conditioned to permit facile penetration of the sludge by moisture, will enhance encapsulation while taking advantage of the alkaline nature of the sludge to the greatest extent possible. In contrast, more random distribution of the sludge will greatly dilute the localized neutralization capabilities of the sludge and tend to increase mobilization of the metals from the sludge particles. Therefore, the method of placement of metal sludges in a municipal landfill can substantially influence ultimate fate in terms of heavy metal mobility. In addition, leachate recycle, as utilized in these investigations, will significantly enhance both of these processes by increasing contact opportunity between the sludge mass and leachate components not otherwise available under conventional single pass leaching.

428 Pohland, Cross and Gould

4. CONCLUSIONS

The behavior of heavy metals co-disposed with municipal solid wastes in simulated landfill columns was influenced by an array of attenuative mechanisms which limited metal mobility. Chemical mechanisms included direct reduction to an elemental metal (Hg), precipitation by reductively generated sulfide (Cd, Zn, Pb and Ni), and precipitation by sulfate (Pb) during acid pH conditions. Acid phase attenuation of zinc, cadmium and nickel was ascribed to either physical sorptive mechanisms or the impact of localized alkaline regions within the waste mass associated with the co-disposed metal sludge. Since all of these attenuating processes had the effect of reducing metal mobilities, it can be concluded that landfills have an inherent *in situ* capacity for minimizing the mobility of toxic metals co-disposed with municipal refuse from unregulated household or small quantity generators. This capacity, if managed effectively, can greatly reduce, or eliminate, the potentially adverse environmental impact of landfills as a source of heavy metal contamination.

ACKNOWLEDGMENTS

This research was supported by the U.S. Environmental Protection Agency under the terms of Cooperative Agreement CR-812158. The results and conclusions presented herein are not necessarily those of the U.S. Environmental Protection Agency.

REFERENCES

1. APHA.AWWA.WPCF. *Standard Methods for the Examination of Water and Wastewater.* 16th Ed. Amer. Public Health Assn., Washington, D.C. (1985).

2. Pohland, F.G., W.H. Cross, J.P. Gould, and D.R. Reinhart, Attenuation of Priority Pollutants Co-disposed with MSW in Simulated Landfills. *Proceedings 15th Annual EPA Research Symposium,* Cincinnati, OH. (1989).

3. Sillen, L.G. and A.E. Martell, Stability Constants of Metal-Ion Complexes. *Special Publication No. 17 of the Chemical Society,* London, 1,150 pp. (1964).

4. Sillen, L.G. and A.E. Martell, Stability Constants of Metal-Ion Complexes, *Suppl. Special Publication No. 25 of the Chemical Society,* London, 838 pp. (1971)

5. Friberg, L. and J. Vostal, *Mercury in the Environment,* Chap. 3. CRC Press, Cleveland, OH. (1972).

6. Hughes, W.L., A Physicochemical Rationale for the Biological Activity of Mercury and its Compounds. *Ann. New York Acad. Sci.,* 66:454-461 (1957).